A GEOLOGICAL EXCURSION GUIDE TO THE
MOINE
GEOLOGY OF THE NORTHERN HIGHLANDS OF SCOTLAND

Dedicated to the late
Frank May and Val Moorhouse
who both contributed much to our present understanding of the Moine geology of Northern Scotland.

A GEOLOGICAL EXCURSION GUIDE TO THE
MOINE

GEOLOGY OF THE NORTHERN HIGHLANDS OF SCOTLAND

Edited by

Rob Strachan, Clark Friend,
Ian Alsop and Suzanne Miller

Published in 2010 by
Edinburgh Geological Society
Geological Society of Glasgow

in association with
NMS Enterprises Limited – Publishing
a division of NMS Enterprises Limited
National Museums Scotland
Chambers Street
Edinburgh EH1 1JF

Publication format, text and images
© Edinburgh Geological Society 2010
© Geological Society of Glasgow 2010
(unless otherwise credited)

No part of this publication may be reproduced, stored in a retrieval system or transmitted in any form or by any means, electronic, mechanical, photocopying, recording or otherwise, without the prior written permission of the publisher.

The rights of Rob Strachan, Clark Friend, Ian Alsop and Suzanne Miller to be identified as the authors of this book have been asserted by them in accordance with the Copyright, Designs and Patents Act 1988.

ISBN: 978 1 905267 33 0

Publication layout and design by
  NMS Enterprises Limited – Publishing.
Cover artwork by Mark Blackadder; photograph by Anna Bird. Cover image: High-grade Moine gneisses and meta-igneous intrusions of the Naver Nappe, Torrisdale Bay, Sutherland.
Printed and bound in China by Eurasia.

For a full listing of titles and related merchandise, please contact:

www.nms.ac.uk/books
www.edinburghgeolsoc.org
www.geologyglasgow.org.uk

Contents

Contributors vi

Foreword 3

Editorial Introduction 4

Summary of Moine Geology 7

    EXCURSION 1:   Ross of Mull 29

    EXCURSION 2:   Fort William to Glenfinnan 53

    EXCURSION 3:   Glenfinnan to Morar 63

    EXCURSION 4:   Invergarry to Kinloch Hourn 87

    EXCURSION 5:   Glen Moriston and Glen Shiel 113

    EXCURSION 6:   West Glenelg and Loch Hourn 123

    EXCURSION 7:   East Glenelg and Loch Duich 137

    EXCURSION 8:   Glen Strathfarrar and Loch Monar 153

    EXCURSION 9:   Loch a' Bhraoin, Braemore and Loch Broom ... 162

    EXCURSION 10:   South and Central Sutherland 176

    EXCURSION 11:   The Moine Thrust Belt at Loch Eriboll 191

    EXCURSION 12:   Durness and Faraid Head 221

    EXCURSION 13:   North Sutherland 231

    EXCURSION 14:   Great Glen 266

References 282-297

Contributors

Dr G. I. Alsop
Dept of Geology and
Petroleum Geology
Meston Building
King's College
Aberdeen AB24 3UE

Dr D. Barr
British Petroleum
Farburn Industrial Estate
Dyce, Aberdeen AB21 7PB

Dr I. M. Burns
(formerly Oxford Brookes
University)
6 McKenzie Crescent
Bettyhill, By Thurso
Sutherland KW14 7SY

Professor R. W. H. Butler
Dept of Geology and
Petroleum Geology
Meston Building
King's College
Aberdeen AB24 3UE

Dr C. R. L. Friend
45 Stanway Road
Risinghurst, Headington
Oxford OX3 8HU

Dr R. Glendinning
BG Group plc
100 Thames Valley Park Dr
Reading RG6 1PT

Professor A. L. Harris
School of Earth, Ocean
& Planetary Sciences
University of Cardiff
Main Building
Park Place
Cardiff CF10 3YE

Professor R. E. Holdsworth
School of Earth Sciences
The University
South Road
Durham DH1 3LE

Dr S. Kelley
Dept of Earth Sciences
The Open University
Walton Hall
Milton Keynes MK7 6AA

Dr M. Krabbendam
British Geological Survey
Murchison House
West Mains Road
Edinburgh EH9 3LA

Dr A. G. Leslie
British Geological Survey
Murchison House
West Mains Road
Edinburgh EH9 3LA

Dr S. Miller
Director of the South
Australian Museum
North Terrace
Adelaide, South Australia

Dr D. Powell
(formerly Royal Holloway
and Bedford New College)
Brunery House
Kinlochmoidart, Lochailort
Inverness-shire PH38 4ND

Professor J. G. Ramsay
(formerly University of
Zurich)
Cratoule, Issirac
F-30760 St Julien de
Peyrolas, France

Dr A. M. Roberts
Badley Geoscience Ltd
North Beck House
North Beck Lane
Hundleby
Spilsby
Lincs PE23 5NB

Professor N. J. Soper
(formerly University of
Sheffield)
Gams Bank
Threshfield
Skipton BD23 5NP

Dr M. Stewart
Department of Learning,
Innovation & Development
Liverpool John Moores
University
2 Maryland Street
Liverpool L1 9DG

Dr C. Storey
School of Earth &
Environmental Sciences
University of Portsmouth
Burnaby Road
Portsmouth PO1 3QL

Dr R. A. Strachan
School of Earth &
Environmental Sciences
University of Portsmouth
Burnaby Road
Portsmouth PO1 3QL

Foreword

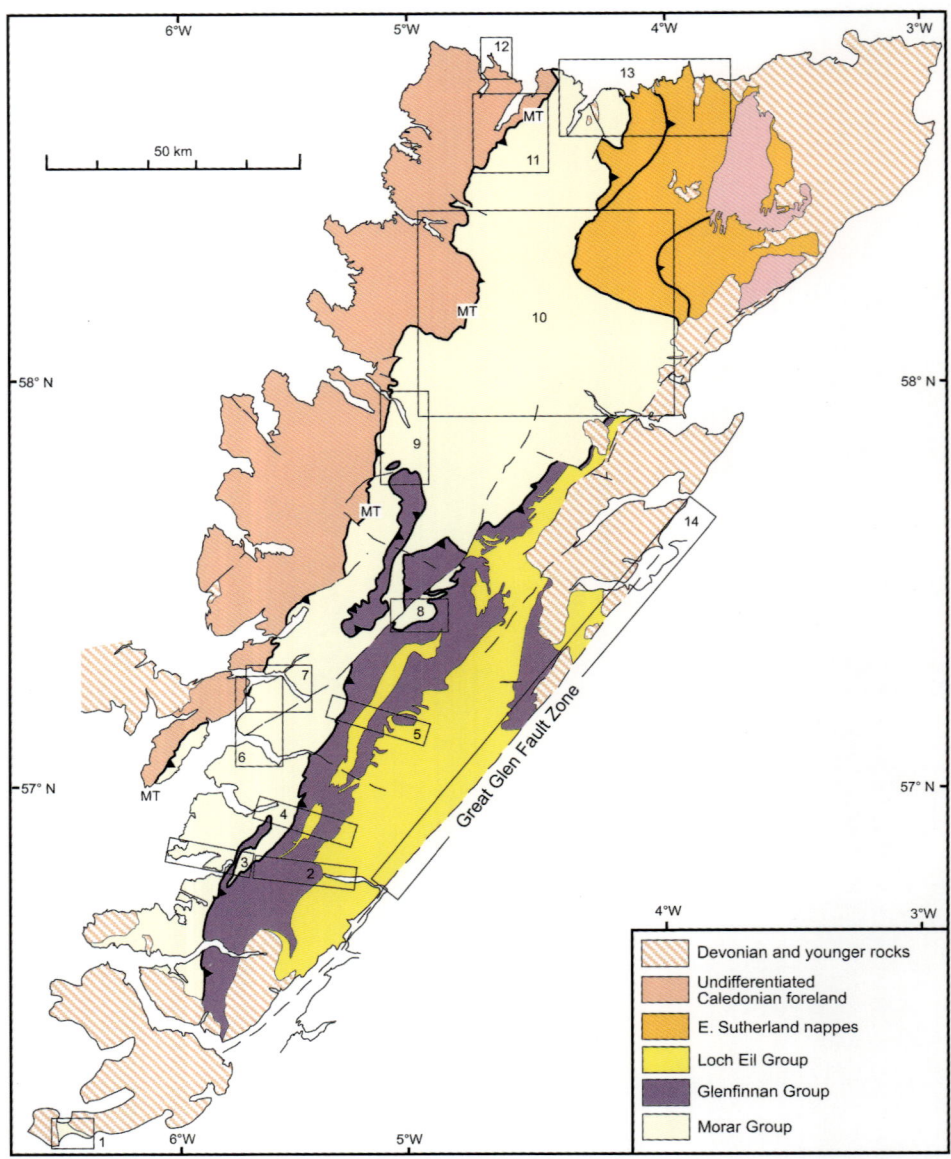

Figure F.1 Locations of the excursions on a generalized geological map of the Northern Highlands of Scotland.

Foreword

The Neoproterozoic rocks of the Moine Supergroup underlie an extensive tract of the Lower Palaeozoic Caledonian mountain belt in NW Scotland. The region contains numerous classic geological localities that have been illustrated in geology textbooks for many years. The superb geology of the region continues to attract field parties of amateur groups, undergraduate students and international scientists. This guide is a new edition of the first 'Moine fieldguide' that was published by Scottish Academic Press in 1988 on behalf of the Edinburgh and Glasgow geological societies, and is now more or less unavailable. As was the case with the first guide, the aim is to provide an up-to-date summary of the geological evolution of the Moine Supergroup, illustrated by the field evidence on which it is based. Owners of the first fieldguide will see that a number of excursions have survived more or less intact, although at a minimum all have been updated to take account of new geological information, as well as any new outcrops and/or additional constraints on access. Other excursions have been more or less completely rewritten. A key feature of this second edition is the inclusion of new excursions to the Ross of Mull, West Glenelg and Loch Hourn, East Glenelg and Loch Duich, Glen Strathfarrar and Loch Monar, South and Central Sutherland, Durness, and the Great Glen (Fig. F.1).

The editors acknowledge the substantial contributions made by Iain Allison and the late Frank May who co-edited the first 'Moine fieldguide'. The authors of the various excursions acknowledge discussions with colleagues too numerous to mention, and also the role of the Natural Environment Research Council who funded studentships which allowed much of the research reported here to be carried out.

Editorial Introduction

The aim of this excursion guide is to allow geological field parties to see the wide variety of rocks and structures that occur within the outcrop of the Moine Supergroup, as well as the Moine Thrust Zone that separates these rocks from those of the Caledonian foreland to the NW. The guide has been written for those who have some previous knowledge of geology: informed amateurs, undergraduate students and professional geologists. Books that provide useful background reading include *The Mapping of Geological Structures* by Ken McClay, and *The Field Description of Metamorphic Rocks* by Norman Fry, which are both published by John Wiley & Sons as part of their 'Geological Field Guide Series'. Two other publications that provide much useful background information are the 2002 edition of the *Geology of Scotland*, published by the Geological Society of London and edited by N. Trewin, and the British Geological Survey *Northern Highlands Regional Guide* published in 1995.

The excursions are mostly easily accessible from the various roads that cross the Moine outcrop. Statutory rights of public access were established over most land through the Land Reform (Scotland) Act 2003. Nonetheless, stalking of red deer occurs from early August and shooting of grouse from 12 August, and field parties should take account of reasonable requests to minimise disturbance at these times. A guide to access rights is published by the Ramblers' Association Scotland. Field parties are also reminded that many of the excursions include localities that have Site of Special Scientific Interest (SSSI) status and hammering and collection of material at these sites is prohibited without permission. Details of SSSIs can be obtained from Scottish Natural Heritage.

It is assumed that all geological field parties will adhere to the codes of practice for safety published by the Geological Society of London and/or the Geologists' Association. Visitors to the Scottish Highlands should be aware that the weather can be highly unpredictable, even in summer. Stout footwear, warm clothing and waterproofs are all necessary, even if the weather looks set fair. Generations of Highland geologists will testify to the need to carry insect repellent during the summer months!

Since the publication of the first 'Moine fieldguide' in 1988, a number of new geological maps of the Moine Supergroup have been produced by the British Geological Survey (BGS). Additionally, the application of modern geochronological techniques has placed important constraints on the timing of major metamorphic and structural events. Despite these significant advances, there still remains a lack of consensus concerning the correlations of certain tectonostratigraphic units and structures, and the nature of the Neoproterozoic evolution. In this guide, no attempt has been made to force a single view; individual authors present the evidence on which they base their views and the reader is invited to follow the excursion guide, to study the rocks and their relationships in the field and to form his or her own conclusions.

Geologists have shown that the Moine Supergroup has been affected by several phases of deformation. These phases, giving rise to recognisable sets of structures, may all be part of one mountain-building event spanning some tens of millions of years, or they may be related to different orogenic events perhaps hundreds of millions of years apart. Some structures, formed during a single phase of deformation, can be correlated over large areas, while others are quite local phenomena. One cannot assume, therefore, that structures with certain labels in one excursion are the same as those with the same label in another excursion. The shorthand terms are D for phase of deformation, S for planar fabric (surface), L for a linear fabric and F for folds. Subscripts (e.g. D_2) are added to denote which phase is being referred to. Thus S_2 is a planar fabric formed during the second (local) phase of deformation (i.e. D_2). The term S_0 may be used to indicate original sedimentary bedding

Summary of Geology

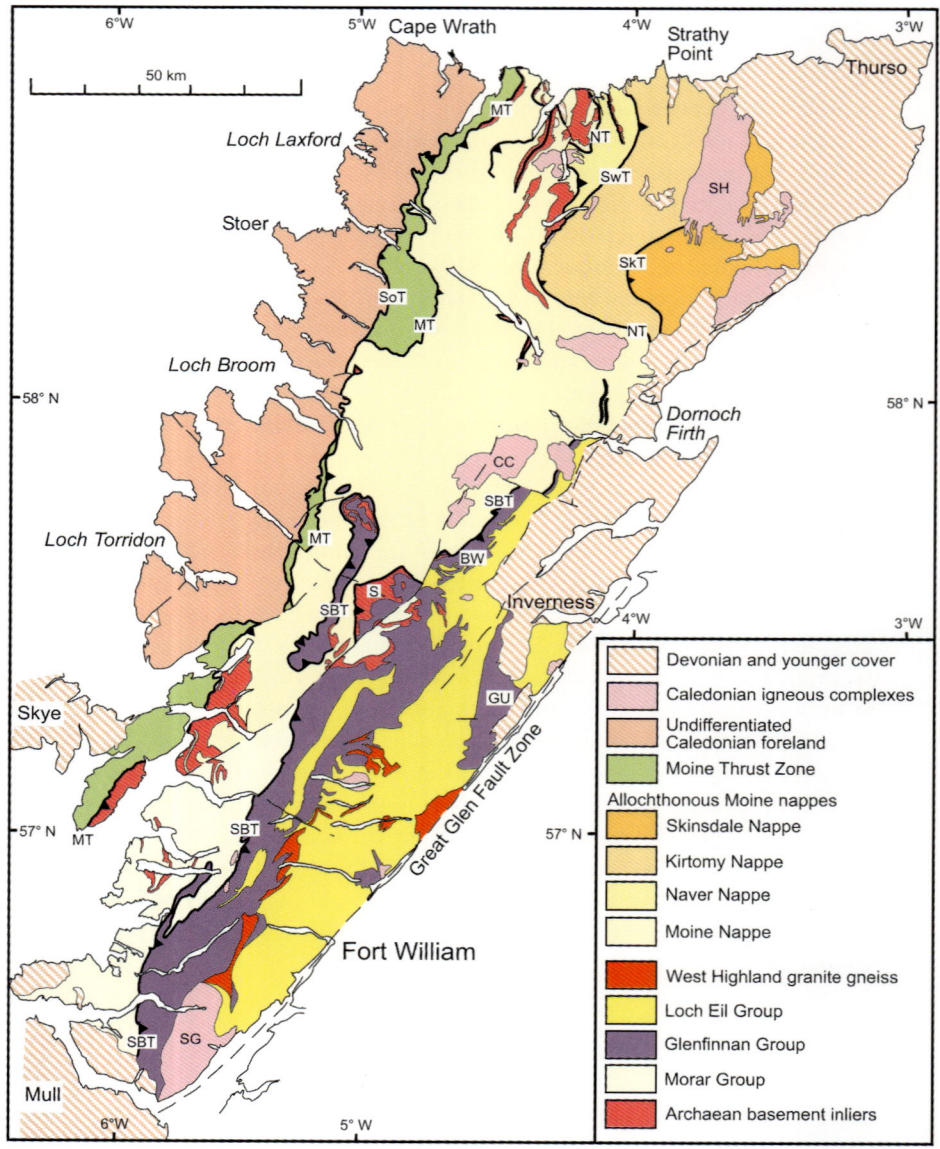

Figure S.1 Geological map of the Northern Highlands of Scotland. Abbreviations as follows:

MT = Moine Thrust; NT = Naver Thrust; SwT = Swordly Thrust; SkT = Skinsdale Thrust; SoT = Sole Thrust; SH = Strath Halladale Granite; CC = Carn Chuinneag-Inchbae Granite; GU = Glen Urquhart; S = Scardroy; SBT = Sgurr Beag Thrust; BW = Ben Wyvis; SG = Strontian Granite

Summary of Moine Geology

Rob Strachan, Bob Holdsworth and Ian Alsop

The Moine Supergroup is a sequence of Precambrian metasedimentary rocks that outcrops in the Northern Highlands of Scotland (Fig. S.1). These rocks, also known informally as 'the Moine', have excited the interest of geologists for well over a hundred years since the pioneering studies of the Geological Survey in the late 1800s. The Moine represents classic ground in the history of structural geology, because it is here that some of the first and most influential studies were carried out on the nature of basement-cover relationships and polyphase fold patterns within an orogenic belt. The Moine Thrust Zone that bounds the Moine to the northwest is one of the best known and most accessible examples worldwide of basement-involved thrusting along the margin of an orogenic belt. Since the first edition of this fieldguide was published, there have been many advances in the understanding of the Moine. These are principally the result of the integration of structural and metamorphic studies with modern isotopic dating techniques, the systematic remapping of selected areas by academic groups in collaboration with the British Geological Survey, and the refinement of Neoproterozoic and Lower Palaeozoic plate reconstructions. Nevertheless, despite the considerable amount of research that has been carried out on the Moine in recent years, various aspects of its geological evolution as well as the nature of its relationship with other Precambrian rock units in the region are still controversial matters that continue to attract geologists to the Scottish Highlands. The aim of this new edition of the fieldguide is to present an updated view of Moine geology as a series of excursions through classic as well as hitherto less well known ground. These excursions are designed to be intelligible and interesting to the casual or amateur visitor, or undergraduate field parties, whilst still providing sufficient detail for the professional enthusiast. Short itineraries are suggested for all excursions, for those with limited time. The summary that now follows is intended as a brief statement of the current understanding of Moine geology; those who require more detail are referred to the syntheses provided by Holdsworth *et al.* (1994), Strachan *et al.* (2002, 2010), and Mendum *et al* (2009).

Summary of Geology

Geological background

The Moine Supergroup of the Northern Highlands of Scotland is a sequence of Neoproterozoic metasedimentary rocks that was involved in the Ordovician-Silurian Caledonian Orogeny. The Moine rocks comprise thick formations of psammites, semi-pelites and pelites, as well as striped or banded units characterized by rapid alternations of lithologies. Sedimentary structures are often present in areas of low tectonic strain and provide the evidence on which the original way up of local successions can be established. Caledonian deformation and metamorphism has long been recognized within the Moine, but in recent years significant isotopic evidence has accumulated suggesting that these rocks were also affected by Precambrian orogenesis at $c.820$-730 Ma.

The Moine rocks are separated from the Hebridean foreland to the NW by the Moine Thrust Zone (Fig. S.1). The oldest component of the foreland is the Archaean-Palaeoproterozoic Lewisian Gneiss Complex (Park *et al.*, 2002; Kinny *et al.*, 2005 and references therein). These basement rocks are overlain unconformably by Proterozoic Torridonian sedimentary rocks (Stewart, 2002) that are in turn overstepped by a Cambrian-Ordovician shelf sequence of quartz arenites, limestones and dolomites (Swett, 1969; Park *et al.*, 2002). Within the Moine Thrust Zone, the cover and basement rocks are imbricated, folded and stacked up in a complex sequence of thrusts (e.g. Peach *et al.*, 1907; Elliott & Johnson, 1980). The Moine rocks rest on the roof thrust to this belt, the Moine Thrust. The Moine Supergroup is limited to the southeast by the Great Glen Fault Zone (Fig. S.1). Possible equivalents of the Moine in the Central Highlands are represented by the Badenoch Group (formerly the Dava and Glen Banchor successions, Smith *et al.*, 1999). These are of generally higher metamorphic grade and greater structural complexity than the overlying metasedimentary rocks of the mid- to late-Neoproterozoic Dalradian Supergroup, although the inferred unconformity between the two has been largely obscured by tectonism.

Various suites of igneous intrusions cut the Moine, including pre- to synmetamorphic amphibolites, granites and pegmatites that were emplaced during the Neoproterozoic, and Caledonian granite plutons and minor intrusions (Stephenson *et al.*, 1999). The Moine is overlain unconformably by Lower Devonian Old Red Sandstone rocks (Fig. S.1) and is cut by regional basic dyke swarms of Permo-Carboniferous and Tertiary age. Table S.1 summarizes the timing of the main geological events in the Northern Highlands.

Summary of Geology

Table S.1	
Summary of the Caledonian and pre-Caledonian history of the Moine rocks of the Northern Highlands of Scotland. Timing based on isotopic dates quoted in the text.	
430-390 Ma	Emplacement of the 'Newer Granite' suite and sinistral displacements along the Great Glen fault system.
435-425 Ma	**Scandian** orogenic event – mid to low amphibolite facies metamorphism, widespread ductile thrusting and folding, culminating in development of the Moine Thrust Zone.
470-460 Ma	**Grampian** orogenic event – mid to upper amphibolite facies metamorphism and deformation of the eastern Moines above Sgurr Beag and Naver thrusts.
600-590 Ma	Intrusion of augen granites (e.g. Inchbae) during continental rifting and development of the Iapetus Ocean.
820-725 Ma	**Knoydartian** orogenic event(s) – garnet grade metamorphism, isoclinal folding, intrusion of pegmatites, early displacement on the Sgurr Beag Thrust.
870 Ma	Intrusion of amphibolites and the granitic protoliths of the West Highland Granitic Gneiss – during an orogenic event or during crustal extension and development of the Moine sedimentary basin?
1000-870(?) Ma	Deposition of Moine sediments.

Age and tectonic setting of Moine sedimentation

The Moine Supergroup is unfossiliferous and its age is constrained only by isotopic data. Detrital zircon grains obtained from Moine rocks, as well as inherited zircons within migmatites and granites that were formed by the melting of Moine sources, have given ages that mostly range between $c.1800$ and $c.1000$ Ma (Friend *et al.*, 1997, 2003; Kinny *et al.*, 1999; Cawood *et al.*, 2004, 2007; Kirkland *et al.*, 2008). The Moine rocks were therefore probably derived in part from the erosion of the $c.1.1$-1.0 Ga Grenville orogenic belt that formed during the assembly of the super-

Summary of Geology

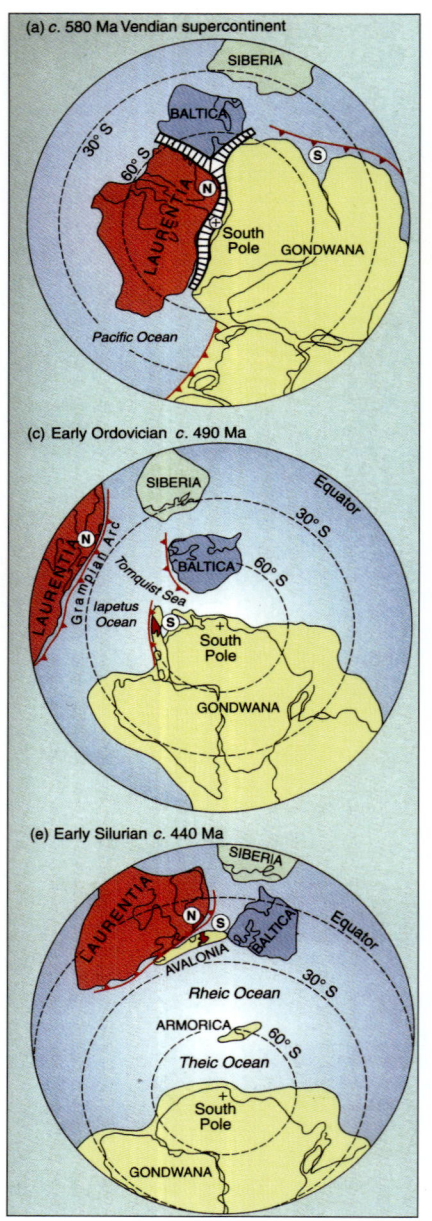
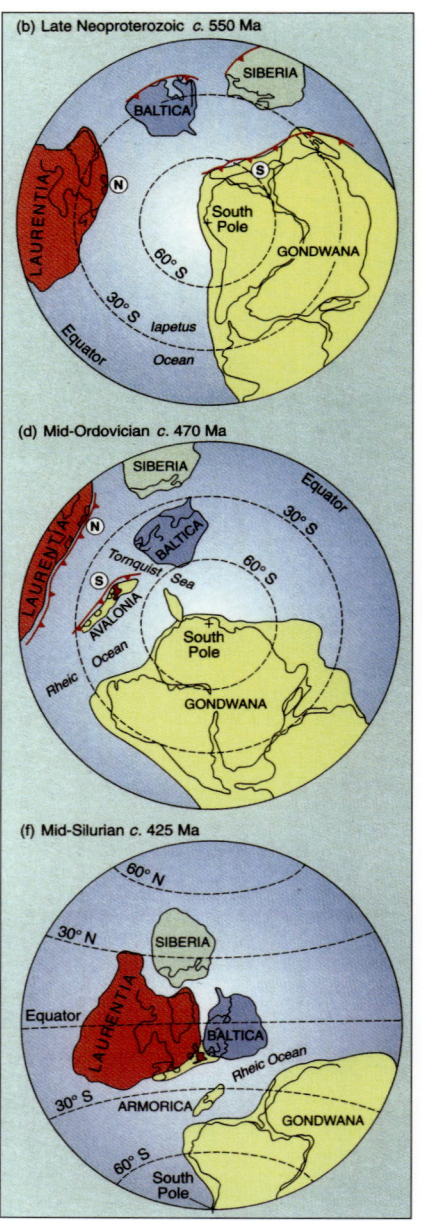

Summary of Geology

Opposite page:
Fig. S.2 Generalized reconstructions of the positions of the main continents from the late Neoproterozoic to late Silurian times (from Open University, 2003). The configuration of Laurentia, Baltica and Gondwana depicted here in **S.2a** remained stable throughout the Neoproterozoic from the formation of the supercontinent Rodinia to the rifting at c.600 Ma which led to the formation of the Iapetus Ocean.

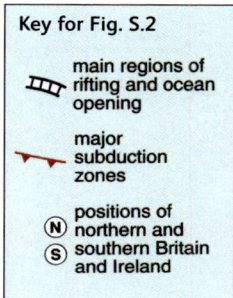

Key for Fig. S.2

- main regions of rifting and ocean opening
- major subduction zones
- (N) (S) positions of northern and southern Britain and Ireland

continent Rodinia (see Fig. S.2a), as well as other basement sources located along the eastern margin of Laurentia. A lower limit for Moine sedimentation is provided by ages of c.870 Ma for igneous rocks that intrude the supergroup (see below). Moine sedimentation is therefore constrained to the period between about 1000 and 870 Ma. The general consensus is that the Moine sedimentary basin was located within the Rodinia supercontinent, near to the junction between three major continental blocks, Laurentia, Baltica and Amazonia (Fig. S.2a; Dalziel & Soper, 2001; Friend et al., 2003; Cawood et al., 2004; see, however, Cawood et al., 2010). The Moine rocks may have been deposited in an aborted zone of crustal extension and rifting that developed along the eastern margin of Laurentia at the same time as various crustal blocks separated from west Laurentia to form the Pacific Ocean (Dalziel & Soper, 2001).

Regional framework

The regional framework that has been developed arises in part from the recognition of regional ductile thrusts, principally the Sgurr Beag and Naver thrusts (Fig. S.1). This enabled the Moine to be considered in terms of a series of thrust nappes, each with a distinctive lithostratigraphy and structural sequence (Tanner et al., 1970; Barr et al. 1986; Holdsworth et al., 1994). In addition, refinement of local Moine successions within thrust nappes resulted in part from the identification of inliers of Archaean orthogneisses that are thought to represent fragments of the basement upon which the Moine sediments were deposited (e.g. Flett, 1905; Peach et al., 1907, 1910; Read, 1931; Ramsay, 1958; Holdsworth, 1989a).

The basement inliers consistently lie at the lowest structural levels in local successions where the effects of thrusting and/or folding are removed (e.g. Richey & Kennedy, 1939; Ramsay & Spring, 1962; Holdsworth, 1989a). In the Glenelg and Fannich areas, as well as Sutherland, many basement inliers lie in the cores of isoclinal folds (Excursions 6, 9, 10 and 13), whereas others are carried as allochthonous slices along Caledonian ductile thrusts, notably the Sgurr Beag Thrust (Excursions 5 and 8). Marbles and pelites within some inliers (e.g. Loch Shin inlier, Excursion 10) appear to be integral parts of the basement rather than infolds or tectonic slices of Moine lithologies. Although ductile deformation has mainly obliterated any sedimentary or structural discordance across the basement-Moine contacts, the present consensus is that the relationship is one of basement and cover (e.g. Peach *et al*., 1910; Ramsay, 1958; Holdsworth, 1989a). Correlation of the basement inliers with the Lewisian Gneiss Complex of the Caledonian foreland has been generally accepted on the basis of lithological and geochemical similarities (e.g. Ramsay, 1958; Winchester & Lambert, 1970; Rathbone & Harris, 1979; Moorhouse & Moorhouse, 1988; Strachan & Holdsworth, 1988). U-Pb zircon dating of some basement inliers has yielded late Archaean protolith ages similar to those of the Lewisian Gneiss Complex (Friend *et al*., 2008). Sm-Nd and U-Pb mineral ages of c.1100-1000 Ma obtained from the eclogite-bearing eastern Glenelg inlier (Excursion 7; Sanders *et al*., 1984; Brewer *et al*., 2003) imply that at least some of the basement inliers were reworked during the Grenville orogeny.

Tectonostratigraphy of the Moine Supergroup

The Moine rocks of West Inverness-shire comprise three lithostratigraphic units – the Morar, Glenfinnan and Loch Eil groups (Fig. S.1; Holdsworth *et al*., 1987, 1994; Roberts *et al*., 1987). Although the Morar and Glenfinnan groups are thought on the mainland to be separated by the Sgurr Beag Thrust, stratigraphic passage between the two has been proposed on the Ross of Mull (Holdsworth *et al*., 1987, Excursion 1). The boundary between the Glenfinnan and Loch Eil groups is also transitional (Roberts & Harris, 1983; Roberts *et al*., 1984, Excursion 4). The Morar Group stratigraphy in its type area is dominated by a tripartite psammite-pelite-psammite succession that is up to 5km thick (Fig. S.3; Richey & Kennedy, 1939; Ramsay & Spring, 1962; Johnstone *et al*., 1969; Brown *et al*. 1970). A discontinuous basal pelite comprises a tectonic melange of Moine semi-pelite and retrogressed

Summary of Geology

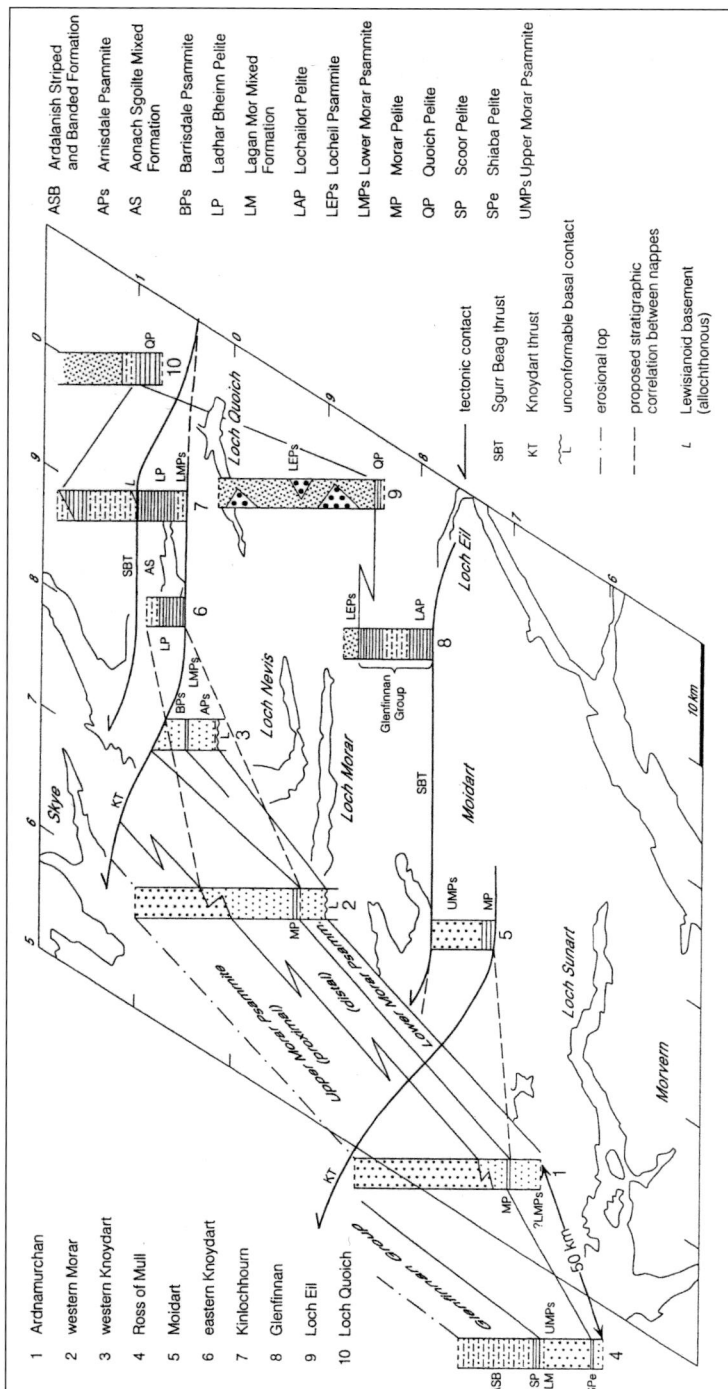

Fig. S.3 A fence diagram, viewed from the south, of different stratigraphic columns within the southern Moine (from Soper et al. 1998). Vertical exaggeration x2.5. Dots represent psammitic lithologies of varying proximality, but coarse dots in column 9 signify quartzite. Close fine dots are Loch Eil Group psammites. Lines represent pelite or semi-pelite (undistinguished). Dot-dash ornament represents striped and banded pelitic, semi-pelitic and psammitic units.

gneisses derived from the underlying basement that occupies the core of an early isoclinal fold. The Knoydart Thrust (Fig. S.3) is the only structure that disrupts the sequence significantly, although a common succession is recognized in its foot-wall and hanging-wall. The thick psammites within the Morar Group are commonly weakly strained and therefore preserve sedimentary structures (Excursions 1 and 3). The Glenfinnan Group mainly comprises striped units of thinly interbanded psammites, semi-pelites, quartzites and pelites, together with thick pelitic formations (Fig. S.3; Excursion 3). Tectonic strain is commonly high and sedimentary structures are therefore rare. Estimates of original thickness vary from 1-4km (Holdsworth *et al.*, 1994). Allochthonous slices of basement present along the trace of the Sgurr Beag Thrust north of Loch Hourn are assumed to lie at the stratigraphical base of the Glenfinnan Group. The Loch Eil Group (Excursions 2 and 5) is a monotonous sequence of psammites, although local quartzite and striped formations are recognized in the type area (Fig. S.3; Stoker, 1983; Strachan, 1985). Sedimentary structures are locally common and the succession may be up to 5km thick. Outliers of the Loch Eil Group occur as synformal infolds within the Glenfinnan Group, and migmatitic gneisses adjacent to the Great Glen Fault probably represent upfolds of the Glenfinnan Group (Fig. S.1).

Following recognition that the Sgurr Beag Thrust can be traced at least as far north as the Dornoch Firth (Fig. S.1; Wilson & Shepherd, 1979; Kelley & Powell, 1985; Strachan & Holdsworth, 1988, Excursion 10), the type Moine succession has been extended into Ross-shire. Further north, there seems little doubt that the psammite-dominated succession of western Sutherland equates with the Morar Group (Fig. S.1; Holdsworth *et al.*, 1994). Further east, however, correlations are less certain because the Sgurr Beag Thrust cannot be linked unambiguously with any of the major ductile thrusts recognised in central and east Sutherland (Friend *et al.*, 2003; Kocks *et al.*, 2006). The Sgurr Beag Thrust has been conventionally linked with the Naver/Swordly thrust system (e.g. Soper & Barber, 1982; Butler & Coward, 1984; Barr *et al.*, 1986; Strachan & Holdsworth, 1988). However, there are significant differences between the metamorphic history of the Loch Coire migmatites that occur above the Naver Thrust in Sutherland and the Glenfinnan Group rocks above the Sgurr Beag Thrust south of the Dornoch Firth. It is possible that the Skinsdale Thrust in SE Sutherland (Fig. S.1) is a more plausible correlative of the Sgurr Beag Thrust (Kocks *et al.*, 2006).

Sedimentological studies of parts of the Moine Supergroup are possible in areas of low strain. However, the interpretation of Precambrian sandstone-dominated sequences is difficult because they contain none of the fossils that might, for example, distinguish between marine and non-marine sequences. Glendinning (1988) interpreted the Upper Morar Psammite between the Ross of Mull and Mallaig as a predominantly tidal shelf deposit (Excursion 3). Complex sand waves, bipolar cross-bedding and gravel lag deposits were thought to compare closely with those found in modern shelf environments. A shallow marine environment is also indicated for the Loch Eil Group in its type area by bipolar cross-bedding, wave ripples and possible lenticular and flaser bedding (Strachan 1986). In contrast, recent analysis of the Morar Group psammites of west Sutherland suggests that they represent fluvial deposits that may correlate with the Applecross Formation of the Torridon Group on the Caledonian foreland (Krabbendam et al., 2008). Similarly, the Upper Morar Psammite of Ardnamurchan has been reinterpreted as an alluvial braidplain deposit (Bonsor & Prave, 2008). Further work is clearly necessary to resolve the depositional environments and basin evolution of the Moine Supergroup. It seems likely that the great thickness of Moine sediments must have accumulated in a basin formed by crustal extension, and both localized rifts (Soper et al., 1998) and larger-scale basins (Cawood et al., 2004) have been proposed.

Regional metamorphism

Metamorphic grade within the Moine is often difficult to establish because of the aluminium-poor nature of pelitic rocks that has inhibited the formation of Barrovian index minerals. Metamorphic grade has therefore been in part defined in terms of mineral assemblages in calc-silicates that have been correlated with Barrovian metamorphic facies (Kennedy, 1949; Johnstone et al., 1969; Winchester, 1974; Tanner, 1976; Powell et al., 1981; Fettes et al., 1985). Metamorphic grade within the Morar Group increases rapidly from the greenschist facies in the west, through the epidote-amphibolite facies and into a broad belt of low amphibolite facies metamorphism where rare kyanite appears in pelites and calc-silicates show hornblende ± plagioclase assemblages. A central area of high-grade rocks occupies a narrow NNE-trending belt, broadly corresponding to the outcrops of the migmatitic rocks of the Glenfinnan Group and the Loch Coire migmatites in Sutherland. These contain hornblende ± pyroxene ± bytownite assemblages in calc-silicates. The western margin of the high-grade belt is broadly coincident with the Sgurr Beag and Naver thrusts, consistent with field evidence that migmatization preceded ductile displacements along both structures (Powell *et al.*, 1981; Strachan & Holdsworth, 1988). The eastward decrease in grade into the low amphibolite facies of the Loch Eil Group is probably the result of the late folding of gently-dipping isograds into a broad regional synform, because high-grade Glenfinnan-type migmatites re-emerge locally adjacent to the Great Glen. The apparent simplicity of the regional metamorphic zonation is, however, illusory since the implication of isotopic studies is that it is composite and polymetamorphic (see below).

Early (*c.*870 Ma) igneous activity in the Moine Supergroup

The West Highland Granitic Gneiss (Johnstone 1975) is a series of separate bodies that mainly outcrop close to the boundary between the Glenfinnan and Loch Eil groups between Strontian and Glen Doe (Fig. S.1; Excursions 2, 4 and 5). Other bodies occur to the east within the Loch Eil Group. Barr *et al.* (1985) interpreted the granite gneisses as magmatic intrusions that were formed by anatexis of Moine rocks at a deeper structural level. Dating of zircons has shown that the granitic protolith of the Ardgour body, and its enclosed segregation pegmatites, formed at 873 ± 7 Ma (Friend *et al.*, 1997). A similar age of 870 ± 30 Ma has been obtained for the igneous protolith of the Fort Augustus granitic gneiss (Rogers *et al.*, 2001).

Sill-like metabasic bodies are common within the Glenfinnan and Loch Eil groups but rare in the Morar Group except in west Sutherland (e.g. Moorhouse & Moorhouse, 1979; Smith, 1979; Roberts & Harris, 1983; Winchester & Floyd, 1983; Winchester, 1984; Rock *et al.*, 1985; Strachan, 1985; Holdsworth, 1989a; Millar, 1999). They also cut members of the West Highland Granite Gneiss (Excursions 2 and 5). Most are foliated amphibolites or hornblende schists, although metagabbros with relict igneous textures are present locally. The metabasic intrusions display a tholeiitic chemistry comparable with modern mid-ocean ridge basalts. Although it is clear that the metabasic rocks were not emplaced within an oceanic setting *sensu stricto*, since the host Moine rocks were apparently deposited on Archaean basement, the chemistry is also consistent with intrusion into continental crust that had been thinned during extension. It therefore seems likely that these early metabasic bodies were intruded during crustal extension and development of the Moine sedimentary basin(s). A U-Pb zircon age of 873 ± 6 Ma obtained from a metagabbro at Glen Doe (Millar, 1999) is thought to date its magmatic crystallization and it is assumed that the rest of the suite is of broadly the same age.

Barr *et al.* (1985) argued that the West Highland Granite Gneiss was syn-orogenic and formed during regional migmatization and D_1 isoclinal folding of the Moine rocks. In contrast, Soper & Harris (1997). Millar (1999) and Dalziel & Soper (2001) have suggested that the granitic protolith of the gneiss was formed during crustal extension, formation of the Moine sedimentary basin and emplacement of the regional metabasic suite. Ryan & Soper (2001) envisage that the metabasic intrusions provided sufficient heat to locally melt both the underlying basement and Moine sediments to produce granitic melts that migrated up through the sedimentary pile to their present location. However, in the absence of reliable pressure-temperature data to constrain the conditions of melting, the origin of the granitic protoliths, the age of their gneissification, and hence the nature of the *c.*870 Ma tectonothermal event remain equivocal.

Evidence for Neoproterozoic Knoydartian orogenic activity at *c.*820-730 Ma

Rather firmer evidence exists for younger orogenic events in the period *c.*820-725 Ma. The first indications that the Moine rocks were metamorphosed during the Precambrian were provided by the Rb-Sr dating of muscovites from deformed pegmatites within the Morar Group (Giletti *et al.*, 1961). Ages of 750-690 Ma were interpreted as the likely age of pegmatite segregation during an early high-grade metamorphic event that was later termed the Knoydartian (Bowes, 1968) or Morarian (Lambert 1969) orogeny. Further isotopic dating of these pegmatites and others has yielded Rb-Sr muscovite and U-Pb zircon and monazite ages of *c.*830-730 Ma (van Breemen *et al.*, 1974, 1978; Piasecki & van Breemen, 1983; Powell *et al.*, 1983; Piasecki, 1984; Rogers *et al.*, 1998). The Loch Eilt pegmatite (Excursion 3) is a classic example of one of these deformed early intrusions. Much debate has focused on the tectonic significance of these pegmatites. Their field relations with host Moine rocks are commonly difficult to evaluate because of the high degree of superimposed Caledonian strain and metamorphic recrystallization (Powell *et al.*, 1983). An alternative interpretation is that the pegmatites are entirely pre-tectonic and were produced during crustal extension and episodic melting of the Moine sedimentary pile (Soper & Anderton, 1984; Soper & Harris, 1997; Dalziel & Soper, 2001), thus challenging the very existence of a Precambrian orogenic event.

Recent studies that have linked modern geochronological techniques and pressure-temperature data have provided firmer evidence for Neoproterozoic orogenesis. Sm-Nd ages of $c.820$-790 Ma obtained from post-D_1 garnets in the Morar Group date early prograde metamorphism during crustal thickening (Vance *et al.*, 1998; Cutts *et al.*, 2009). In Inverness-shire, this event is thought to have been associated with nappe-scale interleaving of Moine rocks and the basement rocks of Glenelg and Morar (Excursions 3 and 6; Ramsay, 1958; Powell, 1974). A U-Pb age of 737 ± 5 Ma has been obtained from prograde titanites that developed after initial displacement along the Sgurr Beag Thrust in the Loch Eilt area of west Inverness-shire (Excursion 3; Tanner & Evans, 2003). Similar U-Pb ages of c.730-725 Ma are recorded by zircons and monazites that formed during high-grade metamorphism of the Moine rocks in Glen Urquhart (Fig. S.1; Cutts *et al.*, 2010). Hyslop (1992) has confirmed that most of the early pegmatites formed *in situ* in zones of high strain and melt generation during metamorphism at garnet grade and higher. Further complexity is provided by Storey *et al.* (2004) who have obtained a U-Pb age of 669 ± 31 Ma from syn-kinematic titanites within a contractional shear zone in the Glenelg area (Excursion 7). It therefore seems possible that the Moine rocks were affected by a number of orogenic events in the mid-Neoproterozoic. The term 'Knoydartian' is probably best employed with reference to this overall period of orogenic activity rather than to any individual phase.

The present consensus is that the earliest prograde metamorphic events and associated foliations and isoclinal folds within the Moine are likely to be of Precambrian age. The tectonic setting of Knoydartian orogenic events is presently uncertain. The lack of any Neoproterozoic calc-alkaline igneous rocks within the Moine means that it is unlikely that orogenic activity occurred near to an active plate margin. Neoproterozoic plate reconstructions mostly depict the continents in close proximity (Fig. S.2a) and there is little scope in the segment of Rodinia where Scotland is thought to have been located for the opening and closure of large ocean basins. Cawood *et al.* (2004) have suggested that Knoydartian orogenic activity resulted from the episodic closure of a Moine intracratonic basin, perhaps driven by far-field stresses arising from terrane accretion on the periphery of Rodinia.

Late Neoproterozoic magmatism

The Moine rocks of Ross-shire and East Sutherland were intruded by granites during the late Neoproterozoic. These include the Carn Chuinneag-Inchbae granite within the Morar Group of Ross-shire (Fig. S.1) and minor granite sheets within the East Sutherland Moine. These have yielded U-Pb zircon ages of 594 ± 11 Ma (Inchbae granite, Oliver *et al.*, 2008) and 599 ± 9 Ma and 588 ± 8 Ma (East Sutherland granites, Kinny *et al.*, 2003a). The preservation within the contact aureole of the Carn Chuinneag granite of delicate sedimentary structures appears to rule out a pervasive pre-granite deformation in the Morar Group country rocks (Peach *et al.*, 1912; Soper & Harris, 1997). The Carn Chuinneag-Inchbae granite is thus presumed to lie within an area of low Knoydartian strain. Contemporaneous magmatism in the Dalradian Supergroup, and in the Appalachians and the Norwegian Caledonides, has been attributed to the break-up of Rodinia and development of the Iapetus Ocean (Figs S.2a and b); Bingen *et al.*, 1998). In this context, the late Neoproterozoic granites in Northern Scotland probably resulted from processes related to continental rifting (Kinny *et al.*, 2003a).

Ordovician (Grampian) structures and metamorphism

Following the breakup of Rodinia in the late Neoproterozoic, the Iapetus Ocean widened through the Cambrian and into the early Ordovician (Figs. S.2b and c; Cocks & Torsvik, 2002). The Moine is thought to have been located on the margin of Laurentia and during the Cambrian and early Ordovician was probably overlain unconformably by shelf sediments that passed southeastwards into deep marine turbidite basins of the Upper Dalradian (Anderton, 1985). Sedimentation was halted in the early to mid-Ordovician by the Grampian orogenic event (Lambert & McKerrow, 1976; Soper *et al.*, 1999). A possible model for the Grampian orogenic event in Scotland and Ireland involves the collision of the Laurentian continental margin with an intra-oceanic subduction zone and a volcanic arc that developed during closure of Iapetus (Figs. S.2c and S.4). This is thought to have resulted in overthrusting of the Laurentian margin by an exotic ophiolite nappe and regional deformation and metamorphism of the Dalradian and Moine rocks (Dewey & Shackleton, 1984; Dewey & Ryan, 1990). Remnants of this nappe may be represented by the ophiolitic rocks exposed on the island of Unst in Shetland and intermittently along the Highland Boundary Fault.

Summary of Geology

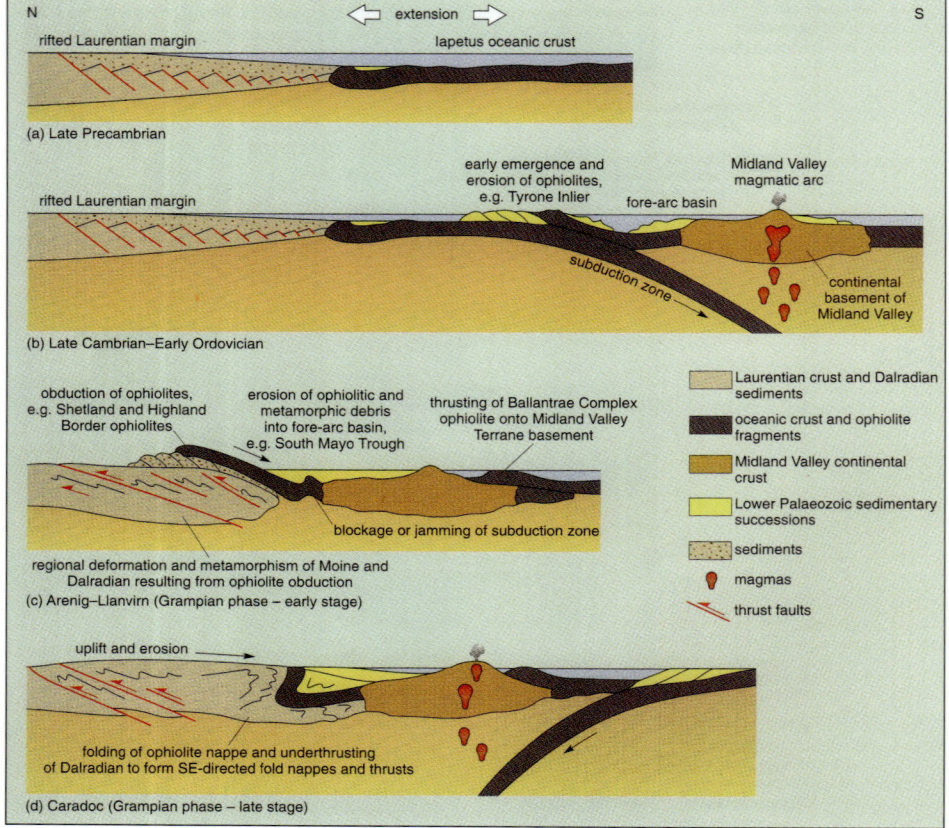

Fig. S.4 Tectonic model for Grampian phase of the Caledonian orogeny (from Open University 2003, with acknowledgement to Woodcock & Strachan, 2000).

(a) A pre-Grampian stage showing the opening of the Iapetus Ocean and sedimentation of the Dalradian Supergroup on the rifted Laurentian crust.

(b) S-directed subduction formed an island arc on continental crust.

(c) Initial collision led to narrowing of ocean tracts, the obduction of ophiolites, and widespread deformation of the Moine and Dalradian supergroups on the Laurentian margin.

(d) Continued collision resulted in folding and underthrusting of the Laurentian margin, and a N-dipping subduction zone.

Various lines of evidence indicate that the eastern Moine was affected by Grampian deformation and metamorphism. In Sutherland, formation of the Loch Coire migmatite complex and its associated N-S-trending lineations and isoclinal folds has been dated at c.470-460 Ma (U-Pb zircon; Kinny et al., 1999; Kocks et al., 2006). Relict garnet-pyroxene assemblages preserved within metabasic sheets in the Naver Nappe are thought to result from the same high-grade Grampian metamorphic event (Friend *et al.*, 2000, Excursion 13). In Inverness-shire U-Pb titanite and monazite ages of *c*.470 Ma record Grampian metamorphism in this part of the Moine (Rogers *et al.*, 2001; Cutts *et al.*, 2010). Recumbent, tight to isoclinal folds that are curvilinear about a N-S mineral lineation are widespread in the Glenfinnan and Loch Eil groups and probably represent the effects of the Grampian event in the eastern Moine (Rogers *et al.*, 2001, Excursion 4). These folds predate intrusion of the Glen Dessary syenite at 456 ± 5 Ma (van Breemen *et al.*, 1979a; Roberts *et al.*, 1984). U-Pb monazite ages of 455 ± 3 Ma obtained from the Ardgour Granitic Gneiss and its host Moine psammites at Glenfinnan provide additional evidence for Grampian metamorphism (Aftalion & van Breemen, 1980). A major concentration of variably deformed pegmatites within the Glenfinnan Group in Inverness-shire and Ross-shire is also arguably late Grampian in age as two members of the suite have yielded Rb-Sr and U-Pb mineral ages of *c*.455-445 Ma (van Breemen *et al.* 1974). As yet, there is no isotopic evidence that the Morar Group was affected by Grampian metamorphism. One solution to this conundrum is that a western 'front' to Grampian orogenic activity is buried beneath younger ductile thrusts (Fig. S.5; Dallmeyer *et al.*, 2001).

Silurian (Scandian) deformation and metamorphism

It is believed that following the Grampian orogenic event, continued closure of the Iapetus Ocean was achieved by a reversal in the polarity of subduction and the development of the Southern Uplands accretionary prism (Fig. S.4; Dewey & Ryan, 1990). The final orogenic events in the Scottish Highlands are the result of the oblique collision in the Silurian of three continental blocks, Laurentia, Baltic and Avalonia (Figs S.2d-f; Soper & Hutton, 1984; Pickering *et al.*, 1988; Soper *et al.*, 1992). Baltica is thought to have collided with the segment of the Laurentian margin that incorporated the Northern Highlands, to result in the Scandian orogenic event (Fig. S.5; Coward, 1990; Dewey & Mange, 1999; Dallmeyer *et al.*, 2001; Dewey & Strachan, 2003; Kinny *et al.*, 2003b). This resulted in regionally significant metamorphism and ductile thrusting and folding of the Moine, culminating in development of the Moine Thrust Zone.

The best place to study the early stages of Scandian thrust-related deformation is Sutherland where the effects of later upright folding are minimal

(Excursions 10 and 13). Within the Morar Group and lowermost parts of the Loch Coire migmatites, widespread tight to isoclinal folding of the Moine accompanied NW-directed ductile displacements along the Swordly, Naver and Ben Hope thrusts (Strachan & Holdsworth, 1988; Holdsworth, 1989a; Holdsworth *et al.*, 2001a). The syn-kinematic growth of garnet and hornblende shows that deformation occurred within the amphibolite facies (Strachan & Holdsworth, 1988; Holdsworth, 1989a; Holdsworth *et al.*, 2001a). Above the Swordly Thrust, the intensity of Scandian deformation dies out and reworking of the Loch Coire migmatites is apparently restricted to the Skinsdale Thrust (Kocks *et al.*, 2006). Syn-kinematic granite sheets in the vicinity of the Naver Thrust have yielded U-Pb zircon ages of $c.435$-420 Ma (Kinny *et al.*, 2003b; Excursion 10). Further east, late stages of displacement along the Skinsdale Thrust were accompanied by intrusion of the Strath Halladale Granite at 426 ± 2 Ma (U-Pb monazite, Kocks *et al.*, 2006). This phase of thrust-related deformation was responsible for the major interleaving of Moine rocks with basement gneisses in Sutherland. Within the Morar Group, many inliers occupy the cores of major sheath folds; thin allochthonous slices of highly strained basement lie along the Ben Hope, Naver and Swordly thrusts (Holdsworth, 1989a). The folding of ductile thrusts by folds developed in their footwalls demonstrates that thrust-related deformation propagated towards the foreland.

Extensive tracts of the Morar Group in Ross-shire as far south as Loch Duich are dominated by NW-trending lineations and associated tight to isoclinal folds that are geometrically and kinematically identical to those described above from west Sutherland (Kinny *et al.*, 2003b). There may have been significant displacement along the Sgurr Beag Thrust during the Scandian event. The total displacement along the thrust is unknown, but likely to be at least tens of kilometres and conceivably > 100km (Powell *et al.*, 1981). However, a wholly Caledonian age for the Sgurr Beag Thrust has been questioned by Tanner & Evans (2003) who argue that in west Inverness-shire it is fundamentally a Knoydartian structure.

Subsequent tight upright folding along NNE-trending axes resulted in the formation of the Northern Highland Steep Belt (Excursions 3, 4 and 5). Outliers of the Loch Eil Group occur within the steep belt along the axial trace of a major curvilinear synform (Roberts *et al.*, 1984, 1987). These folds gradually become less intense northwards and in Ross-shire they deform NW-trending thrust-related lineations that are correlated with the Scandian structures identified in west Sutherland (Kinny *et al.*, 2003b). This implies that the steep belt folds are most probably of Scandian age.

Moine Thrust Zone

The Moine Thrust Zone is the westernmost and youngest of the system of Scandian thrusts on the mainland of Northern Scotland. The thrust zone at Loch Eriboll and Durness (Excursions 11 and 12) is historically important ground in structural geology as it is here the existence of large-scale thrusts was first demonstrated by Lapworth (1883, 1885). In a classic publication, Peach *et al.* (1907) recognized that the Moine rocks had been displaced along the Moine Thrust at a low angle to the WNW across the Hebridean foreland. Between the Moine Thrust and the undeformed foreland lay a 'belt of complication', up to 11km-wide, within which cover and basement rocks were interleaved by folding and thrusting: this is the Moine Thrust Zone. Peach *et al.* (1907) recognized major, low-angle thrusts, that delimited nappes within the thrust belt, and also families of small-scale thrusts or reverse faults that repeated the stratigraphy many times over in a series of imbricate slices. These faults generally rooted down onto one of the major thrusts.

Structurally above the Moine Thrust is an extensive belt of mylonites, best exposed at Loch Eriboll and Durness (Excursions 11 and 12). These formed from the intense ductile shear and recrystallization of Moine rocks, associated slices of Lewisian(oid) basement, and Cambrian quartzites at temperatures of $c.350$-$400°C$ (Holdsworth *et al.*, 2007). In the Morar Group of Sutherland, the regional Scandian lineation is parallel with the main mineral and stretching lineation within the mylonite belt (Soper & Brown, 1971; Soper & Wilkinson, 1975; Holdsworth *et al.*, 2006, 2007). This suggests that the internal ductile thrusting and folding of the Morar Group is linked kinematically with development of the mylonite belt during the same Scandian event. Within the underlying Moine Thrust Zone, slices of Lewisian, Torridonian and Cambrian units are limited by sharp, brittle thrusts that lack much mylonite. This part of the thrust zone clearly developed at higher crustal levels than the mylonite belt, and the Cambrian-Ordovician rocks record peak temperatures of only about $275°C$ (Johnson *et al.*, 1985). Notable examples of major overturned and recumbent folds occur within the thrust zone in the Assynt region and between Loch Carron and Skye. Structural analysis has shown that the thrusts generally developed in a foreland-propagating sequence, with successively younger and lower thrusts transporting older and higher thrusts to the WNW in 'piggyback' fashion (Elliott & Johnson, 1980; McClay & Coward, 1981; Butler, 1982, 2009). This is complicated in some areas by later, low-angle 'out-of-sequence' faults that cut through previously thrust and folded strata. Such

Summary of Geology

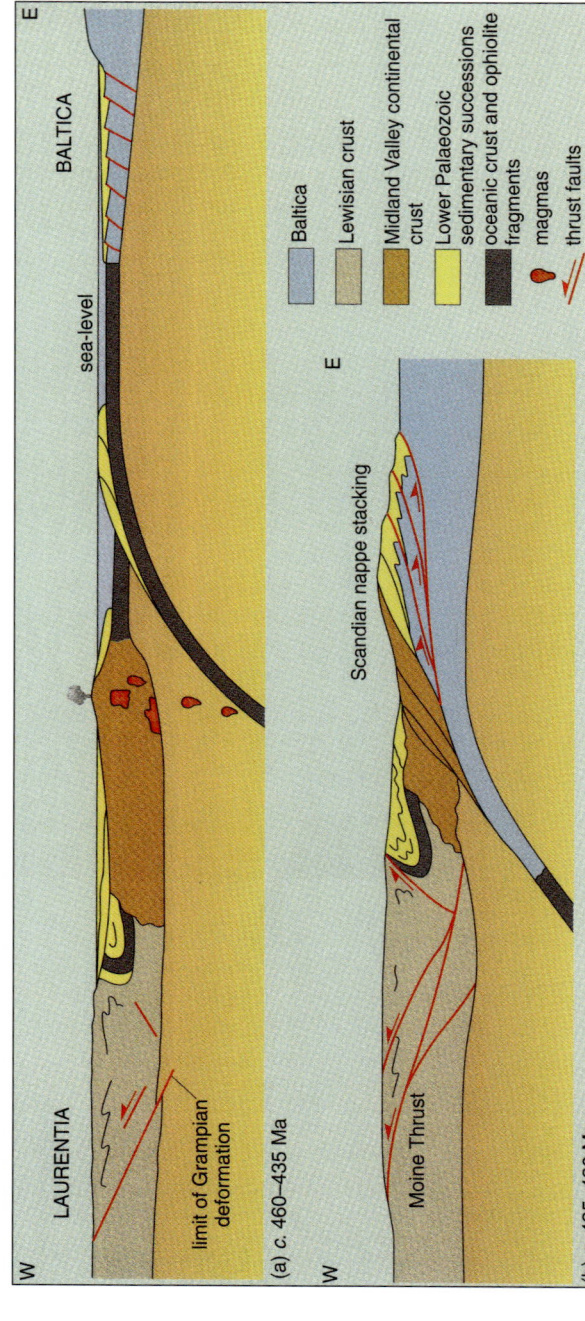

Fig. S.5 A simplified tectonic model for the Scandian phase of the Caledonian orogeny at c.435–424 Ma in Northern Scotland (from Open University, 2003).

(a) Development of an active margin in post-Grampian times by westwards subduction of the northern arm of Iapetus beneath Laurentia.

(b) Underthrusting of Baltica beneath Laurentia and the development of both E- and W-directed thrust systems. In the Northern Scotland sector of the collision zone, this culminated in the Moine Thrust. This was followed by strike-slip displacement along the Great Glen Fault (omitted here for clarity) that brought the Northern Highlands to its present position relative to the Grampian Highlands which was unaffected by Scandian phase deformation and metamorphism.

structures may either be late thrusts or extensional faults that developed due to gravitational instability of the evolving thrust zone (Coward, 1985; Holdsworth et al., 2006).

Isotopic dating of recrystallized micas within Moine mylonites suggests that thrusting occurred at c.435-430 Ma (Johnson et al., 1985; Kelley, 1988; Freeman et al., 1998; Dallmeyer et al., 2001). This is consistent with the U-Pb zircon age of 430 ± 4 Ma obtained from the syn-tectonic Loch Borralan syenite complex within the thrust zone in the Assynt area (van Breemen et al., 1979b). Isotopic ages as young as c.408 Ma obtained from some mylonites suggest that at least locally thrusting continued into the Early Devonian (Freeman et al., 1998). Minimum displacements across the thrust zone are c.50-80km (Elliott & Johnson, 1980; Butler, 1982; Butler & Coward, 1984). It is difficult to assess the displacement on the Moine Thrust itself, but its association with a thick mylonite belt suggests a minimum offset of several tens of kilometres. A total minimum displacement for the Moine Thrust Zone of c.100km is therefore likely.

Deep seismic reflection profiling carried out in the Pentland Firth was designed to determine the sub-surface profile of the Moine and Outer Isles thrusts (hence the acronymn MOIST). Brewer & Smythe (1984) identified several mid-crustal, east-dipping reflectors, either of which could represent the down-dip continuation of the Moine Thrust. It is difficult, however, to reconcile either solution with Butler & Coward's (1984) interpretation that the Cambrian-Ordovician foreland succession originally extended c.54 km to the ESE. This implies that the Moine Thrust must follow a shallow trajectory (c.3°) within the upper crust over this distance before any steep ramp occurs. The geometry and deep structure of the margin of the Caledonian orogen is thus still problematical.

Late Caledonian strike-slip faulting and plutonism

The main phase of Scandian ductile thrusting and folding was followed by sinistral strike-slip displacements along the Great Glen Fault and associated structures (Fig. S.1). The development of these faults most likely occurred during the terminal stages of the oblique collision between Laurentia, Baltica and Avalonia in the late Silurian to Early Devonian (Fig. S.2f; Soper et al., 1992; Dewey & Strachan, 2003). The final stages of the Caledonian orogeny in Scotland were also marked by a major phase of subduction-related plutonism, to form the 'Newer Granite' suite (e.g. Read, 1961; Stephens &

Halliday, 1984; Watson, 1984; Soper, 1986; Thirlwall, 1988; Stephenson *et al.*, 1999; Oliver, 2001; Atherton & Ghani, 2002). The emplacement mechanisms of many intrusions were structurally controlled. As indicated above, some granites were emplaced during Scandian thrusting. However, the main phase of plutonism accompanied displacements along strike-slip faults that appear to have acted as ascent pathways for magmas (Jacques & Reavy, 1994).

The Great Glen Fault (e.g. Kennedy, 1946; Smith & Watson, 1983) has been linked with the Walls Boundary Fault in Shetland (Flinn, 1961; McBride, 1994) and to the southwest with the Loch Gruinart-Leannan Fault in Islay and Ireland (Pitcher *et al.*, 1964; Alsop, 1992). Seismic reflection studies show that it is coincident with a subvertical structure that extends to at least 40km depth (Hall *et al.*, 1984). Mantle-derived, late Caledonian lamprophyre dykes appear to have different isotopic signatures either side of the fault, suggesting that it has some expression in the upper mantle (Canning *et al.*, 1996, 1998). On the Scottish mainland, the Great Glen Fault comprises a *c.*3km-wide belt of fracturing and intense cataclasis of Moine and Dalradian protoliths (Stewart *et al.*, 1997, 1999, 2000; Excursion 14). Kinematic indicators demonstrate a consistent sinistral sense of displacement with a minor southeasterly component of downthrow. Related minor faults in the Northern Highlands include the Strathconon Fault, and possibly also the Strath Glass and Helmsdale faults, all of which trend NE, subparallel to the Great Glen Fault. A subordinate set of NW-trending faults, such as the Strath Fleet Fault, may represent anti-Reidel shears to the main Great Glen fault system.

Caledonian sinistral displacements along the Great Glen fault system probably occurred between *c.*430 Ma and *c.*400-390 Ma (Stewart *et al.*, 1999). Evidence for Silurian displacement is indicated by the U-Pb zircon ages of structurally controlled plutons located along or adjacent to major faults. These include the Clunes Tonalite (428 ± 2 Ma, Stewart *et al.*, 2001), the Strontian Granite (425 ± 3 Ma, Rogers & Dunning, 1991) and the Ratagain pluton (425 ± 3 Ma, Rogers & Dunning, 1991; Hutton & McErlean, 1991). A lower age limit of *c.*400-390 Ma is indicated by the low strain nature of Old Red Sandstone (latest Emsian?) sedimentary rocks within the fault zone. These are relatively undeformed compared with the metamorphic basement, and the deformation fabrics described above clearly predate Old Red Sandstone deposition (Stewart *et al.*, 1999, see also Mykura, 1982; Stoker, 1982). The magnitude of early sinistral displacement along the Great Glen Fault has been controversial because there is no unambiguous correlation of pre-Devonian features across the fault. The general consensus has been that sinistral displacements are unlikely to have exceeded 200-300 km, consistent with the available palaeomagnetic evidence (Briden *et al.*, 1984). A rather larger displacement of *c.*500-700km is implied, however, by tectonic reconstructions that place the Northern Highlands opposite Baltica during the Scandian collision (Dewey & Strachan, 2003; Kinny *et al.*, 2003b).

Excursion 1
Ross of Mull

Tony Harris

Purpose: A general excursion across the Moine rocks of the Ross of Mull.

Aspects covered: Metasedimentary lithologies typical of the Morar and Glenfinnan groups, amphibolites, polyphase folds and fabrics, the margin of the (Caledonian) Ross of Mull Granite, regional and contact metamorphism, Tertiary minor intrusions.

Useful addresses: Hotel and B&B accommodation are available at Bunessan and Fionnphort, and also at Scoor House SW of Loch Assapol.

Maps: OS: Explorer 1:25,000 Sheet 373, Iona, Staffa and Ross of Mull; BGS: 1:50,000 sheet 43S Ross of Mull.

Type of terrain: Rocky coastline (some scrambling required), beaches and hillside.

Short itineraries: East limb of Assapol Synform from Scoor House: Localities 1.1, 1.3, 1.6 (+/-1.13), 1.10, 1.8, 1.9 and 1.14; distance is 5km, taking a whole day. West limb of Assapol Synform from Ardalanish: Localities 1.16, 1.17, 1.19, 1.20, 1.23; distance is 4km, taking 3-4 hours. Granite relationships from Bunessan and Ardalanish: Localities 1.15, 1.24 and 1.25; distance is 4km, taking 3-4 hours. Whole area (very abbreviated) starting from Scoor House: Localities 1.13, 1.10, 1.9, 1.14, 1.19 or 1.20, and 1.23; distance is 7km, taking a whole day.

The Moine rocks of the Ross of Mull (Figs 1.1, 1.2) constitute the south-westernmost occurrence of the Supergroup. The inlier contains evidence that is critical in assessing the stratigraphic relationships between the Morar and Glenfinnan groups which are, respectively, thought to be laterally equivalent to the Shiaba and Assapol groups of Mull (Holdsworth *et al.*, 1987). The Moine outcrop is terminated to the west by the Ross of Mull Granite which was intruded towards the end of the Caledonian orogeny at 421 ± 5 Ma (Oliver *et al.*, 2008). Lewisian basement and Iona Group cover rocks of the Caledonian foreland occur on the island of Iona directly west

EXCURSION 1

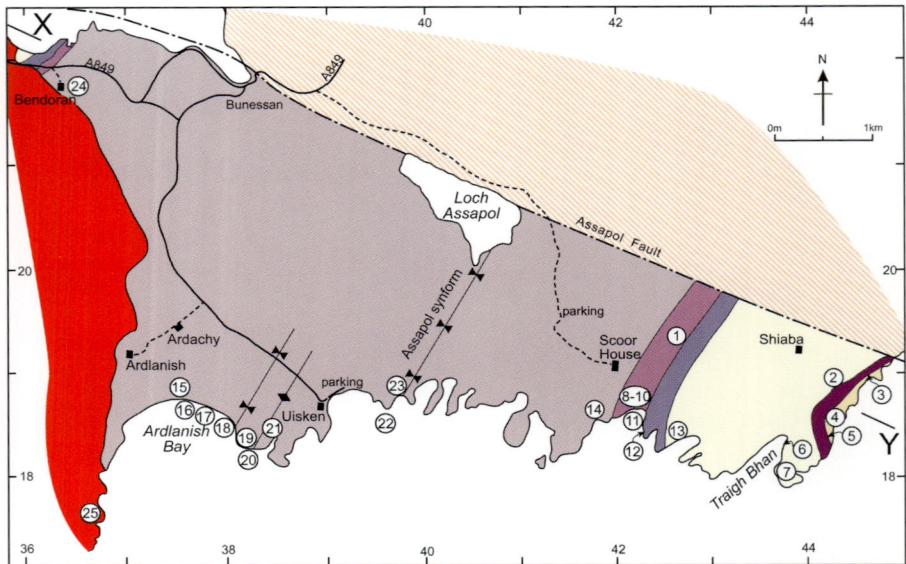

Fig. 1.1 Simplified geology map of the Ross of Mull, showing the localities described in the text.

of the Ross of Mull. The mid-amphibolite facies metamorphic grade of the Moine rocks is anomalously high for rocks adjacent to the Moine Thrust and the Caledonide foreland (Holdsworth *et al*., 1987). The former had been inferred to lie in the Sound of Iona, and Potts *et al*. (1996) concluded that it had been displaced by a major normal fault. This structure had brought Moine rocks of the Knoydart Nappe into contact with the rocks of the foreland before emplacement of the Ross of Mull Granite.

Almost all the localities lie on the southern coast of the Ross (Fig. 1.1). Access is most suited to cars and minibuses. Coaches should not be used without local advice. Itineraries A and B focus on Localities 1.1-1.13, and Itineraries C and D describe Localities 1.15-1.23 (Fig. 1.1). Itineraries A and B describe a traverse through the units of the Shiaba (= Morar) Group and the younger Assapol (= Glenfinnan) Group. Itineraries C and D traverse the Assapol Group on the western limb, and within the core, of the regional F_3 Assapol Synform (Fig. 1.2) and includes various lithologies not seen in itineraries A and B. Two additional localities (24 and 25) focus on relationships between the Ross of Mull Granite and the Moine rocks in its envelope.

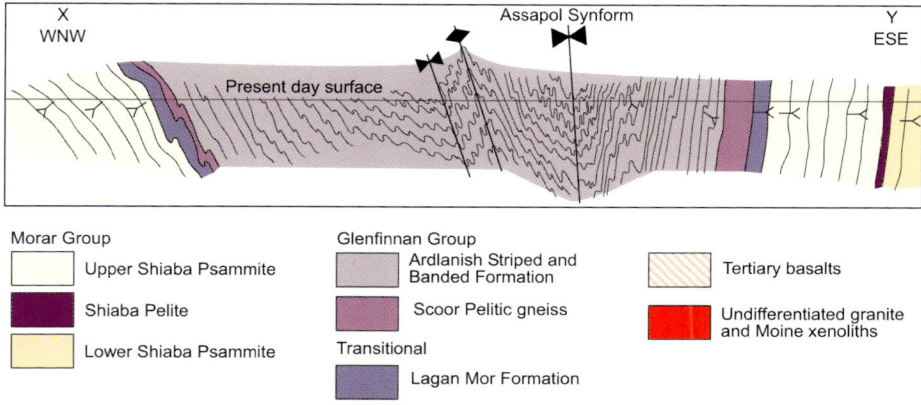

Fig. 1.2 NW-SE cross-section across the Moine rocks of the Ross of Mull (modified from Holdsworth *et al.*, 1987).

Itinerary A

Eastern limb of the Assapol Synform (1); access via Scoor House.
Total distance is *c.* 7 km from Scoor House and return, taking 6-8 hours.

Follow the A849 from Bunessan towards Pennyghael for about 0.5 km and turn right up the very minor road leading southeast towards Scoor House (signposted Scoor 3.5 km). The road initially tarmacked, becomes a partly metalled track which is liable to potholes. Public parking is available about 3 km from the main road beside the track (Fig. 1.1; [NM 419 196]), above a graveyard with a ruined church (parking for about eight cars). Coaches should not be used on these roads without local advice. Continue along the track on foot as far as Scoor House [NM 419 191]. Passing the cottage and farm steadings on your right immediately before you reach Scoor House, you will note the low cliff cut into a dolerite sheet. This is a very extensive and persistent intrusion which you will come across on many places on this excursion, because it overlies or underlies much of the eastern part of the Moine inlier.

Locality 1.1 [NM 4268 1931]

Viewpoint of the regional and local geology (Fig. 1.1).

Follow the track that leads up the hill from Scoor House for about 300m as far as a gate into a field on the north side of which there is a sheep fank. Cross the field to a gate on the far side leaving the fank to your left. Passing through this gate, leave the track that heads off to the ENE and instead, walk up to the nearby (~100m) smooth glaciated crag from the summit of which [NM 4268 1931] there is a fine view that enables a perspective of the local and regional geology. BGS Sheet 43S and the BGS 1:625,000 North Sheet would be useful to refer to here. Looking to the south and southeast to the other side of the Great Glen Fault, the course of which runs several kilometres offshore, can be seen the low-lying islands of Islay and Colonsay, the latter comprising the enigmatic Colonsay Group rocks of uncertain affinity. In contrast, the island of Jura is dominated by prominent pale mountains of Dalradian Islay Quartzite. Further east on a clear day can be made out the low-lying Garvellach islands underlain by the Dalradian Port Askaig tillites. Looking west over the end of the Ross of Mull, both Coll and Tiree can be seen, both comprising Archaean Lewisian gneisses of the Caledonian foreland. Closer, Iona, itself largely comprising Lewisian gneiss, and probably also part of the foreland, rises up beyond the rocky exposures of the Ross of Mull Granite. To the northwest, the Treshnish Islands and Staffa and to the north and east the trap topography of the Wilderness of Burgh and of the eastern part of the Ross itself, are all formed of Tertiary basalts. These are separated from the Caledonian rocks by the Assapol Fault, the course of which can be easily seen trending west-northwest along the far shore of Loch Assapol. Nearer to hand, the position of the eastern margin of the Ross of Mull Granite can be made out with the help of Sheet 43S where rising ground to the west becomes much more rocky than that to the east, underlain by the Moine.

Locality 1.2 [NM 4425 1893]

Upper Shiaba Psammite with cross-bedding and heavy mineral bands.

From Locality 1.1, proceed N40°E for about 400m along the side of the An Crosan ridge, staying above the extensive peaty area to the southeast which can be treacherously boggy. A rough track leads to Shiaba from approximately [NM 4302 1967] and should be followed east-southeast for about 800m. This track crosses some boggy ground and rocky knolls that expose Upper Shiaba Psammite that display poorly preserved cross-bedding that youngs and dips ~70° northwest. Crossing the shallow burn Allt Cnoc na Fearnaige, a well preserved though roofless croft house is conspicuous enough to serve as a landmark for positioning in an otherwise rather featureless landscape. Locality 1.2 lies some 500m southeast of the point where the track crosses the burn (Fig. 1.1). Here a number of crags comprise typical micaceous, laminated Upper Shiaba Psammite with cross-bedding and heavy mineral bands.

Locality 1.3   [NM 4452 1898]

Shiaba Group lithologies; Tertiary intrusions and faulting.

From Locality 1.2, walk south-southeast towards the shore via the slope just east of a gully and then via a rough scramble down a gully cut in steeply dipping Upper Shiaba Psammite starting at [NM 4425 1898] at about 35-40m OD down to about 20m OD. A line of somewhat rounded crags, trending east-northeast (055°) comes into view, on the other (south) side of an open hollow. These expose coarse garnetiferous Shiaba Pelite (e.g. [NM 4432 1888]). There are many exposures where this can be examined and searched for sparse occurrences of staurolite and kyanite, but it is recommended that the Shiaba Pelite is also studied at Locality 1.3 in a narrow fault gully where the contact with the Lower Shiaba Psammite can also be examined. Access is possible at all states of the tide, but several points of interest will be missed if the visit takes place above middle water.

Once on the foreshore, the high water mark (HWM) should be followed to Locality 1.3. Here, a few metres southwest of an old metal deer fence running across the Upper Shiaba Psammite on the foreshore, the gully that is Locality 1.3 trends 345° exposing garnetiferous Shiaba Pelite, locally containing granitic pegmatite, in both walls. There is a rapid transition of the steeply dipping Lower Shiaba Psammite contact into the Shiaba Pelite which is well displayed on the west-southwest side of the entrance to the gully and above and beyond the head of the gully. It can be demonstrated from cross-bedding in nearby exposures of the Upper Shiaba Psammite that it is younger than the Shiaba Pelite.

The whole width of the outcrop of Lower Shiaba Psammite is very well exposed to the south and west on the wave-cut platform and in seastacks for a distance of 600-700m. These exposures display well the coarse feldspathic, somewhat migmatitic nature of the Lower Shiaba Psammite, its heavy-mineral seams, lack of obvious clastic grains and of any obvious sedimentation structures; nevertheless, dubious cross-bedding in places does hint at younging towards the Shiaba Pelite. These contrasts with the Upper Shiaba Psammite are strikingly reminiscent of the distinctions between the Upper and Lower Morar psammites further north in the type localities.

Locality 1.3 is also an excellent place for examining the complex timing and displacement history of the fault within the gully:

(i) the contact between the psammite below and the pelite above, dipping 60° towards N30°W is displaced, apparently sinistrally, across the fault to appear in the east-northeast wall of the gully some 25m from the HWM;
(ii) a steep, apparently concordant dolerite dyke trending 060° across the line of the gully and a few metres from its mouth on the seaward side is neither brecciated nor displaced;
(iii) an irregular, flat-lying basaltic sheet cutting the pelite/psammite boundary, and seen high up in the west-southwest wall of the gully, has been displaced down to the east to appear in its east-southeast wall;
(iv) on the rocky foreshore on the line of the gully, there is brecciated basalt and psammite.

Thus, this is an excellent locality to discuss the Tertiary history of intrusion and faulting in the area, and it may well have some regional significance because the fault in the gully is probably a splay off the Tertiary Assapol Fault. The psammite/pelite contact can be followed along near the HWM, east-northeast from here for ~300m to the point where the Assapol Fault crosses the foreshore causing some brecciation of both pelite and basalt. From here and from the wave-cut platform at Locality 1.3, there are fine views of basaltic cliffs and spectacular waterfalls to the north-northeast.

Locality 1.4 [NM 442 185]

Folds within the Lower Shiaba Psammite.

En route southwestwards to Locality 1.4 along or near HWM, try to walk as close as possible to the Lower Shiaba Psammite/Shiaba Pelite boundary. Exposures frequently show elliptical doubly-plunging outcrop patterns of the psammitic banding. This is taken to imply the presence of highly curvilinear folds reflecting unusually high tectonic strains in this contact zone. At the western end of the wave-cut platform at Locality 1.4, large raised slabs of Lower Shiaba Psammite near HWM display many steeply plunging small-to-intermediate scale folds of M- and W- geometry. These probably lie in the hinge zone of the open fold around which the Shiaba Pelite passes, changing its strike from northeast-southwest to approximately north-northwest-south-southeast (Fig. 1.1).

Locality 1.5 [NM 4410 1841]

F_2 sheath folds within the Lower Shiaba Psammite.

Southwards from Locality 1.4, the coastline becomes more difficult and dangerous to access, but the nimble-footed and stout-hearted are rewarded by the sight of remarkable F_2 sheath folds, suggestive of very high strains in the Lower Shiaba Psammite adjacent to the Shiaba Pelite. To reach the most spectacular of these, on a falling tide below mid-tide, follow the Lower Shiaba Psammite/Shiaba Pelite boundary, trending S10°E towards the headland. Proceed a few metres within the psammite (the pelite generally occurs in the adjacent cliffs). Having crossed a small bay filled with large boulders, largely of garnetiferous pelite, follow a concordant Tertiary dyke (0.3m wide) southwards for some 25-30m. Locality 1.5 exposes metre-scale oval-shaped patterns of bedding/banding as a result of F_2 sheath folding (see Holdsworth et al., 1987, figure 1.1). It is on a rock shelf sloping gently southwards above the narrow cleft, occupied by the dyke. Note that the psammite near the contact with the pelite has a generally flaggy habit, dipping at ~70° towards ~250°. The flags carry a down-dip lineation.

Return to Shiaba by walking from Locality 1.4 as far as the course of the Allt Cnoc na Fearnaige stream, and follow the general line of the stream uphill as far as the track leading back to Scoor House. The paths leading up to Shiaba from the sea can be very obscured by vegetation in the summer.

Locality 1.6 [NM 436 183]

Sedimentary and tectonic structures within the Upper Shiaba Psammite.

If time permits, instead of following the burn up the hill, it can be followed down to Traigh Bhan which is Locality 1.6 (Fig. 1.1). On the south-east side of the bay, at mid-tide and below, there are good exposures of Upper Shiaba Psammite preserving northwest-younging cross-bedding and showing thin (~5mm) layers of dark semipelitic schist. These schists carry the regional S_2 foliation and skeletal porphyroblasts of white mica. Restored to the horizontal, S_2 cleavage/bedding relationships imply originally west-facing, very tight F_2 folds that pre-dated the regional-scale F_3 synform. Westwards across the bay can be seen the crags forming the lowest parts of Torr na Stallachhdach [NM 4325 1827], where they have been rendered very unstable and dangerous by the presence of an easily eroded sheet of Tertiary basalt.

Locality 1.7 [NM 4365 1805]

Upper Shiaba Psammite cut by Tertiary dolerite.

A difficult scramble to the south-southeast through and across very wet gullies, leads to Port na Eglaise [NM 437 181] where there are ruins, possibly of a small church. A small rocky bay comprises the roof of the basaltic sheet seen in the unstable crags on the northwest side of Traigh Bhan. This locality is the frequent haunt of a herd of feral goats that graze this part of the coast. On the headland beyond and to the southwest, a stack that comprises Locality 1.7 (Fig. 1.1) is formed of the dolerite sheet and from this vantage point can be seen that the same sheet on the opposite side of the bay [NM 4325 1827] cuts a camptonite dyke, probably of Permo-Carboniferous age. The country rocks all comprise Upper Shiaba Psammite.

> *Note*: If time permits, it would be possible to move up-section, i.e. in the direction in which younger rocks come on, following Itinerary B, but in reverse. To do this, proceed west-northwest from the head of Traigh Bhan, curving around and passing to the northeast of the generally craggy area shown on OS maps as Torr nan Stallachdach. It is strongly recomended that parties do not try to follow the coast to the southwest of Torr nan Stallachdach, because long, deep fissures, descending vertically to sea level, penetrate far inland where they are almost completely obscured by vegetation. Skirt the head of Slochd a Mhuilt Bay and the steep cliffs on its northwest side, descending to the coast at approximately [NM 425 184]; this is the approximate position of Locality 1.13 of Itinerary B.

Itinerary B

Eastern limb of the Assapol Synform (2); access via Scoor House.
Distance is either 3 or 4.5 km, depending upon inclusion of Localities 1.13 and 1.14, taking 4-6 hours.

Access is via Scoor House and Farm (for the route to the house see Itinerary A). Turn right (south) in front of the house, passing the old steading on your right; following the path around you arrive at a gate, adjacent to a cottage, that leads on to the hill. This area is very wet in bad weather. The adjacent rocks and smooth rock surface that are passed on the way to the gate form the top of the same 2-3 m-thick dolerite sheet on which Scoor House stands and which you will have encountered in several places on Itinerary A.

Locality 1.8 [NM 4208 1884]

Geological viewpoint.

Pass through the gate and turn left, skirting the side of the hill towards the fence that runs south-southeast up the hill from a large modern (2002) shed. Follow the fence on its western side until it reaches the top (~100 m O.D.) of the very steep slope running down to Traigh na Sgurra – a beautiful sandy bay, flanked by large rock slabs (Fig. 1.1). From a suitable vantage point to the west of the bay (Locality 1.8) and looking east along the coast, the control of the configuration of the headlands by the strike of the subvertical Moine rocks can be appreciated. Note also the brown colouring and distinctive columnar jointing of the same dolerite sheet that crops out at Scoor.

Locality 1.9 [NM 4215 1882]

S_0/S_2 relationships within the Ardalanish Striped and Banded Formation.

In high crags some 50m east of the fence [NM 4215 1882] alternating subvertical stripes and bands of somewhat gneissose pelite and semipelitic micaceous psammitic schist can be seen striking approximately 50° east of north and dipping 70° west. This is Locality 1.9. S_2 is marked by a preferred orientation of micas, producing a foliation dipping westwards at ~60°. Looking northeast at the subvertical surfaces, the foliation appears to lie a few degrees clockwise of the bedding/banding; this implies that the F_2 folds are very tight and may have an interlimb angle as low as 20°. If the foliation is axial planar to the F_2 folds and the line of bedding/foliation intersection is not very steeply plunging, it also implies that there is an F_2 synform to the east and an F_2 antiform to the west. Although the distance to these inferred structures is unknown, it may be significant that the relationship is the same as that in the psammites at Locality 1.6, nearly a kilometre to the east.

Locality 1.10 [NM 4214 1876]

Scoor Pelitic Gneiss.

Return to the fence and follow it down towards the beach to the point where it bears round to the right [NM 4214 1876] and walk eastwards to Locality 1.10 (Fig. 1.1) where a large crag displays very well the characteristics of the Scoor Pelitic Gneiss. Here the psammitic bands that had been a feature of Locality 1.9 are absent and the exposure consists of alternating stripes and bands of pelitic and semipelitic somewhat gneissose material, showing clear indications of original compositional layering, probably bedding. The mineral assemblage comprises quartz, plagioclase feldspar, biotite, muscovite and almandine garnet. The S_2 foliation dips generally west and looking northeast at steep faces, appears to lie slightly clockwise of the subvertical banding, i.e. as at Locality 1.9. Here the relationship, so clear at Locality 1.9, has been complicated by the overprinting by D_3 open crenulations of the S_2 planes and the planes of bedding/banding. The axial surfaces of the crenulations dip gently eastwards, while the hinges of the crenulations plunge shallowly seawards.

Locality 1.11 [NM 4231 1873]

S_0/S_2 relationships within the Scoor Pelitic Gneiss; Tertiary intrusion.

Looking back up the hill and eastwards across the Blown Sand on the narrow low raised beach, the relationships of the major dolerite sheet with the Moine can be appreciated. At the eastern end of the bay, to the west of a fault gully trending 060° [NM 4231 1873], there is an exposure of inter-banded pelitic gneiss and micaceous psammite which is Locality 1.11 (Fig. 1.1). Looking north-northeast, the S_2 foliation again dips clockwise of the dip of bedding and is weakly crenulated. The dolerite sheet above and to the north-northeast of this locality carries xenoliths of a psammite different from that in the adjacent country rocks. These xenoliths were probably derived from the Upper Shiaba Psammite to the east, implying an east to west direction of flow of the magma.

Locality 1.12 [NM 4229 1854]

Top of the Lagan Mor Formation.

From the east end of the beach move south to the headland, a difficult scramble and, especially when wet, the fallen slabs of pelite and dolerite are extremely slippery. Near the headland at the traverse encounter a band of white pure quartzite, about a metre thick. This bed is Locality 1.12 (Fig. 1.1) and marks the stratigraphic top of the Lagan Mor Formation, the main outcrop of which is in the Lagan Mhor Bay to the east. At the headland there are excellent exposures of pelitic gneiss, of quartzite and of the dolerite sheet. A second, higher, dolerite sheet marks the top of the headland. Cross-bedding younging westwards has been recorded in the quartzites. These exposures are at the top of the transitional sequence between the Shiaba Group (= Morar Group) psammites to the east and the Shiaba Pelite of the Assapol Group (= Glenfinnan Group) to the west; thus they constitute part of the evidence for a stratigraphic transition from the Morar Group to the Glenfinnan Group (Holdsworth *et al.*, 1987, figures 3 and 4).

Locality 1.13 [NM 4251 1834]

Transition from the Lagan Mor Formation into the Upper Shiaba Psammite.

Lagan Mor Bay is Locality 1.13, and a traverse across the bay for some 60 m (Fig. 1.1) shows the intensely flaggy nature of the interbanded, subvertical N10°E-striking pelite/quartzite of the Lagan Mor Formation. Towards the east, quartzite passes into siliceous psammite, locally with cross-bedding in which the angles between sets of lamination have been severely reduced. At the headland on the eastern side of the bay [NM 424 183] the rocks have passed by transition into the rather flaggy, highly strained psammites with poorly preserved cross-bedding that have been referred to as the Upper Shiaba Psammite by Holdsworth *et al.* (1987). These psammites pass, also by transition, stratigraphically downwards into weakly strained psammites of fluviatile or shallow marine origin (see also Glendinning, 1988). They carry complex sedimentary structures, including beautifully preserved cross-bedding with somewhat over-steepened cross-lamination, slumps and dewatering structures (Fig. 1.3). The best localities are around [NM 4251 1834]. *Please photograph, not collect.* Younging is unequivocally towards the west across bedding that strikes just west of north and is subvertical. Access to and within the immediate vicinity of this locality is difficult with strong fissuring that, on the landward side, can be treacherously covered by thick vegetation.

Fig. 1.3 Cross-bedding and cross-lamination within the Upper Shiaba Psammite at Locality 1.13.

One locality (14) is especially valuable in demonstrating the complex structural history of the rocks on the eastern limb of the Assapol Synform; this is described below. Being accessible from Scoor House, the locality could be visited in conjunction with either Itinerary A or B.

> *Note*: The note accompanying Locality 1.7 indicated that Itinerary B could be run in reverse, if time permitted at the end of Itinerary A. This would have the advantage of passing through the whole sequence from oldest to youngest. That note recommends joining Itinerary B at [NM 425 184]; this is only a few tens of metres east along the coast from Locality 1.13 described above. This precise location can be identified by the presence of a subvertical ~1m basaltic dyke approximately concordant with the bedding/banding in the psammite; the continuation of the dyke to the south is clearly visible across the Slochd a Mhuilt inlet. The dyke itself is remarkable for the spinifex texture involving feathery pyroxene crystals at its margins.

Locality 1.14 [NM 4165 1847]

F_2/F_3 relationships within the Ardalanish Striped and Banded Formation on the eastern limb of the F_3 Assapol Synform.

This can be accessed from Scoor House using the gate at [NM 4190 1905] mentioned at the start of Itinerary B. From the gate, walk southwest for about 200m, crossing a marshy burn/drain (do not follow this to the sea) and a low ridge trending north-south towards the sea. Using a gap in the substantial stone dyke, pass along the western side of the ridge, at the eastern edge of an extensive marshy area formerly cut for peat. This marshy area is drained by a small burn that plunges over a fall [NM 4169 1862] and flows via a narrow alluvial strip almost due south to the sea. (Please note that although the burn itself is not shown on either the 1:10,000 map or 1:25,000 map, the V-shaped contours defining its valley are shown.) The mouth of this valley, here some 25m wide, and in close proximity to the foreshore is Locality 1.14 (Fig. 1.1). Two parts of this locality are of particular interest:

(i) on the east side in the subvertical wall of the valley, looking towards N25°E, is a pair of almost upright F_2 folds verging towards a synform to the east-southeast. The plunge of a D_2 lineation on the folded foliation surfaces is 30°NNE, unequivocally steeper than the crenulation lineations caused by the crumpling of the foliation surfaces by small-scale F_3 folds. Looking north-northeast, it can be seen that the S_2 foliation transects the banding clockwise, while the axial planes of the D_3 crenulations transect the banding and S_2 anticlockwise;

(ii) the rocky knoll in the centre of the valley mouth [NM 41655 18476], just above HWM, can be used to demonstrate most of the structural history of the area. Looking at the seaward side of the knoll, a pair of tight F_2 folds, with S-shaped vergence, is disposed generally horizontally. These are transected by axial planes of F_3 crenulation cleavage dipping 70°E. A tight S_2 crenulation cleavage related to the F_2 fold pair is shallowly inclined/subhorizontal; these crenulations have crumpled an earlier penetrative mica fabric (S_1).

(iii) on the western side of the valley, the banding and early foliation (S_2) have been bent down to become subvertical, and thus complete an F_3 fold pair that verges westwards to a synform, the distant Assapol Syncline.

Itinerary C

Structures and lithologies within the Ardalanish Striped and Banded Formation on the western limb of the Assapol Synform; access from Ardalanish.
Distance from Ardalanish and return is c. 3 km, taking 4-5 hours.

In Bunessan, leave the A849 and turn up the steep hill that passes behind the Argyll Arms Hotel. After about 600m, turn left (south) following for about 2km the narrow metalled road signposted to Uisken. A narrow metalled road joins the Uisken road from the right (southwest). Follow this past the Ardachy Hotel towards Ardalanish Farm. About 0.5km along this road it is joined from the left by a track running alongside some ruined outbuildings. Parking for minibuses or cars is provided beside the track near its junction with the metalled road [NM 3731 1930], shown with 'P' on the 1:25,000 Ordnance Survey topographic map.

Access to several of the localities, notably 1.15 to 1.18, is easy and involves no more than walking along a fairly rough track and thereafter along the extensive sands interspersed with rocky knolls and reefs that is Ardalanish beach. Beyond the eastern limit of the beach, however, access to Localities 1.19 and 1.20 requires quite strenuous scrambling over rocky and bouldery terrain which becomes more difficult eastwards, especially under wet conditions produced either by rain or an outgoing tide. The access to exposures on the main beach (Localities 1.15-1.18) is, to a small degree, tide-dependent. Ideally they would be visited at or below mid-water on an ebbing tide; this would also reduce problems of access further to the east (Localities 1.19 and 1.20). Access to Locality 1.19 is quite easy at low water, but Locality 1.20, a tidal island, and a very rewarding, highly photogenic locality, should be attempted only by fit people capable of strenuous scrambling and minor rock climbing, and then only on an ebbing tide.

Locality 1.15 [NM 3753 1884]

Regional and contact metamorphism of the Assapol Group Moine.

From the car park, pass through the gate and walk south-southeast along the track, noting the extensive flat very gently sloping fields, especially to the northeast of the track. These surfaces comprise higher (older) and lower (younger) raised beaches. The higher beach at 15-20m O.D. has coarse cobble/pebble deposits, and the lower at ~10m O.D. has much finer silty/sandy deposits. Both have excellent back (fossil cliff) features and the lower beach has raised stacks and sea caves. The lower beach is extensively covered by Blown Sand.

Locality 1.15 is Dun Fuinn which would have been an offshore rocky island at the time the lower raised beach was cut. The exposure, originally described by MacKenzie (1949), comprises largely Assapol Group pelitic and semipelitic schists, remarkable for their metamorphic mineral assemblage. In addition to garnet and micas, some 3-4cm-wide bands contain blades of kyanite 1-2cm long. The high-pressure Al_2SiO_5 polymorph kyanite formed during the early period of Barrovian regional metamorphism that also produced the garnets. Because the exposure lies within the aureole of the Caledonian Ross of Mull Granite (Zone 2 of Wheeler *et al.*, 2004, figure 1.1), rims of pink andalusite, the low-pressure polymorph, are common around the kyanite blades (Wheeler *et al.*, 2004, figures 1, 4).

Abundant tourmaline also occurs at the head of a shallow gully within the exposure at its western end. These may also be linked to the effects of the granite, the exposed margin of which lies some 500m to the west.

Locality 1.16   [NM 3772 1876]

Migmatitic fabrics and polyphase folds.

Locality 1.16 is a low elongate crag that emerges from the beach sand and extends south-southwest from HWM for some 100m down the beach. The crag consists of interbanded psammite, semipelite and pelite; the last is commonly garnetiferous and carrying quartzofeldspathic *lits* appears somewhat migmatized. The migmatitic segregations have been shredded by the apparently penetrative fabric that is interpreted as S_2, axial planar to sub-isoclinal folds, and well displayed at the northern end of the exposure. Calc-silicate stripes and lenticles up to a few centimetres thick are an important feature of this exposure. These are characteristically pale-weathering and comprise amphibole, garnet, plagioclase feldspar, quartz and epidote. In several parts of the exposure they have clearly been folded into sub-isoclinal (F_2?) folds and subsequently refolded by more open (F_3?) structures.

Locality 1.17   [NM 3778 1870]

Garnetiferous amphibolites.

Locality 1.17 lies some 60m east of Locality 1.16 (Fig. 1.1). It is best examined on its eastern side where it exposes large pods of metabasic garnetiferous amphibolite. The red almandine garnets, commonly a centimetre in diameter, contrast vividly with the dark green groundmass consisting of hornblende, plagioclase quartz, epidote and titanite. Contacts between the metasediments and amphibolite at this locality appear to be concordant with the metasedimentary layering. The protolith of the meta-amphibolite is basaltic with MORB ('mid-ocean ridge basalt') chemistry. Its concordant pod-like form is not unique to this locality, but it is unusual and elsewhere the forms are sheet-like and commonly cut bedding.

Locality 1.18 [NM 3805 1855]

Relationships between amphibolites and host metasediments.

Locality 1.18 lies ~250m east of Locality 1.17 and is situated about half way down the beach at low tide (Fig. 1.1). It is important for establishing the original intrusive relationships of the metabasic amphibolites and the metasediments. The margin of the 3.5m-wide (not thick) amphibolite at its eastern side strikes N37°E and dips 70°NW. It unambiguously cuts the bedding in the psammitic country rocks which strike N44°E and dip 50° NW. The marginal zone of the amphibolite carries abundant large (~1cm) garnets that rapidly decrease in size while increasing in abundance away from the margin.

Localities 1.19-1.21 demonstrate the nature of the tectonic structures which have been imposed on Assapol Group rocks during polyphase deformation.

Locality 1.19 [NM 3816 1837]

East-verging F_3 folds.

Locality 1.19 can be reached at all states of the tide, but is best approached at low water (Fig. 1.1). It comprises a ~2m-wide gully with subvertical sides, running approximately east-west for about 15m and ending at its eastern end with a shingly more open space surrounded by low crags to north and south, but a much higher cliff to the east. Access to this gully is possible at low water directly from the eastern end of Ardalanish beach, following a sandy inlet that leads off to the east. Otherwise the gully should be approached by scrambling eastwards from higher up the beach, across and around crags surrounded in most part by shingle, turf and boulder-strewn terrain.

    Several structural features are worth noting:

(i)   at the southern side of the western entrance to the gully, on a low, sand-abraded rock face are well displayed ptygmatic folds of thin (~1cm) pegmatitic veins as well as stripes of siliceous psammite, set in a more micaceous matrix. The exposure can be used to demonstrate the minimum shortening accomplished by the D_3 deformation and the variable wavelengths of buckle folds where the competent layers are of different thickness;

(ii) along both sides of the gully F_3 folds and related crenulations and crenulation cleavage (S_3) in pelitic bands are well displayed, while a lineation (L_2) on the folded banding lies obliquely to the F_3 fold hinges. The F_3 folds verge eastwards to an F_3 antiform;

(iii) at the eastern end of the gully [NM 3816 1837] a fold pair, antiform/synform, at least an order of magnitude larger than those displayed on the gully walls, similarly verge towards an antiform to the east;

(iv) in the southern wall at the end of the gully, one limb of the antiform is brecciated and, at first sight, appears faulted. However, there is little sign of displacement and psammite flags can be traced, essentially continuously through the breccia;

(v) the psammitic flags carry regularly spaced, quartz-filled tension gashes.

Locality 1.20   [NM 3824 1832]

F_3 folds.

Locality 1.20 can be reached from Locality 1.19 by climbing out of the east end of the gully by climbing up the lowest part of the vertical face and walking/scrambling south-southwest down the fold hinge zone as far as the next east-northeast-trending gully some 20m to the south (Fig. 1.1). Alternatively, return to Ardalanish beach and, if the tides permit, walk down the beach to a gain access to the gully via a sandy inlet. This gully marks the line of a strike-slip fault. Walk eastwards along the gully noting the many subvertical garnetiferous amphibolite bodies, many locally cross-cutting, exposed in the gully walls. Because of the lateral-slip displacement of the fault these cannot be matched in detail across it.

Passing along the gully eastwards, into the next bay, a very rocky boulder strewn inlet, note in its north wall in particular, good examples of 'necking' and boudinage extension of the psammite layers. These are the result of the extension of the competent psammitic layers and are accompanied by tension gashes, similar to those at Locality 1.19.

In the bay itself, scramble across the boulders, which are exceptionally slippery when wet, to the tidal island situated opposite a cliff some 30m high.

The cliff, the face of which is probably the fault along which you walked to gain access to the locality, displays spectacular intermediate-scale folds that verge towards a major antiform the hinge zone of which lies some 100m to the east. The lithologies involved in these structures are striped and banded pelites and psammites with abundant calc-silicates. The crest of the major antiform which forms the roof of a cave marked on the 1:25,000 Ordnance Survey topographic map is not normally accessible from the west, but can be reached, with difficulty, from Uisken to the east (see Itinerary D, below). It forms the lowest structural level of a complex antiformal zone that extends westwards for some 300m from the inlet known as Slochd nam Ba [NM 385 184].

The island itself displays Assapol Group lithologies very well, on clean surfaces. Some psammitic bands, several centimetres thick, have very sharp contacts with extremely micaceous material on one side but pass by rapid transition into micaceous material on the other as the pelitic component of the metasediment increases. These occurrences can be readily, but perhaps not safely, interpreted as graded bedding, indicating the original way-up of the sediments. If this interpretation is to be believed, however, it would imply that the F_3 folds at this locality at least, are downward-facing, and if widely applied would be at variance with more convincing younging evidence elsewhere on the Ross.

Leaving the island to return to Ardalanish, glance up the scree slope to the north, to where the structural features displayed on a very large fallen and tilted block are worth a detour. These include excellent examples of ptygmatically folded pegmatitic veins and of the 'cusp and lobe' mode of buckle folding of alternating psammite (competent) and pelite (incompetent) layers.

Itinerary D

Western limb, continued, and core of the Assapol Synform; access from Uisken. Distance from Uisken and return is c. 3 km, taking 3-4 hours.

For access to Uisken, proceed as for Ardalanish, but instead of turning right towards the Ardochy Hotel and Ardalanish Farm, continue on the road from Bunessan, past the telephone kiosk to the coast. Here, in the small scattered settlement of Uisken, at the head of a sandy bay with scattered rocky reefs and islands, there is limited parking (out-of-season!) for about eight cars and minibuses.

Locality 1.21 [NM 384 183 to NM 398 188]

Hinge and eastern limb of F_3 antiform.

This locality requires mid- and falling tide. It is recommended that you do not follow the coast from Uisken, but instead walk cross-country over, in places, somewhat boggy terrain and largely raised beaches. To do this proceed southwest from the parking area (bearing approximately 235°) reaching the head of Slochd nam Ba [NM 385 184] after about 600m. If possible, it would be courteous to call at the house with outbuildings immediately above the road that led down to Uisken.

The northwest coast of Slochd nam Ba exposes, along the top of the crag, the most eastern of the intermediate-scale antiforms that comprise the major antiformal area mentioned in Itinerary C. Access is strictly for the physically capable. On the southeast side of the Slochd, walk ~150m south-southwest to the point where a substantial east-west-trending hollow feature cuts across the strike. En route there are small-scale paired folds that plunge shallowly south-southwest giving a clear indication of a larger scale synform to the east and an antiform to the west. Proceed eastwards along the hollow feature, noting the regular ~60° east-southeast dip of the strata and the more steeply inclined S_3 foliation giving clear indications of a synform to the east.

Return to Uisken Bay and walk eastwards across the sands, using a small stile [NM 3945 1890] to gain access to the rocks to the east. The section along the coast from here to Slochd Mhi Chriscain [NM 3975 1880] requires some difficult scrambling and minor climbing. It would be easier to leave the shore area, walking away from the coast across the scrubby moorland, only returning to the coast to reach Localities 1.22 and 1.23.

Locality 1.22 [NM 3965 1864]

Discordant amphibolite and F_2/F_3 structures.

Locality 1.22 exposes a thick strongly cross-cutting garnetiferous amphibolite (Fig. 1.4). There is considerable structural complexity whereby garnetiferous amphibolite is repeated by an F_3 antiform that intervenes between the two amphibolite outcrops; the complexity is compounded by a further

Fig. 1.4 (left) Garnetiferous amphibolite cutting obliquely across bedding within host Moine psammite at Locality 1.22.

Fig. 1.5 (above) F_3 folds within the core of the Assapol Synform at Locality 1.23.

antiform, the hinge of which can be clearly seen to plunge steeply (~60°) seawards by looking over the edge and to the east of a 10-20m cliff at the headland of the Slochd. Because the latter and the F_3 antiform that repeats the amphibolite are adjacent, i.e. without an intervening synform, it is inferred that the inaccessible structure must be a F_2 fold.

Locality 1.23   [NM 3980 1885 to NM 4100 1880]

Hinge zone of the F_3 Assapol Synform (Fig. 1.2).

Locality 1.23 comprises the coast between [NM 3980 1885] and [NM 4100 1880] (Fig. 1.1) The coast between Localities 1.22 and 1.23 at the head of Slochd Mhi Chriscain is very jagged and can be difficult to traverse especially at high water. It may be advisable to walk away from Locality 1.22 via the moorland and to drop down to the head of the bay to visit Locality 1.23 which comprises a set of low crags above and below HWM. This exposure lies in the core of the Assapol Synform (Figs 1.2, 1.5) and displays abundant upright minor folds with overall W-shaped vergence;

they plunge shallowly south-southwest, although having somewhat curvilinear hinge lines. A rodding fabric, possibly a product of D_2, passes around the hinges of these folds. Eastwards from this locality, the F_3 minor folds and the S_3 bedding/banding relationships consistently indicate a F_3 synform core to the west. It is noteworthy that this persists at least as far east as Locality 1.14 where relationships between D_1, D_2 and D_3 structures and fabrics can be established.

Once east of the hinge zone of the Assapol Synform, the vergence of the F_2 folds and the F_3 folds is in the opposite direction, while the F_3 fold plunge becomes shallowly inland to the north-northeast. The F_3 folds are pointing to a synform to the west, while the F_2 indicate an antiform to the west. The F_2 and F_3 are normally easily distinguished; the axial planar fabric S_2 of the F_2 folds commonly appears penetrative, while the S_3 is unambiguously a crenulation fabric. The interference of the two sets of folds and fabrics is well seen in many localities along the coast section designated as Locality 1.23. Notably, these relationships can be studied on Eilean Dubh (e.g. [NM 4010 1875]) and on the unnamed headland [NM 403 187] to the east across Port Bheathain. Interference between the two sets of structures is common, e.g. on a north-facing crag [NM 4041 1886].

Two further localities are recommended to point the contrast between the phenomena in pelitic and psammitic rocks associated with the emplacement of the Ross of Mull Granite.

Locality 1.24 [NM 365 217]

Contact metamorphic phenomena within the aureole of the Ross of Mull Granite. Distance is less than 1 km, taking c.1 hour.

The locality is reached by means of the A849 towards Fionnphort from Bunessan, and lies ~2 km west of Bunessan. The track leading to Bendoran Cottage has been gated near the roadside, and parking is limited. Walk to the cottage and turn left following paths up the craggy hill Torr na h-Annaid [NM 365 217], to the east of the cottage. This is Locality 1.24 (Fig. 1.1). Here you are in the contact zone of the granite where it has intruded striped and banded pelitic/semipelitic gneiss, reminiscent of the Scoor Pelite. Zones and veins of granite with rather diffuse margins, possibly the result of partial melting *in situ*, transect the banding. These commonly contain cm-scale xenoliths. The quartzofeldspathic gneissic lenticles are preserved in the country rocks. Sillimanite knots, possibly after kyanite, are locally well displayed. *These should not be hammered.* The rocks lie within Zone IV of Wheeler *et al.*, 2004 (figures 1, 6, 7) who recorded cordierite coronas around garnet, euhedral garnet, K-feldspar and prismatic sillimanite at this locality. The Moine country rocks here are strongly hornfelsed, but there is little or no sign of the forceful wedging of magma into planes of pre-existing fissility. This is in strong contrast to the granite relationships to the flaggy Assapol Group psammites well displayed at Locality 1.25.

Locality 1.25 [NM 3680 1765]

Margin of the Ross of Mull Granite. Distance from Ardalanish and return is c.4km, taking 3 hours.

It would be advisable to seek permission from the farmers at Ardalanish Farm before visiting this locality. The locality is accessed by the same route as that leading to Itinerary C. However, from Locality 1.15 the excursion should turn westwards, making towards the western end of Ardalanish beach, while passing through rocks of Zones II and III of Wheeler *et al.* (2004, figure 1). At the end of the beach a path should be followed southwards across the low raised beach to Slochd na Beiste [NM 3683 1827] where a low stone dyke can be easily crossed, and a steep path can be followed a few tens of metres up the slochd, a gully with rocky walls, as far as a col on its west side (you are here at the granite margin). Drop down from here on to the low raised beach below and walk down to the shore, to which the approach can be very boggy in wet weather. Proceed along the coast southwards past Carraig Mhor with its offshore rocky island, and continue to the next headland to the south [NM 3680 1765]. This is Locality 1.25 (Fig. 1.1).

A great range of igneous features can be studied at this locality, although the spectacular wedging of the granite between the psammitic flags is the most obvious (see also Zaniewski *et al.*, 2006). The precise contact of the pluton is not easily determined, because there is a gradual transition from granite with xenoliths and psammite essentially *in situ* with abundant granite wedges. This is in marked contrast to the contact with the pelite at Bendoran (Locality 1.24). Additional features of interest at this locality include an excellent example of a graphic granitic pegmatite dyke cutting a large hornfelsed metabasic body, a porphyritic felsite sheet about 4m thick, dipping 15°-20°N and displaying feldspar phenocrysts having cores of fresh microcline and red, heavily altered sodic plagioclase rims, as well as rounded, heavily corroded quartz ?xenocrysts, and a northwest-trending Tertiary dolerite dyke.

Excursion 2

Fort William to Glenfinnan

Rob Strachan, Dave Barr and Alan Roberts

Purpose: A general excursion to examine the main aspects of the geology of the Loch Eil Group and easternmost parts of the Glenfinnan Group.

Aspects covered: Psammites, sedimentary structures and calc-silicates in the Loch Eil Group; migmatitic gneisses in the Glenfinnan Group; the Ardgour Granite Gneiss; amphibolites and micro-diorites; steep belt and flat belt.

Maps: OS: 1:25,000 sheets 391 Ardgour and Strontian and 392 Ben Nevis and Fort William; BGS: 1:50,000 sheet 62W Loch Quoich.

Type of terrain: Roadside exposures.

Distance and time: By car 31 km (19 miles), about 45 minutes; on foot 6 km, c. 7 hours.

Short itinerary: Localities 2.1, 2.2, 2.9 and 2.10 cover the main rock types, tectonic structures and minor intrusions present within the metasedimentary rocks and the granitic gneiss. Time may be reduced to 3-3½ hours.

The Moine rocks of this area comprise psammites and quartzites of the Loch Eil Group that are underlain by interbanded pelites and psammites of the Glenfinnan Group. Both have been subjected to a common history of deformation and metamorphism. The structural sequence involves two phases of recumbent folding (D_1 and D_2) and three phases of upright folding (D_3-D_5) (Strachan, 1985). D_1 and D_2 folds have been recognized at least as far north as Loch Quoich and Glen Garry (Holdsworth & Roberts, 1984). D_3 folds, however, are only well developed in the southern part of the Loch Eil Group and are apparently absent in the Loch Quoich area where the structural sequence is generally simpler (Strachan, 1985). D_3 folds as defined above are absent west of Locality 2.5, so Barr (1983), Roberts (1984) and Barr *et al.* (1985) applied their Loch Quoich structural sequence in that area, i.e. D_3 of these authors (see Excursion 4) corresponds to D_4 of

EXCURSION 2

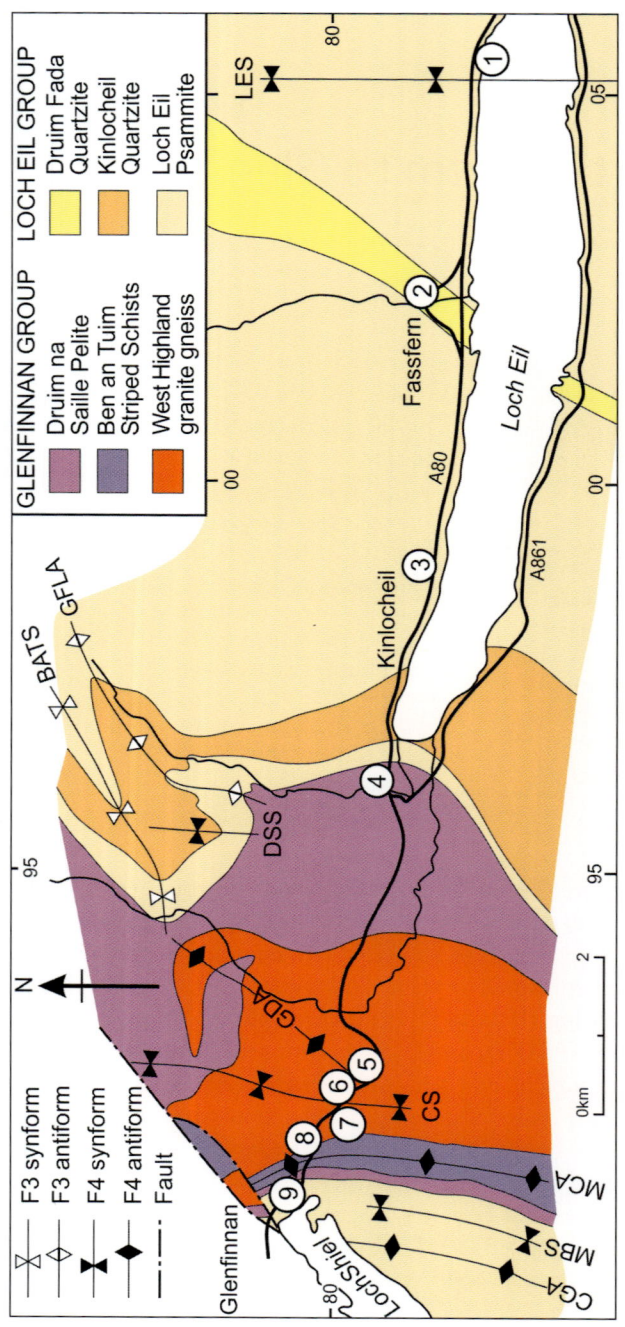

Fig. 2.1 Map of the Loch Eil-Glenfinnan area showing the generalized geology and the localities for the excursion.

BATS = Ben an Tuim Synform, D_3
GFLA = Glen Fionne Lighe Antiform, D_3
LES = Loch Eil Synform, D_4

DSS = Druim na Saille Synform, D_4
GDA = Glen Dubh Lighe Antiform, D_4
CS = Callop Synform, D_4
MCA = Meall na Cuartaige Antiform, D_4
MBS = Meall a Bhainne Synform, D_4
CGA = Coire Ghuibhsachain Antiform, D_4

Strachan (1985) and the present account. The most important set of upright folds on a regional scale are the NNE-SSW-trending D_4 structures (= D_3 at Loch Quoich) that are responsible for the formation of the steep belt and the Loch Quoich Line. D_1 is most likely Knoydartian in age, and D_2-D_5 are assigned to the Caledonian orogeny (Roberts *et al.*, 1984; Rogers *et al.*, 2001; Strachan & Evans, 2008).

The first locality is reached by driving north from Fort William on the A82 road. Three kilometres north of Fort William, turn left at Lochy Bridge onto the A830. Drive through Corpach to the first locality near the Loch Eil Outward Bound Centre [NN 057 784]. Good views of the Ben Nevis range and the Mamores may be obtained by looking eastwards from this point.

Locality 2.1 [NN 057 784]

Loch Eil shore section (Fig. 2.1). Loch Eil Psammite intruded by amphibolites.

Parking for two to three cars is available opposite the boat sheds on the north side of the road at the entrance to a small forest track. Cross the rail track at the level crossing to access the shore of the loch. Walk west for *c.*200 m to low-lying outcrops of quartzo-feldspathic psammites and concordant amphibolite sheets. The amphibolites carry a mineralogy of hornblende + quartz + andesine and represent basic intrusions probably emplaced prior to D_1 (Smith, 1979; Strachan, 1985). The gently-dipping to subhorizontal schistosity is interpreted as a composite $S_0/S_1/S_2$ fabric. Both lithologies carry a N-S-trending L_2 mineral and extension lineation defined by aligned amphiboles and quartz-feldspar aggregates. Lithological banding and the lineation are folded by upright, open folds in the core of the D_4 Loch Eil Synform (Strachan, 1985). Thin cross-cutting granitic veins are presumed to be of late Caledonian age.

Locality 2.2 [NN 0215 7895]

Fassfern (Fig. 2.1). Sedimentary structures within the Druim Fada Quartzite; microdiorite intrusions within the Loch Eil Psammite.

Due to the narrowness of the access road, coach parties are advised not to visit this locality.

From Locality 2.1, drive $c.3.5$ km further west and turn right onto the minor road signposted to Fassfern. Continue for 0.5 km until reaching the small bridge over the Allt Suileag. Cars and minibuses may be parked west of the bridge on the north side of the road adjacent to a white farm building. Walk over the bridge and climb over the stile north of the road. Descend the grassy slope down to the stream with care, and examine the exposures underneath the bridge. These comprise well-bedded quartzite and siliceous psammite assignable to the Druim Fada Quartzite, dipping gently to the east. Symmetrical straight-crested ripples are common on bedding surfaces. These are considered to result from the reworking of the sediments by wave activity. The quartzites are underlain upstream by grey micaceous Loch Eil Psammite, which locally preserves small-scale cross-laminations that young to the east. Three hundred metres north of the bridge [NN 0220 7925], amphibolite sheets are cut by pegmatite veins that are in turn cut by a weakly foliated microdiorite sheet. Retrace your steps to the parking place.

Locality 2.3 [NM 988 788]

Loch Eil road section (Fig. 2.1). Calc-silicates in Loch Eil Psammite; late granite veins.

From Locality 2.3, drive westwards out of Fassfern and turn right onto the A830. Drive 2.5 km further west and park in a large lay-by on the south side of the road. Carefully cross the road to the cutting where dark, micaceous psammites dip gently to the west. The locality is notable for the development of calc-silicate rocks. These are commonly white or pale grey and show every variation from large elliptical pods to small lenticles and calcareous wisps. They are considered to represent concretions formed during the diagenesis of the enclosing sediments (Tanner, 1976; Strachan, 1986). The calc-silicates contain a prograde mineral assemblage of hornblende +

andesine + quartz + garnet ± biotite, broadly indicative of low-amphibolite facies metamorphism. At the west end of the locality, lithological banding (S_0/S_1) dips gently to the west and is noticeably oblique to subhorizontal S_2, suggesting proximity to an F_2 fold hinge. The psammites are cut by numerous sheets of granite and aplite that belong to one of the late Caledonian granite vein complexes (Fettes & MacDonald, 1978).

Locality 2.4 [NM 962 793]

Kinlocheil (Fig. 2.1). Psammites and hornblende schists of the Loch Eil Group; pelitic gneisses of the Glenfinnan Group.

Due to the narrowness of the access road, coach parties are advised not to visit this locality. From Locality 2.3, drive c.4km further west. Slightly east of the junction with the A861 Strontian road, turn right and drive over the old road bridge over the Allt Fionne Lighe. Cars and minibuses may be parked on the hard-standing east of the bridge. A small tree-covered hillock 300m east of the bridge exposes the lowermost levels of the Loch Eil Group, represented here by gently-dipping micaceous psammites and hornblende schists. The upper unit of the Glenfinnan Group, the Druim na Saille Pelite, is exposed 100m west of the hillock and beneath the old road bridge. Here this unit comprises medium to coarse-grained pelitic gneisses with concordant migmatitic quartzofeldspathic segregations. The gneisses are composed of biotite + muscovite + quartz + feldspar + garnet ± fibrolite. The actual contact between the two divisions is unexposed; however, the general absence of widely developed high strain fabrics at any of these localities precludes the presence of a major tectonic break such as that which separates the Morar and Glenfinnan groups further west. The conformable nature of the Glenfinnan-Loch Eil contact is more clearly demonstrated at Loch Quoich (see Excursion 4). Return to the vehicles.

Locality 2.5 [NM 925 794]

Callop road section (Fig. 2.1). Flat-lying Ardgour Granite Gneiss with S_2 fabric and quartzofeldspathic *lits*; microdiorite sheet; D_2 folds in flat belt.

Drive westwards along the A830 for $c.3$ km and park in the Callop track (300 m past the railway bridge, on the left). The low-lying area between Loch Shiel and Loch Eil is commonly known as the 'Callop gap' and only attains a height of $c.20$ m above sea level. It is a good example of a low level glacial breach formed by intense glacial erosion and breaching of a pre-existing watershed during the movement of ice sheets from west to east. On the way to the parking place, passengers can note in roadside exposures the gentle, undulating dip of the granitic gneiss, locally interrupted by steep, NNE-SSW-trending D_4 fold limbs. The wide outcrop of the granitic gneiss is a consequence of its gentle sheet dip. Most exposures lie in the upper part of the body, which is essentially sheet-like and less than 1 km thick.

The cutting on the north side of the A830 exposes the gently-dipping Ardgour Granite Gneiss. The gneiss is readily distinguished from the metasedimentary country rocks by its strictly granitic composition (subequal quartz, K-feldspar and oligoclase with lesser biotite). It is homogeneous on the scale of an exposure, but a strong foliation is defined by discontinuous biotite laminae and light-coloured veins or *lits*. This corresponds to the S_2 fabric described in the Loch Eil Group rocks (Strachan, 1985). The veins or *lits* formed by segregation from biotite-bearing granitic gneiss and often have a prominent biotite-rich selvedge (restite). The gently-dipping gneissic fabric is folded by open, upright D_4 folds, but at the eastern end of the exposure is axial-planar to recumbent, isoclinal D_2 folds of the quartzofeldspathic *lits*. Quartz in the *lits* carries an S_2 shape fabric and other *lits* form augen within the foliation. The Loch Quoich Line is defined further west by an increase in the tightness of the upright D_4 folds. This exposure demonstrates that the granitic gneiss possessed a strong fabric and had been migmatized prior to D_4, so a genetic link between granitic gneiss and the Loch Quoich Line (e.g. Dalziel, 1966) is unlikely.

An ESE-dipping foliated microdiorite sheet has contacts steeper than banding in the granitic gneiss. In the steep belt (e.g. Locality 2.9), microdiorites have similar dips but with contacts *shallower* than banding, showing that they post-date the initiation of the steep belt and the Loch Quoich Line. If more than one vehicle is available, it is worthwhile leaving one here and another at Locality 2.10, and walking the *c.*1.5 km of road section without interruptions. Otherwise, drive *c.*400 m west and park opposite the track leading to Craigag Lodge, in the process crossing the open D_4 Glen Dubh Lighe Antiform of Dalziel (1966).

Locality 2.6 [NM 922 796]

Craigag road section (Fig. 2.1). Amphibolites and ptygmatic folds in granitic gneiss.

Cross the A830, and in cuttings observe gently-dipping granitic gneiss containing early *lits* that define intrafolial D_2 folds and rootless fold cores (Fig. 2.2). A number of metabasic early Moine amphibolites, concordant with the gneissic fabric, are present within the granitic gneiss. Compare these highly foliated and lineated rocks with the late Caledonian microdiorite at Locality 2.5. To the west, a series of broadly north-south-trending, open, upright D_4 folds deform the S_2 fabric and the amphibolites, and several cross-cutting late Caledonian pegmatites are present. Walk NW along the road towards the small stream, distinguishing between early *lits* that define intrafolial folds, and later pegmatites that cut S_2 but in some cases are ptygmatically folded. D_4 structures show no overall vergence in this area which lies in the core of the Callop Synform and thus corresponds to the Loch Quoich Line *sensu* Barr *et al.* (1985). Return to the vehicles and proceed to Locality 2.7.

Fig. 2.2 Typical exposure of the Ardgour Granite Gneiss at Locality 2.6, showing abundant concordant quartz-feldspathic segregations parallel to gneissic banding.

Locality 2.7 [NM 919 799]

Allt na Criche road section (Fig. 2.1). Low-strain zone in centre of granitic gneiss sheet.

From Locality 2.6, drive *c*.400m west to the next lay-by. Walk east along the A830 for *c*.100m, to where, in a cutting on the north side of the road, a 3m-thick pegmatite sill with vertical feeders disrupts the granitic gneiss. The variably-dipping gneissic foliation is axial-planar to rootless D_2 isoclines, and is locally steepened by D_4 folds. The pegmatite cuts these folds, and also intrudes a folded hornblende schist which is broadly concordant but has locally cross-cutting apophyses. Walk *c*.100m west along the road to view the outcrops opposite the lay-by, noting that the S_2 foliation has an average easterly dip and that open, NNE-plunging D_4 folds verge towards an antiform to the west. In the vicinity of the stream, S_2 is subvertical and strikes NNE-SSW; 100m beyond the stream, gentle easterly dips return. This part of the section defines an east-facing D_4 monocline that brings up the central part of the granitic gneiss sheet.

Walk further west for another 100m, noting that within this central part of the granitic gneiss, D_2 strain is low and the gently undulating but east-dipping S_2 fabric is axial-planar to open or close, ENE-plunging folds of early *lits* (Barr *et al.*, 1985, Fig. 4a). Traces of an earlier S_1 gneissic fabric can be recognized, sub-parallel to the *lits*. This fabric is also defined by discontinuous biotite-rich laminae. Locally, the S_2 fabric becomes extremely weak and sub-recumbent folds of S_1 are preserved (Barr *et al.*, 1985, figure 4b). The *lits* and their biotite selvedges cut the S_1 fabric, confirming their post-D_1, pre-D_2 age. Both D_1 and D_2 structures are cut by a microdiorite sheet. In this lowest-strain region in the centre of the granite gneiss, the origin of the granitic gneiss is clear: it represents a deformed granite that was itself affected by segregation. U-Pb dating of zircons from samples collected at this locality of the granitic gneiss and its quartzofeldspathic *lits* has yielded an age of 870 ± 7 Ma that is thought to date Neoproterozoic high-grade metamorphism and anatexis of this part of the Moine Supergroup during an orogenic event (Friend *et al.*, 1997). An alternative interpretation is that the protolith of the granite gneiss is entirely pre-tectonic and was generated by crustal melting during extensional rifting and development of the Moine sedimentary basin (Millar, 1999; Dalziel & Soper, 2001). Return to the vehicles and proceed to the next lay-by adjacent to the abandoned roadway.

Locality 2.8 [NM 916 802]

Glenfinnan road section (Fig. 2.1). Loch Quoich Line; minor intrusions; western margin of the granitic gneiss; steep belt.

Cross to granitic gneiss exposed in the adjacent road cut, where an east-dipping gneissic foliation steepens westwards about a D_4 fold hinge plunging at 45° towards 030°. It is axial-planar to tight, ENE-plunging folds of early *lits* that can be seen in various stages of transposition into this S_2 fabric. At this locality the sheet-dip is steep towards the east, reflecting a general increase in the intensity of D_4 strain and upright folding towards the west. However, no transposition into an S_4 fabric has occurred, and a strong S_2 fabric is preserved in the granitic gneiss (Fig. 2.2). An increase in D_2 strain towards the base of the granitic gneiss sheet is inferred on the basis that no S_1 fabric is apparent. This locality also displays a ESE-dipping, foliated microdiorite sheet and a later discordant pegmatite.

Walk westwards along the north side of the A830 for *c*.200m, noting that abundant upright D_4 folds in the granitic gneiss are cut by late Caledonian pegmatites. These folds verge towards an antiform to the west and their tightness, steep NNE plunge and attenuated long limbs are typical of the steep belt, west of the Loch Quoich Line. Note the uniformly steep dip of the gneissic foliation. Approaching the sign 'Glenfinnan Monument ½ mile', the gneiss remains steeply-dipping and planar-banded. One hundred metres past the sign, a road cut exposes typical Glenfinnan Group rocks of the steep belt. Migmatitic psammites, semi-pelites and pelites are intensely deformed and most leucosomes are transposed into the steep fabric, although some reclined, isoclinal S-folds are still preserved. These are related to the very tight D_4 Meall na Cuartaige Antiform (Barr *et al.*, 1985, figures 2, 3), although throughout most of this area, intense D_4 strain makes systematic mapping of D_4 fold vergence difficult. A large amphibolite at this locality carries an internal fabric that is deformed by the reclined D_4 folds. Return to the vehicles.

Locality 2.9 [NM 909 804]

Glenfinnan road section (Fig. 2.1). Loch Eil Group psammites within the steep belt.

From Locality 2.8, drive *c.*600m west and park in the disused road immediately before a road cut through large crags north and south of the A830. The crags expose Loch Eil Group psammites similar to those seen at Localities 2.1-2.4, but here folded into a steep attitude within the regional steep belt. The psammites occupy the core of a major D_4 fold, the Sgurr Ghiubhsachain Synform (Dalziel, 1966; Barr *et al.*, 1985), complementary to the Meall na Cuartaige Antiform. High strain associated with the D_4 folding has obliterated any sedimentary way-up evidence at this locality, and minor folds are also rare. Most foliation surfaces carry a steeply north-plunging intersection lineation that is likely to approximate to the plunge of the Sgurr Ghiubhschain Synform in this area. The steeply-inclined foliation of the psammites is cut by numerous steeply-inclined, post-tectonic pegmatites. The westward continuation of this road section is described in the following excursion. Parties who do not intend to continue any further may, however, like to drive to the Glenfinnan Visitor Centre where there are splendid views of the Glenfinnan Jacobite Monument and Loch Shiel. The monument was erected to commemorate the 1745 Highland uprising, led by Bonnie Prince Charlie who erected his standard at Glenfinnan.

Excursion 3

Glenfinnan to Morar

Derek Powell and Clark Friend
with additions by Robert Glendinning

Purpose: A general traverse across the structurally complex, high grade rocks of the eastern 'vertical' belt in the southwest Moine (= Northern Highland Steep Belt) and some of the lower grade rocks of the type area of the Morar Group to the west.

Aspects covered: Various metasedimentary lithologies and aspects of their sedimentology; tectonic structures and fabrics; metamorphic minerals and migmatites; igneous rocks such as microdiorites and pegmatites; features relating to the development of ductile thrusts.

Useful information: Hotel, B&B accommodation and camping are available in Fort William, or alternatively, sporadically along the road between Lochailort and Mallaig.

Maps: OS: 1:25,000 Explorer sheet 397 Loch Morar & Mallaig; BGS: 1:50,000 sheets 52 Tobermory, 61 Arisaig and 62W Loch Quoich.

Type of terrain: Many roadside exposures that require extreme caution because of the speed of the traffic. Moderately rough hillsides and moorland.

Distance and time: The road route is ~50km long and on foot an additional 10km. 3-4 days are recommended for the whole of this excursion. It is best followed from the east. See each locality for suggested times.

Short itinerary: Visits to Localities 3.1 (A & B), 3.3, 3.4b and 3.6 (B & C) will provide an appreciation of the geology of the area during a full, one-day excursion.

High-grade rocks of the 'vertical' belt

Despite many years of research, the structural and metamorphic development of the 'vertical' belt of the southwest Moine (= Northern Highland Steep Belt) has proved difficult to unravel because of the high levels of strain and a complex deformation history (Dalziel, 1966; Brown *et al.*, 1970; Powell, 1974; Baird, 1982, 1985). Equally, attempts to date events affecting

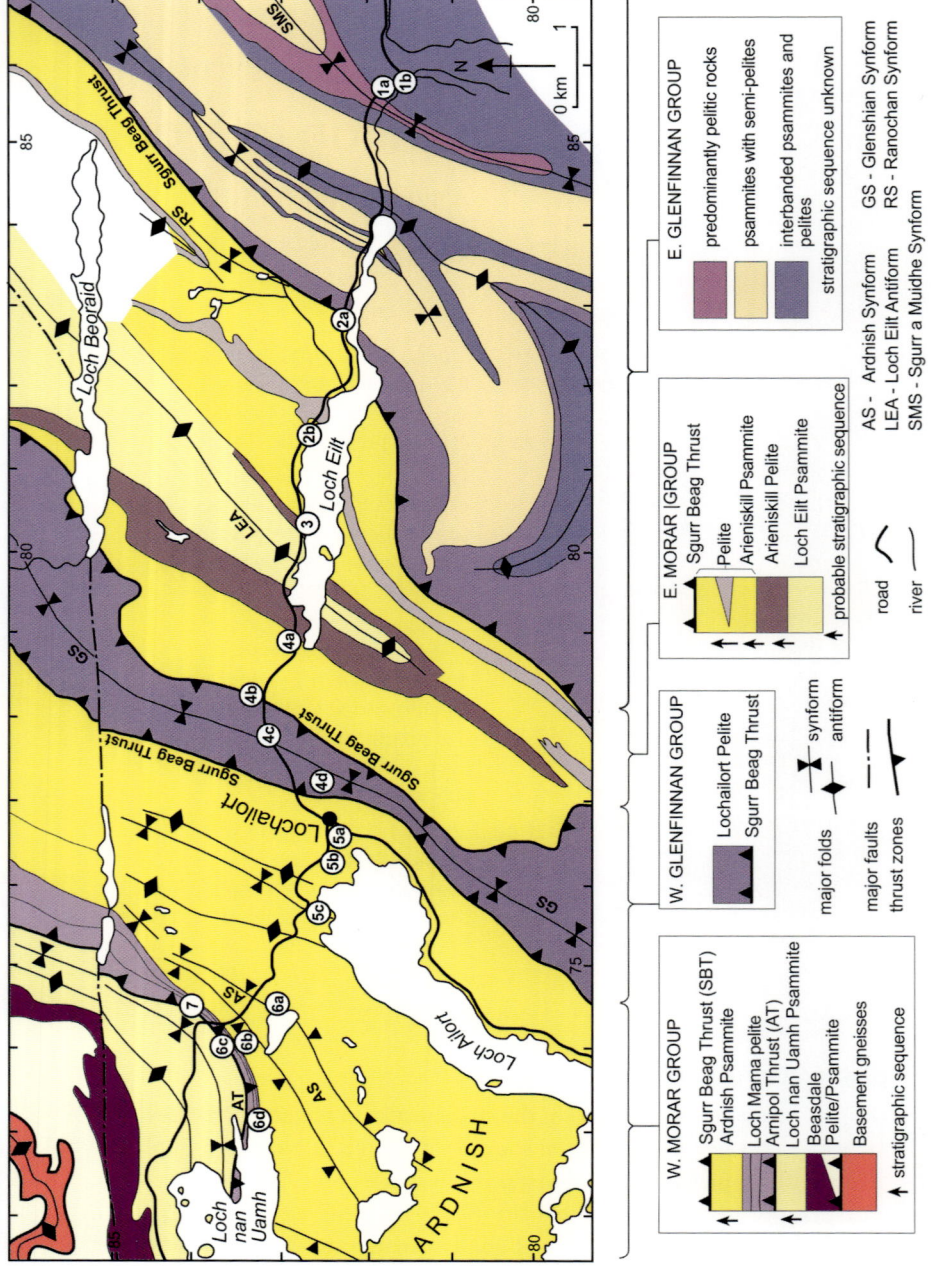

the belt have proved confusing and controversial (see Summary of Geology). However, recent geochronological work vindicates the earlier view of a complex Neoproterozoic history for the Moine rocks (e.g. Friend *et al.*, 1997; Rogers *et al.*, 1998; Vance *et al.*, 1998; Millar, 1999; Tanner & Evans, 2003). In the 'vertical' belt, at least three sets of tight to isoclinal folds can be detected (Baird, 1982) which pre-date intrusion of a suite of transgressive but folded, Caledonian pegmatites. A later suite of microdiorite intrusive sheets is deformed by open to moderately tight folds (Smith, 1979; Talbot, 1983). Whilst it would appear that many of the fabrics might have originally been Neoproterozoic, the development of the steep belt has been ascribed to crustal reworking during the Caledonian orogeny (Roberts & Harris, 1983).

Locality 3.1 [NM 857 815]

The Muidhe (Fig. 3.1). Glenfinnan Group psammites, pelites and migmatites; pegmatites; microdiorite sheets.

Parking is available in a lay-by on the south side of the A830 at [NM 8605 8142], 4.25 km west of Glenfinnan. The traverse involves a moderate climb to $c.50$ m above road level – 1-2 hours should be allowed. Climb uphill northwestwards for about 250 m to reach a gently inclined glacial pavement at [NM 8575 8150]. Here, at Locality 3.1A, large flat exposures comprise banded metasediments of the Glenfinnan Group. These are interbanded white psammites (originally impure quartzites); light grey psammites (arkoses); dark grey, generally homogeneous semi-pelites (silty mud rocks); and schistose, migmatitic pelites (mud rocks). Though as elsewhere in the Moine, metamorphic index minerals are only rarely preserved (Winchester, 1974), these rocks are at sillimanite grade (amphibolite facies). The pelitic rocks contain garnet (almandine) + biotite + muscovite + oligoclase + quartz, rarely sillimanite (e.g. MacQueen & Powell, 1977). In many places copious, often randomly oriented, late retrogressive muscovite flakes probably

Opposite page:
Fig. 3.1 Geological map of part of the SW Moine between Glenfinnan and Arisaig, showing locations of Localities 3.1-3.7.

represent breakdown products of an aluminium silicate, presumably kyanite or sillimanite, which are both known to be present in the area. The pelites often develop segregations of plagioclase and quartz (migmatitic *lits*), indicating that, in places, partial melting may have taken place, a phenomenon consistent with amphibolite facies conditions.

All exposures lie on the SE limb of the D_3 Sgurr a' Muidhe Synform (Baird, 1982) and many of the open to moderately tight, asymmetric minor folds that verge towards the SW are parasitic to the major synform (Fig. 3.1). The minor folds plunge at moderate to steep angles to the NE and are accompanied by crenulations that fold the migmatitic planar fabric in the pelitic rocks. The folds display considerable variation in style, which is largely attributable to variations in rock type and scale of interbanding, and thus ductility and ductility contrasts. They fold previously boudinaged layers and tight to isoclinal folds. Occasionally 'eye' folds occur wherein folded layers close both to the NE and SW; the fold hinges plunge to the northeast and southwest and are thus curved, and probably have a sheath-like morphology (Fig. 3.2).

At least three generations of intrusive veins are distinguishable: early quartz veins which cut the early isoclinal folds and may relate to the phase of layer extension that produced the boudinage; thin foliated pegmatites with diffuse margins which in places are sub-parallel to the axial planes of the D_3 minor folds; and usually thick, discrete, coarse-grained pegmatites trending northwest-southeast, which cut all minor folds, but are themselves folded. The latter, transgressive Caledonian pegmatites, are discussed further below at Locality 3.1B.

Fig. 3.2. 'Eye'-folds deforming interbanded psammites and migmatitic pelites of the Glenfinnan Group at Locality 3.1A. Lighter is 8cm long.

Return to the road by following the cliff edge near the road to the SE until a small valley is reached giving easy downhill passage. Along the top of this cliff, many exposures display fine examples of sheath folds. Some 350m to the W, roadside exposures, Locality 3.1B, occur on the NE side of the road opposite the lay-by at [NM 8570 8127] (Fig. 3.1), and provide sections that lie normal to those of the previous locality. At the SE end of the exposure, an E-W-trending vertical dyke belonging to the Permo-Carboniferous camptonite-olivine basalt suite cuts a 2m-wide, steeply dipping pegmatite, which exhibits a strong crystal growth fabric lying at a high angle to its contacts. The pegmatite is one of the transgressive early Caledonian suite characterised by their coarse grain-size and the presence of both muscovite and biotite with plagioclase + K-feldspar + quartz ± garnet. Similar bodies have given isotopic ages of about 450 ± 10 Ma (van Breemen *et al.*, 1974), *i.e.* late Ordovician.

A few metres to the NW, a partly foliated, intrusive microdiorite sheet crops out. At its NW end, the sheet dips moderately to the east, thins rapidly upwards and contains in places an oblique, internal, metamorphic fabric that lies clockwise of the sheet margins. The schistosity is most strongly developed adjacent to the margins but progressively decreases in intensity, and increases in dip, towards the centre. In the lower part of the exposure, the microdiorite cuts a coarse-grained pegmatite and appears to contain a xenolith of the same pegmatite plus country rock. At the bottom of the exposure, what may be the same pegmatite appears to have been displaced across the microdiorite with an apparent 0.5m shift. Above and to the left of the xenolith, the internal schistosity is seen to be axial planar to the folds of internal veins, whilst above and to the right the microdiorite margin displays cusp and lobe form indicating that during deformation the microdiorite was less ductile than the country rocks. Traced to the SE, part of the sheet changes attitude to dip westwards and thence it becomes horizontal before thinning and terminating. With these changes of dip the geometry of the internal schistosity becomes more complex but appears to relate to the overall folded form of the sheet; possibly relating to flexural flow on the fold limbs. Within the schistose margins of the sheet an extension mineral lineation in the schistosity relates to the movement across the sheet during its deformation; here NW-SE.

Microdiorite sheets occur throughout the south-western Moine areas and are thought to belong to a late Caledonian suite (Smith, 1979). Their structural development has been discussed by Talbot (1983). Generally sheets dipping to the west contain a schistosity, often sinusoidal, which dips westwards at steeper angles than the sheet margins, those dipping eastwards show the opposite relationships, whilst folded sheets have sub-horizontal enveloping surfaces and show changes in the attitude of the internal schistosity that relate to fold limbs. The sheets cross-cut structures generated during the first three deformation phases affecting the Moine, and the early Caledonian pegmatites. According to Smith (1979) they are earlier than the Strontian and Ross of Mull granites, which have, respectively, given isotopic ages of 425 ± 3 Ma (Rogers & Dunning, 1991) and 421 ± 5 Ma (Oliver *et al.*, 2008). Thus intrusion of the microdiorites together with syn-metamorphic regional deformation of both the sheets and their host rocks, would appear to have taken place between *c.*450 and 425 Ma, *i.e.* during the late Ordovician or early Silurian. Post-orogenic cooling of the southwestern Moine is dated by common K-Ar muscovite and biotite cooling ages between 420 and 410 Ma (Brook & Powell, unpublished data).

Locality 3.2 [NM 828 821 to 817 824]

The Loch Eilt Antiform eastern limb (Fig. 3.1). Glenfinnan Group pelites; Morar Group psammites and pelites; Sgurr Beag Thrust Zone.

Proceed west from Locality 3.1B for 3.5km and park in a lay-by on the south side of the road at an emergency telephone [NM 8290 8210]. Exposures along the north side of the road to the east of the lay-by are Locality 3.2A, and display at their western end strongly schistose pelites of the Glenfinnan Group which contain lenses, knots and stringers of migmatitic, plagioclase-quartz *lits*. Characteristic small, claret-coloured garnets are common and chemically homogeneous in contrast to the larger red-brown garnets of low-grade Morar Group rocks to the west (Anderson & Olympio, 1977; MacQueen & Powell, 1977; see below). The exposure lies within the Sgurr Beag Thrust Zone (Ranochan Slide of Baird, 1982) where this is brought to ground surface on the eastern limb of the Loch Eilt Antiform (Figs 3.1, 3.3). The junction, not here exposed, between the Morar and Glenfinnan groups lies some 50m to the west. Note the occasional ribs of blue-grey psammite that lie parallel to the predominant, near vertical schistosity, and the development of small sinistral (extensional?) shear bands dipping eastwards.

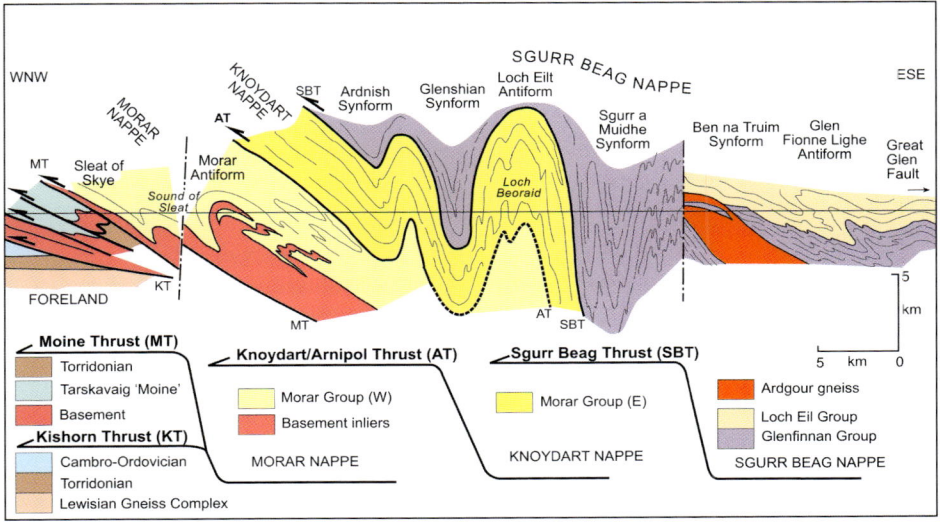

Fig. 3.3 Geological section across the SW Moine.

To the east, along the exposure, more varied rock types come in, with the proportion of psammitic layers increasing. Note the thin (Caledonian?), cross-cutting pegmatitic veins containing garnet + muscovite + feldspar + quartz, and the tight to isoclinal folds of early migmatitic veins and lithological layering. One hundred metres to the west of the emergency telephone, exposures on the north side of the road comprise disturbed and considerably altered rocks of the Morar Group. At the extreme western end of the road cut, interbanded psammitic and pelitic rocks are folded by open to tight minor folds plunging to the SE that are crossed by open folds and coaxial crenulations plunging SW. The pelitic layers do not contain migmatitic segregations; rather they are highly schistose and muscovite-rich. A concordant biotite + feldspar + quartz pegmatite is boudinaged and a folded cross-cutting microdiorite sheet, in places, contains a strong, internal schistosity that is oblique to the sheet margins on fold limbs, but axial planar in the axial zones of the folds. These exposures lie on the western limb of the D_3 Ranochan Synform (Baird, 1982) whose western limb is, according to Baird (1982), replaced by the Sgurr Beag Thrust Zone (Fig. 3.3). Because the Sgurr Beag Thrust is folded over the Loch Eilt Antiform to reappear further west (Figs 3.1, 3.3), it follows that formation of the thrust zone and generation of Baird's D_3 folds are earlier.

Displacements on the Sgurr Beag Thrust are a minimum of 15 km in this area (Powell *et al.*, 1981), but 50km elsewhere (Kelley & Powell, 1985), and it is therefore significant that this microdiorite sheet shows no greater degree of deformation than outside the thrust zone. Clearly, thrusting pre-dated emplacement of the microdiorite sheet. Petrographic and geochemical analysis of the microdiorites suggests that they are members of a con-sanguineous, igneous suite (Smith, 1979), thus they provide relative time markers which were here emplaced after Baird's third phase of deformation and formation of the thrust, but before at least one late phase of ductile deformation (see also Baird, 1985; cf. Talbot, 1983).

About 1km along the road to the west, park in a lay-by on the south side at [NM 8175 8238]. Exposures on the north side of the road, some 80m to the west, are Locality 3.2B and show small claret-coloured garnets, and migmatitic segregations parallel to a dominant schistosity. Staurolite and sillimanite (fibrolite) occur sporadically in this unit which forms, together with adjacent rocks, part of the Morar Group succession lying within the Loch Eilt Antiform (Fig. 3.3).

Walking westwards some 130m, the contact of the pelitic unit with a predominantly psammitic group is encountered. The psammite is exposed for a further 200m. Inter-layered striped and banded psammites, blue-grey semi-pelites and occasional pelitic layers show no unequivocal evidence of sedimentary structures. Indeed the presence of tight to isoclinal minor folds of the lithological layering, concordant pegmatites and boudinaged pegmatites, and the platy nature of many of the psammites indicate the high levels of strain that these rocks have undergone. The exposures lie within 1km of the Sgurr Beag Thrust Zone (Fig. 3.1). Microdiorite sheets dipping to the east show a remarkable variation in the intensity of the internal schistosity, but are not folded.

Locality 3.3 [NM 806 827]

Morarian (= Knoydartian) pegmatites at Loch Eilt (Fig. 3.1). Pegmatites, Morar Group psammites and pelites; isoclinal folds; microdiorite sheet.
PLEASE DO NOT HAMMER THESE EXPOSURES.

Some 1.2km along the road to the west, limited parking is available on the south side of the road at [NM 8060 8270]. Allow about an hour here. A further 250m west is a rocky knoll on the south side of the road, partly hidden by a road-sign, that displays near vertical, interbanded psammitic and semi-pelitic rocks with occasional layers of migmatitic pelite. These Morar Group rocks lie near the core of the Loch Eilt Antiform (Figs 3.1, 3.3). In the large vertical rock face, isoclinally folded, boudinaged and apparently concordant pegmatites are composed of garnet + muscovite + microcline + plagioclase + quartz + accessory green apatite. In the boudin cores, the pegmatites are coarse-grained and non-schistose, but at the boudin margins, and where they are thin, the pegmatites are finer grained and schistose. The pegmatites belong to the Knoydartian suite and at this locality have been dated at 730 ± 20 Ma (van Breemen *et al.*, 1974) and 740 ± 30 Ma (van Breemen *et al.*, 1978).

From exposures such as these, several authors have argued that the pegmatites, because they are concordant to the migmatitic foliation, are deformed, and have a metamorphic mineralogy, constitute *in situ* metamorphic segregations. Thus the isotopic ages record regional metamorphism at, or shortly before, 740 Ma (Giletti *et al.*, 1961; Lambert, 1969; van Breemen *et al.*, 1974, 1978). Further, on this basis, these authors postulate a Morarian (= Knoydartian in present usage) orogeny. With this in mind it is pertinent to note that the rocks exposed here have suffered high levels of strain subsequent to formation of the pegmatites, and that the pegmatite bodies lack the continuous biotite selvedges characteristic of segregations and contain microcline. It is difficult at this, and similar high grade/high strain localities, to assess the original relationships of these pegmatites. The only pegmatite of this suite so far discovered in a relatively low grade, low strain environment, is on the Ardnish peninsula [NM 6937 8127], 11 km to the WSW (Powell *et al.*, 1983). Here the relationships of the pegmatite were originally interpreted as indicating pre-pegmatite Precambrian deformation and metamorphism to garnet grade with pegmatite intrusion at or before 776 ± 15 Ma (Powell *et al.*, 1983). It was suggested that this and other similar-aged pegmatites might represent a suite of intrusive bodies generated during crustal extension. However subsequent work suggests intrusion of the Ardnish pegmatite during deformation and metamorphism at 827 ± 2 Ma with the pegmatite suite as a whole being diachronous (Rogers *et al.*, 1998).

In the exposures directly opposite, on the north side of the road, open sinusoidal folds of a microdiorite sheet have vertical axial surfaces and a southerly plunge. Note the obliquity of strike of the country rocks to the fold hinges; crustal shortening which generated the folds of the intrusive sheet has been accommodated by ductile, homogeneous strain in the host rocks. Forty metres to the east, a 35cm-thick, transgressive, garnet + muscovite + K-feldspar + plagioclase + quartz pegmatite, probably of early Caledonian age (450 Ma), cuts isoclinal folds of the lithological layering. The isoclines also fold a lens of Morarian (Knoydartian) pegmatite and both are cut by a thin pegmatite, which is folded by apparent dextral shearing along the lithological layering.

Locality 3.4 [NM 789 828 to NM 777 830]

The Glenshian Synform and the Sgurr Beag Thrust (Fig. 3.1). Glenfinnan Group pelites in the core of the Glenshian Synform and Morar Group psammites and migmatitic pelites on the limbs; foliated microdiorites; strong planar and linear fabrics of the Sgurr Beag Thrust Zone.

This traverse, which will take 3-4 hours, crosses a major, tight to isoclinal fold, the Glenshian Synform (Figs 3.1, 3.3), which regionally can be traced for at least 22km along strike. It has the form of an elongate boat (BGS sheets 52 and 61) and folds the Sgurr Beag Thrust (Powell *et al.*, 1981). The fold core is occupied by rocks of the Glenfinnan Group, the Lochailort Pelitic Group of Powell (1964), whilst Morar Group rocks form the limbs to east and west – the Arieniskill and Ardnish psammitic groups of Powell (1964). Although difficult to assess accurately, the amplitude of the Glenshian Synform is likely to measure several kilometres and thus the repeated outcrops of the Sgurr Beag Thrust Zone would have originally been at different crustal depths (Powell *et al.*, 1981).

From Locality 3.3 drive 1.6km westwards to park in a lay-by adjacent to the railway line at Locality 3.4A [NM 7885 8285]. NB: the road here is extremely dangerous for unwary pedestrians. The road cut opposite the lay-by continues to the east around a sharp bend for some 200m. The exposure comprises near vertical interbanded grey psammites, blue-grey semi-pelites and migmatitic pelites of the Morar Group. *Please do not hammer these exposures.*

Just to the north of the lay-by, predominantly psammitic rocks give way, moving to the south and east, to more varied rock types. Eastward-dipping microdiorite sheets which cross-cut quartz veins and several generations of pegmatite, have internal schistosities indicating a sinistral shear sense across them. The attitude of the schistosities and associated extension lineations suggest NW-SE movement and in one case a displacement of at least 2.5 m can be deduced from the offset of markers in the country rocks. In many cases the dramatic increase in the intensity of the internal schistosity toward the margins of the sheets testifies to the rapid increases in shear strain and it is likely that detachment along the contacts has taken place (Smith, 1979). For accounts of the petrography, and further significance of the micro-diorite suite, the reader is referred to Johnson & Dalziel (1963), Smith (1979) and Talbot (1983) wherein very different opinions are expressed.

Tight to isoclinal minor folds of the lithological layering have very variable plunges and where these fold pale, thin calc-silicate layers and lenses a strong, axial planar, penetrative schistosity is apparent. In many places the obliquity of the dominant schistosity to layering is noticeable and it is deformed by crenulations that plunge moderately to the SW. At the eastern end of the exposure stacks of minor folds, verging towards the east, form cores to large boudin pods some 3 m long. Judging by the intensity of folding of the layering, boudinage of pegmatites and fold stacks, folding of cross-cutting pegmatites, and the deformation of the microdiorites, the country rocks at this locality have suffered several phases of shortening normal to the layering. Thus the 'flattening' strains must be intense, and planar fabrics likely to be composite. Minor and major folds probably have sheath geometry. Along strike in this zone, pelitic rocks contain sillimanite and staurolite, while the calc-silicate ribs and lenses comprise garnet + hornblende + biotite + plagioclase (An_{80-90}) + quartz; the grade of regional metamorphism is therefore high (Powell *et al.*, 1981).

The next locality can be reached by either traversing up and across the hillside to the north then the NW for some 700 m to reach the line of crags that form Locality 3.4B at [NM 7877 8335], or by driving west to park in a lay-by at [NM 7832 8318] from which the crags are accessible via the footpath signposted to Meoble, which leaves the north side of the main road, under the railway, some 200 m to the east. Climb 250 m towards 050° to reach the crags. Here vertical, banded psammites of the Arieniskill Psammitic Group (Morar Group) that lie 350 m east of the Glenfinnan Group-Morar Group contact (the Sgurr Beag Thrust) can be examined. Following the 100 m contour, traverse west across Arieniskill Burn for 350 m towards the WSW. The banded, generally platy, psammitic rocks in the exposures near Allt Dileige, lie adjacent to the major lithological contact (Fig. 3.1) and display characteristic features of high strain psammites within the Sgurr Beag Thrust Zone. Note the regular, small scale, interbanding of quartz-rich psammite and thin mica-rich lamellae, and the apparent structural simplicity. In many exposures concordant and isoclinally folded quartz veins can be found and occasionally isoclinal folds of the lithological layering. A steeply-plunging grain-alignment lineation is in places apparent. The extremely strong planar fabric of these rocks and the lineation are attributed to movement on the Sgurr Beag Thrust. The characteristics of similar thrust zones are discussed by Tanner (1971), Rathbone & Harris (1979), Powell *et al.* (1981), Baird (1982), Rathbone *et al.* (1983) and Kelley & Powell (1985).

Examination of the exposures on either side of Allt Dileige allows the contact of the predominantly migmatitic pelitic rocks of the Glenfinnan Group rocks to the west (Lochailort Pelitic Group) and Morar Group psammites (the Arieniskill Psammitic Group), to be traced up and down the hillside. Within the pelitic rocks, migmatitic *lits* have suffered disruption and a reduction in size, when compared with similar rocks further west. Because of the mineral assemblages developed in pelitic and calc-silicate rocks across the Sgurr Beag Thrust Zone here, it would appear that movement took place under medium to high grade metamorphic conditions and the Glenfinnan Group rocks have been brought from crustal depths of 15-20km, assuming a 30°C/km geothermal gradient (Powell *et al.*, 1981; Powell, unpublished data). Whilst tectonic slivers of gneissic basement rocks, which provide the most obvious evidence for the magnitude of displacement on the thrust further north (Tanner *et al.*, 1970; Tanner, 1971), are absent in the SW Moine, contrasts in metamorphic grade across it at its most westerly outcrop, indicate considerable displacement (Powell *et al.*, 1981). Unfolding the Glenshian Synform and the Loch Eilt Antiform (Fig. 3.3), suggest that the thrust zone, though now vertical, originally had a gentle to moderate easterly dip and assuming a NW-SE displacement, minimum movements of 15-20km can be suggested but they are likely to be much greater.

Throughout the Moine Supergroup generally, but particularly in this area, discontinuous bands and pods of calc-silicate rocks commonly occur. These are interpreted to represent calcareous concretions that probably formed during diagenesis. In the Lochailort area, the calc-silicate layers show a wide range of mineral assemblages and accordingly have been used to define several distinct metamorphic zones (Tanner, 1976; Fettes, 1979; Powell *et al.*, 1981). In the Morar Group below the Sgurr Beag Thrust, on the western side of the Loch Eilt Antiform (Fig. 3.1), assemblages contain quartz + bytownite + garnet + hornblende + titanite + zircon + biotite ± chlorite, assigned to zone 3, equating to kyanite/sillimanite grade metamorphism. Titanite from a calc-silicate assemblage within the Morar Pelite, sampled between Localities 3.4A and 3.4B at [NM 7914 8318], has been dated using conventional U-Pb techniques (Tanner & Evans, 2003). The results indicate that the reactions forming titanite in these layers occurred at 737 ± 5 Ma. This age equates well with the $c.750$ Ma ages on muscovite books from nearby minor shear zones in the same group of rocks (Piasecki & van Breemen, 1983; Piasecki, 1984). This metamorphic event is thus quite separate from the anatectic event that formed the West Highland Granite Gneiss, which is dated at $c.870$ Ma, for example, at Ardgour and Fort Augustus (Friend et al., 1997; Rogers et al., 2001).

These titanite ages force us to examine the time at which the Sgurr Beag Thrust was formed. Currently there are two hypotheses: (1) it is an early Caledonian ductile structure, which carries the already gneissose Glenfinnan Group rocks over lower-grade Morar Group rocks to the west (e.g. Rathbone et al., 1983; Barr et al., 1986); (2) the main movements on the Sgurr Beag Thrust occurred during the regional metamorphism that affected both the Glenfinnan and Morar group rocks for the first time (Powell et al., 1981). In this hypothesis, both the Moine and Sgurr Beag nappes share the same metamorphic history. In a study of micro-fabrics (Grant & Harris, 2000) it was discovered that the Sgurr Beag Thrust has two phases of movement; an earlier phase of compressive movement and a later extensional phase. It is plausible therefore, that the early phase is Neoproterozoic and that the later phase might be Caledonian (Grampian).

Some 850m west along the road, parking is available in the lay-by (north side) at Locality 3.4C [NM 7773 8308], just west of Susan Macallum's memorial cairn. *Please do not hammer these exposures.* The roadside exposures at the cairn lie near the core of the Glenshian Synform (Fig. 3.3) and display typical interbanded migmatitic pelites, impure quartzites and micaceous psammites of the Lochailort Pelitic Group, here vertical in attitude. Examination of both the vertical faces and sub-horizontal surfaces above, reveals the presence of early pegmatites and the development of steeply-plunging, asymmetric, sub-angular folds accompanied by crenulations of the migmatitic fabric, which refold earlier isoclines. Opposite the western end of the exposure, proceed SE from the road to a low ridge running south toward the River Ailort. Here the complexities of the folding

Fig. 3.4 Isoclinally refolded isoclines picked out by vertically dipping sedimentary layering in the Lochailort Pelitic Group at Locality 3.4C. Compass-clinometer is 10 cm long.

Fig. 3.5 Foliated microdiorite sheet dipping eastwards and cutting near vertical interbanded psammites and semi-pelites of the Upper Morar Psammite at Locality 3.5A. Note easterly-dipping internal schistosity oblique to sheet margins. Lighter is 8 cm long.

are picked out by upstanding ribs of white psammite. Isoclinally refolded isoclines (Fig. 3.4) demonstrate the composite nature of the dominant schistosity and many of the folds of both generations have curvilinear hinges and resemble sheath folds. To the east, small boudins of amphibolite have biotite-rich, often highly garnetiferous rims. These amphibolites belong to a regionally developed suite of early basic intrusives which, in this area, only occur in the Glenfinnan Group.

The Lochailort pegmatite, emplaced in migmatitic Glenfinnan Group pelites, can be seen at Locality 3.4D. Parking for cars can be found near the entrance to the electricity substation [NM 7715 8275]. Cross the road, and then walk 170m southwards to reach hummocky exposures of a coarse-grained feldspar + quartz pegmatite containing large muscovite plates, Locality 3.4D [NM 7722 8261]. The pegmatite exhibits little deformation, largely because of its competence and coarse grain size. In contrast, the host rocks comprise isoclinally folded migmatitic pelites of the Glenfinnan Group. Muscovite samples from the pegmatite have been isotopically investigated and the $c.750$ Ma ages for the cores represent either ages of crystallization or the effects of resetting. Pelitic schists at this locality and elsewhere in the Glenfinnan Group, contain abundant, randomly-oriented flakes of late, retrogressive(?) muscovite that cross-cut earlier fabrics. The thermal event causing this late development of muscovite could be responsible for extensive resetting of isotope systems. The muscovites are sometimes intergrown with fibrolite and this may be significant as staurolite, kyanite and sillimanite have been found in high-grade migmatites of the Lochailort Pelite but are very rare. Winchester (1974) suggests that this is because of the overall, inappropriate bulk chemistry of the rocks. However, some of the abundant, retrogressive muscovite flakes may be derived from alumino-silicates.

Western margin of the 'vertical' belt and low grade rocks of the western 'flat' belt

From the western limb of the Glenshian Synform as far as the coast, in the Morar-Arisaig districts some 17km to the west, the Moine rocks whilst folded on a major and minor scale, locally vertical and overturned, form a regional 'flat' belt (Fig. 3.3). Throughout, the metamorphic grade is low, rising from greenschist to lowest amphibolite facies from west to east (Kennedy, 1949; Lambert, 1959; Tanner & Miller, 1980; Powell et al., 1981). It was within this belt, and its extension northwards into Knoydart, that the lithostratigraphic succession of the Morar Group was originally established (Richey & Kennedy, 1939; Kennedy, 1955; Ramsay & Spring, 1962; Powell, 1964, 1974) and a four-phase deformation sequence recognized (Powell, 1964, 1974; Powell & MacQueen, 1976).

A rapid drop in metamorphic grade coincides with the western outcrop of the Sgurr Beag Thrust Zone near Lochailort (Fig. 3.1). However, the change from the vertically disposed rocks of the 'vertical' belt to the more variable attitudes of the 'flat' belt does not occur until 2.5km to the west. Rocks, within which unequivocal and common sedimentary structures are recognizable, crop out within 500m of the contact between the Morar and Glenfinnann groups (Powell, 1964; Rathbone & Harris, 1979) suggesting that the high levels of strain associated with both the 'vertical' belt and the Sgurr Beag Thrust Zone die away quite rapidly. Sedimentary structures are common in many of the psammitic rocks of this western belt and have allowed detailed control of structural mapping. In particular, changes in younging directions within the Ardnish Psammitic Group (Powell, 1964) delimit several late major fold closures (Fig. 3.1) and show that the Ardnish Psammitic Group youngs away from the underlying Loch Mama Pelitic Group, and towards the Lochailort Pelitic Group (Powell, 1964, main map). The major folds are extremely variable in style and geometry (Powell, 1966).

Locality 3.5 [NM 7675 8236 to NM 7570 8260]

The western limb of the Glenshian Synform (Fig. 3.1). Morar Group psammites, pelites and calc-silicate lenses; sedimentary structures.

About an hour should be allowed for these localities which lie to the west of the Lochailort Inn. Parking for cars is available on the verge off the south side of the road west of the junction at [NM 7668 8226]. Coaches may be able to park at the Inn on request.

Opposite the road junction west of the Lochailort Inn, near vertical rocks of the Morar Group (Ardnish Psammitic Group of Powell, 1964) crop out in road cuttings which are Locality 3.5 [NM 7675 8236]. This locality lies 500m west of the Sgurr Beag Thrust but is within the thrust zone. It comprises interbanded white-grey psammites, blue-grey micaceous psammites, and dark semi-pelites, together with thin calc-silicate ribs and lenses. In places, levels of strain are sufficiently low to enable deformed sedimentary structures to be seen (main map in Powell, 1964). However, the apparent structural simplicity of the rocks, their uniform attitude and platiness, may reflect their situation just within the thrust zone. Calc-silicate rocks here comprise garnet + biotite + zoisite + plagioclase (An $_{40\text{-}60}$) + quartz, whilst occasional pelitic horizons in the vicinity, are non-migmatitic garnet + muscovite + biotite + plagioclase + quartz schists (Powell *et al.*, 1981).

On the south side of the road, immediately west of the road junction, further exposures of psammitic rocks at Locality 3.5A reveal sedimentary structures, largely ripples and trough cross-laminations that suggest younging to the east. The cross-sets have been modified by strain relative to their positions on minor folds. Calc-silicate lenses and pods are common. Gently-inclined microdiorite sheets are also well exposed in the southern side of the cutting. The sheets all show brittle, igneous, emplacement features and usually have internal shear fabrics (Fig. 3.5). The sheets commonly display a number of features including en-echelon segments and, where the intervening bridge has been breached, *en bayonet* relations. Frequently along the length of a sheet there are apophyses on either side that have opposing directions of termination. These features are interpreted to be related to either the linking of two segments through the brittle extensional fractures into which the magma was emplaced, or represent side-steps. In most examples the magma has breached the separating bridge to form a semi-continuous sheet. In some dykes elsewhere, isolated xenoliths of the host rock may be found, interpreted to be fragments of the bridges between segments (e.g. Rickwood, 1990).

Locality 3.5B is some 300m to the west, at [NM 7635 8230], where exposures on the both sides of the road display steeply dipping psammites where levels of strain are low. Planar and trough cross-bedding give convincing evidence for younging to the east. A 5m-wide Tertiary basalt dyke is well displayed in the north side cutting.

Locality 3.5C lies a further 800m to the west at [NM 7570 8260], where roadside exposures can be examined. Parking is available on wide grass verges, and outcrops on the north side contain minor folds, which are associated with the major folds (Fig. 3.1). Sedimentary structures give clear evidence of directions of younging – predominantly to the east. A thin microdiorite sheet shows displacements across semi-pelitic layers in the host psammites and, internally, a sinusoidal shear fabric indicating a top to the west sense of movement. Dextral shear along the semi-pelitic layers has caused attenuation and thinning of the microdiorite and successive uplift to the west. Shearing was associated with retrogression allowing the growth of chlorite.

Locality 3.6 [NM 7478 8305 to NM 7401 8330]

The Ardnish Synform – lithostratigraphy of its north-western limb (Fig. 3.1). Well-preserved sedimentary structures and minor folds; lithostratigraphy.

Visiting localities 3.6A, B & C will require half a day, but if 3.6D is included allow a further two hours. Park in the lay-by on the north side of the road 1.3 km west of Locality 3.5C, 2.4km from Lochailort [NM 7478 8305]. Cross the road and go through a small gate giving access to the viewpoint above Loch Dubh. Walk west along the fence line to Locality 3.6A which comprises low exposures above the grass verge of the road. These outcrops lie near the core of the Ardnish Synform (Fig. 3.1) where levels of deformation are so low that original sedimentary features are very well preserved. Differential weathering has etched out several types of cross bedding which clearly demonstrate that the rocks young to the SW. Similar sedimentary features are still recognizable in the immediately adjacent outcrops, but blasting during road widening has ruined their surface expression.

> **Note on sedimentary features contributed by R. Glendinning**
>
> Trough cross-bedding is well displayed in these exposures and in many places oversteepened to overturned foresets suggest either soft-sediment deformation shortly after deposition or a later tectonic effect. The association of oversteepening of the foresets with the development of parasitic folds perhaps suggests modification of the cross-sets by ductile strain.

A further 600 m to the NW, parking is available in a large lay-by on the west side of the road [NM 7420 8356]. Some 50 m to the north, the public footpath giving access to the Ardnish peninsula (signposted), leaves the road to the SW. About 50 m to the north of this new path, low lying exposures adjacent to the south side of the old Ardnish track, Locality 3.6B lies on the NW limb of the Ardnish Synform (Fig. 3.1) and contains very well preserved, though folded, cross-bedding. The psammites dip and young to the SE and are arkosic in composition; truncated cross-sets are common and different orientations of foreset beds suggest trough cross-bedding. Herringbone cross-bedding suggests currents with either a northerly or southerly transport direction. The minor folds plunging steeply to the SW show extreme, though systematic, changes in layer thickness from limb to hinge, and variation in style along their axial surfaces. Evidently, the rocks had a high mean ductility during deformation. The folds are parasitic to the Ardnish Synform and characteristically have quartz veins sub-parallel to their axial surfaces. Elsewhere, folds of this generation refold isoclinal folds and are deformed by SE-plunging open folds and crenulations.

The next part of the excursion can be completed quickly, but less informatively, by returning to the main road and walking northwards along it for 100 m to reach Locality 3.6C. Here outcrops on the west side of the road start with NW-dipping garnetiferous pelites of the Loch Mama pelite. Retrogressive metamorphism, of uncertain age, has caused chloritisation of both garnet and biotite. Further along the exposure the lithology gradually changes with an increase of semi-pelitic layers and eventual loss of pelites. Characteristic of this outcrop and that of the Loch Mama pelite elsewhere, is the platiness of the layering and rarity of minor folds. A further 200 m along the road, on its south side, uniformly banded psammites of the Loch nan Uamh Psammite crop out. Though not obvious here, sedimentary structures in neighbouring psammites indicate younging to the SE, i.e. into the Loch Mama pelite. In this area there is, therefore, evidence for a tri-partite, lithostratigraphic succession starting with the Loch nan Uamh Psammitic Group, passing through the Loch Mama Pelitic Group and up into the Ardnish Psammitic Group. Such a succession is similar to that on the western limb of the Morar Antiform (Kennedy, 1955; Richey & Kennedy, 1939). There is, however, evidence for a thrust running through the lower part of this succession (see below).

If time is available, a more informative traverse across the NW limb of the Ardnish Synform from the Ardnish Psammitic Group into the Loch Mama Pelitic Group (equivalent to the Striped and Garnetiferous groups of the Morar succession) can be made by following the Ardnish path to the SW. On crossing the railway footbridge [NM 7401 8330], walk by way of scattered exposures, across country for 800m WNW, to reach Locality 3.6D at the coast [NM 7320 8340].

The coastal exposures around the small headland immediately to the north contain highly garnetiferous muscovite + biotite + plagioclase + quartz schists. In thin sections the garnet porphyroblasts are seen to contain planar inclusion fabrics, are in places texturally and chemically zoned, and are interpreted as having grown before development of the dominant schistosity (Powell & MacQueen, 1976; Anderson & Olympio, 1977; MacQueen & Powell, 1977). Detailed studies of Sm-Nd isotopes and the chemical zonation of garnets from here, from near Locality 3.6C, and elsewhere (Vance *et al.*, 1998), suggest early garnet growth at 823 ± 5 to 788 ± 5 Ma. These results, taken with the work on the Knoydartian pegmatites and the calc-silicate rock mentioned above (stop 4D), provide very strong isotopic evidence for Neoproterozoic orogenesis. Acceptance of such a conclusion implies that many of the tectono-metamorphic features and fabrics of the south-western Moine are Precambrian rather than Caledonian in age (Brewer *et al.*, 1979; Powell *et al.*, 1983, Holdsworth *et al.*, 1994, Vance *et al.*, 1998, Rogers *et al.*, 1998, 2001; Tanner & Evans, 2003; Cutts *et al.* 2009).

Locality 3.7 [NM7421 8370 to NM 7438 8430]

The Arnipol Slide Zone (Fig. 3.1). Structures associated with a regional ductile thrust.

A traverse, lasting up to three hours, across the lower part of the Ardnish Psammitic Group, the Loch Mama Pelitic Group, into the Loch nan Uamh

Psammitic Group (equivalent to the Lower Psammitic Group of Richey and Kennedy, 1939), and across the Arnipol Slide Zone, can be made by leaving the main road at [NM 7421 8370] and taking the following route: walk for 200 m north-eastwards up to the 90-100 m contour; follow the 90-100 m contour to the north, meeting Arnabol Burn at [NM 7465 8412] and then walk northwards to the sheepfold at [NM 7468 8430]; cross Allt Mama and walk westwards, downhill for 30 m, examining outcrops of psammites belonging to the Loch nan Uamh Psammitic Group.

On this traverse the lithological character and low-grade metamorphic state of these three members of the Morar Group can be appreciated both within and, in the case of the psammites, outside the Arnipol Thrust Zone. Whilst initially the thrust zone here was thought to be of only local significance (Powell, 1964, 1966), the contrasts in the attitudes of early minor fold structures across the zone are dramatic; moving from NW to SE, tight isoclinal folds with east to NE gently-plunging hinges (D_2 of Powell, 1964, 1974) are rotated, within the plane of the lithological layering across the zone, to become steeply SE-plunging reclined structures (Powell, 1966). The development of platy fabrics in many lithologies is also apparent and the thinning of the Loch Mama Pelitic Group from north to south is remarkable (Fig. 3.1). The Arnipol Thrust may be a major structure representing the southward extension of the Knoydart Thrust Zone (Poole & Spring, 1974).

Locality 3.8   [NM 1613 7842]

The Upper Morar Psammite: Rhue peninsula, by R. Glendinning.
Well-preserved sedimentary structures.

There is no access for coaches to this locality; allow about two and a half hours. Drive westwards from Arnipol towards Arisaig (10 km). On the outskirts of Arisaig [NM 665 865], turn left off the new A830 and at [NM 661 864] turn left and follow the minor road along the Rhue peninsula for 5 km to the end of the metalled road (Fig. 3.6). Limited parking is available adjacent to a small boathouse (Fig. 3.7). Follow a track to Rhue House, 1 km to the SW, and thence walk due west to the coast at [NM 1613 7842].

Rocks near the top of the Upper Psammite are well displayed on a raised wave-cut platform where the western limb of a large (locally D_3) minor anticline is exposed (Fig. 3.7). The sequence (Fig. 3.8), consists of lenticular sheets of medium- to coarse-grained, occasionally gravely, arkosic sands (now psammites) interbedded with silts and silty clays (now semi-pelites and pelites). Internally, the sand sheets are dominated by trough cross-bedding with subordinate tabular cross-bedding (Fig. 3.9), flat-bedding, ripple cross-lamination and climbing ripple cross-bedding. Palaeocurrents measured from cross-bedding show unidirectional, northerly flows. Locally these sheets have erosive bases and frequently incorporate rip-up clasts of the underlying silts. Further evidence for erosion is provided by occasional winnowed sand and gravel layers. Titanite-rich heavy-mineral bands, which appear as thin yellow lines on weathered surfaces, are common. The intervening finer grained sediments are rarely massive and usually contain thin layers and lenses of sand.

Locality 3.9 [NM 1668 7933]

The Upper Morar Psammite: Morar Bay, by R. Glendinning.
Soft-sediment deformation structures in meta-sandstones.

From Rhue, drive back towards Arisaig and turn right to rejoin the A830. Drive north towards Mallaig. Alternatively, a more scenic coastal route can be taken by turning left and following the B8008 to Morar village. Vehicles can be parked just off the road immediately north of the Morar Motors Garage (Fig. 3.6, inset). Two hours should be allowed for this locality; walking time is about 20 minutes.

Follow the footpath along the north shore of Morar Bay. This crosses the conformable contact between the Striped and Pelitic schists and the Upper Psammite; both groups are steeply dipping to vertical and young consistently to the west. The former are particularly well exposed around [NM 1668 7933], immediately to the west of a small rocky headland, and consist of finely interbedded sands and silts with abundant calc-silicate pods and lenses. The Upper Psammite is exposed on an extensive wave-cut platform to the north of Sgeir Mhor (Fig. 3.6). The salient features of the sequence are shown in Fig. 3.8. Again, the Upper Psammite consists of laterally extensive sand sheets with intervening silts and silty-clays. The sands contain a variety of sedimentary structures including flat-bedding, trough and tabular cross-bedding and ripple cross-laminations. Of particular interest are thick sand layers, which were subjected to intense soft-sediment

EXCURSION 3

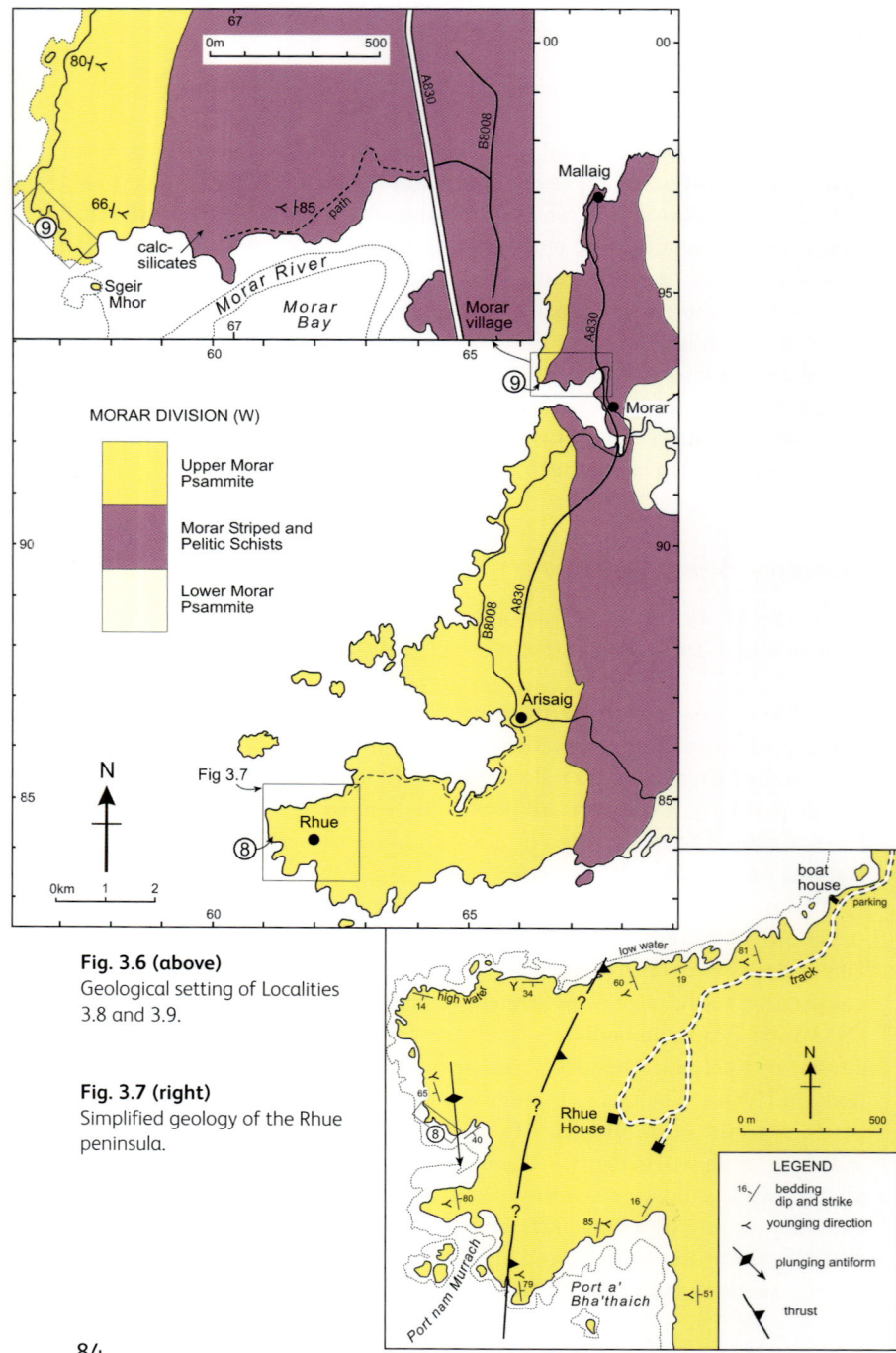

Fig. 3.6 (above)
Geological setting of Localities 3.8 and 3.9.

Fig. 3.7 (right)
Simplified geology of the Rhue peninsula.

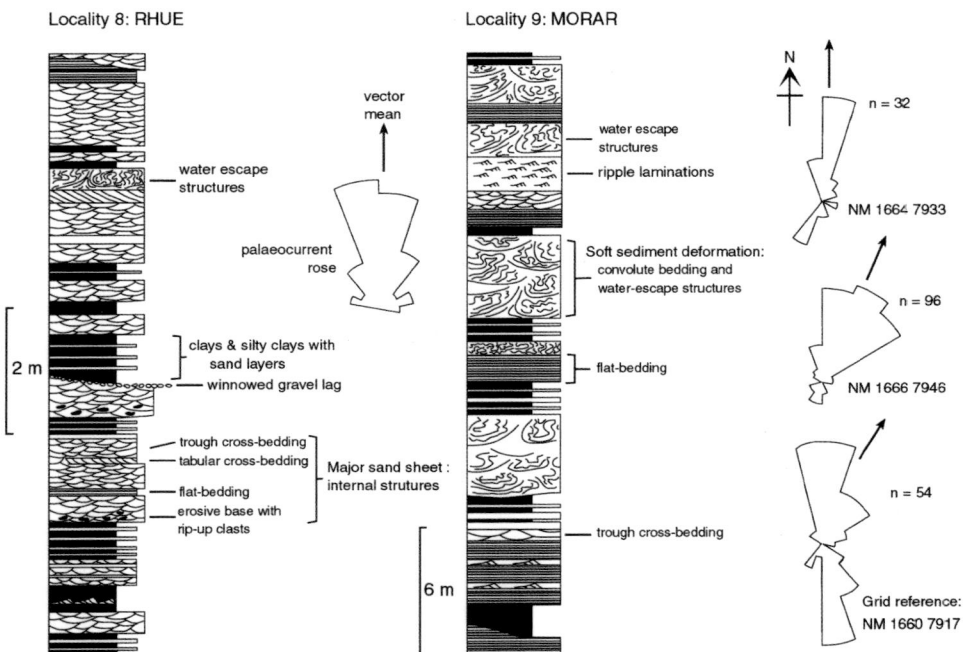

Fig. 3.8 Examples of graphic sedimentological logs from the Upper Morar Psammite.

deformation. Various mechanisms can be invoked to explain such structures: for example, earthquake-induced shock or slumping; however, in this case the contortions are generally associated with trough cross-bedded sands and intervening sediments are undisturbed. This suggests that under shock of rapid deposition the sediment liquefied generating convoluted bedding and water escape structures. This section, in contrast to that described above (Locality 3.8), lies at the base of the Upper Morar Psammite.

Interpretation of the sedimentary environment of the Upper Psammite

Examination of the Upper Psammite at other localities, around Morar and southwards along strike on Ardnamurchan, confirms the dominance of south to north and SW to NE palaeocurrents. These are accompanied by a northwards decrease in both the grain size and scale of the sedimentary structures within the psammites, and a corresponding increase in the proportion of semi-pelite and pelite. Such a transition suggests two broad depositional

Fig. 3.9 Cross-bedding in psammite layer of an interbedded psammite-pelite sequence of the Upper Morar Psammite at Locality 3.8. Truncation of the cross-beds indicates slight inversion and younging to the west. Lighter is 8cm long.

models: either (a) a fluvio-deltaic wedge, or (b) a shallow marine shelf with deposition controlled by tidal and/or storm generated currents. Of these, the former is considered unlikely as there are no recognizable fluvial sequences and channelling is rare. Instead, the sands and silts form laterally extensive sheets. Furthermore, abundant sand lenses and layers within the Striped and Pelitic Schists suggest a starved sediment, rather than a low energy pro-delta environment. A shallow marine shelf model is perhaps more feasible (Glendinning 1988) and Johnson (1978) and Walker (1979) describe similar changes in grain size and sedimentary structures along tidal transport-deposition paths. The palaeocurrent reversals identified around Morar (Fig. 3.8) may represent a weaker opposed tidal flow. Anderton (1976) emphasizes the importance of winnowed gravel lags and describes a tidal shelf model in which a zone of scour is followed by a sandwave zone and finally by silt and mud deposition. When normal tidal flows are strengthened during storms these zones migrate so that, for instance, sandwaves become eroded. This may explain the distribution of winnowed gravel layers within the Upper Psammite, which are occasionally present around Morar but much more common on Ardnamurchan. For further discussion of the sedimentation of the Upper Morar Psammite, see Bonsor & Prave (2008) who prefer to interpret it as an alluvial braidplain deposit.

Excursion 4
Invergarry to Kinloch Hourn

Alan M. Roberts and David Barr

Purpose: An east-west traverse through the Sgurr Beag Nappe to examine the internal structure of the nappe and its basal thrust.

Aspects covered: Curvilinear, recumbent folds in the Loch Eil Group; sedimentary structures in the Glenfinnan/Loch Eil group transition; the Quoich Granitic Gneiss; the Loch Quoich Line and intense Caledonian upright folds; curvilinear folds in the Glenfinnan Group; syn-deformational pegmatites; Lewisianoid inlier in the Sgurr Beag Thrust Zone; reworking of the ductile thrust-related fabric by steep-belt structures; a lower (Knoydart?) ductile thrust within the Morar Group.

Maps: OS: 1:25,000 sheets 399 Loch Arkaig, 400 Loch Lochy & Glen Roy, and 414 Glen Shiel & Kintail Forest; BGS: 1:50,000 sheets 63E Loch Lochy, 62W Loch Quoich and 72W Kintail.

Type of terrain: A mixture of roadside, lochside and off-road exposures involving some hill-walking.

Distance: by car 40km (25 miles), one-way from Invergarry, plus *c.*10km walking.

Time: By car *c.*2 hours return journey plus a long day's fieldwork. Could usefully be split into two days, after visiting the Quoich dam spillway. Coaches can proceed no further than Locality 4.7 (Quoich bridge).

Useful information: This section is best studied while staying at Invergarry, where there are two hotels, two campsites and many B&B establishments. If you wish to take a large party into Coir' an t-Seasgaich during the stalking season (1 September to mid-October), it is best to seek advice first from the keeper (Mr A. L. MacNally, Tel: 01809 511220) in the small village of Kingie, 3km east of the Quoich dam.

Short itinerary: Locality 4.1 (Garry Quarry), Locality 4.3 (Quoich dam spillway), Locality 4.5 (Quoich shore), Locality 4.8 (Coire Shubh Beag, 4.8D and 4.8E only), Locality 4.9 (Kinloch Hourn, omit 4.9A and walk as far as time permits).

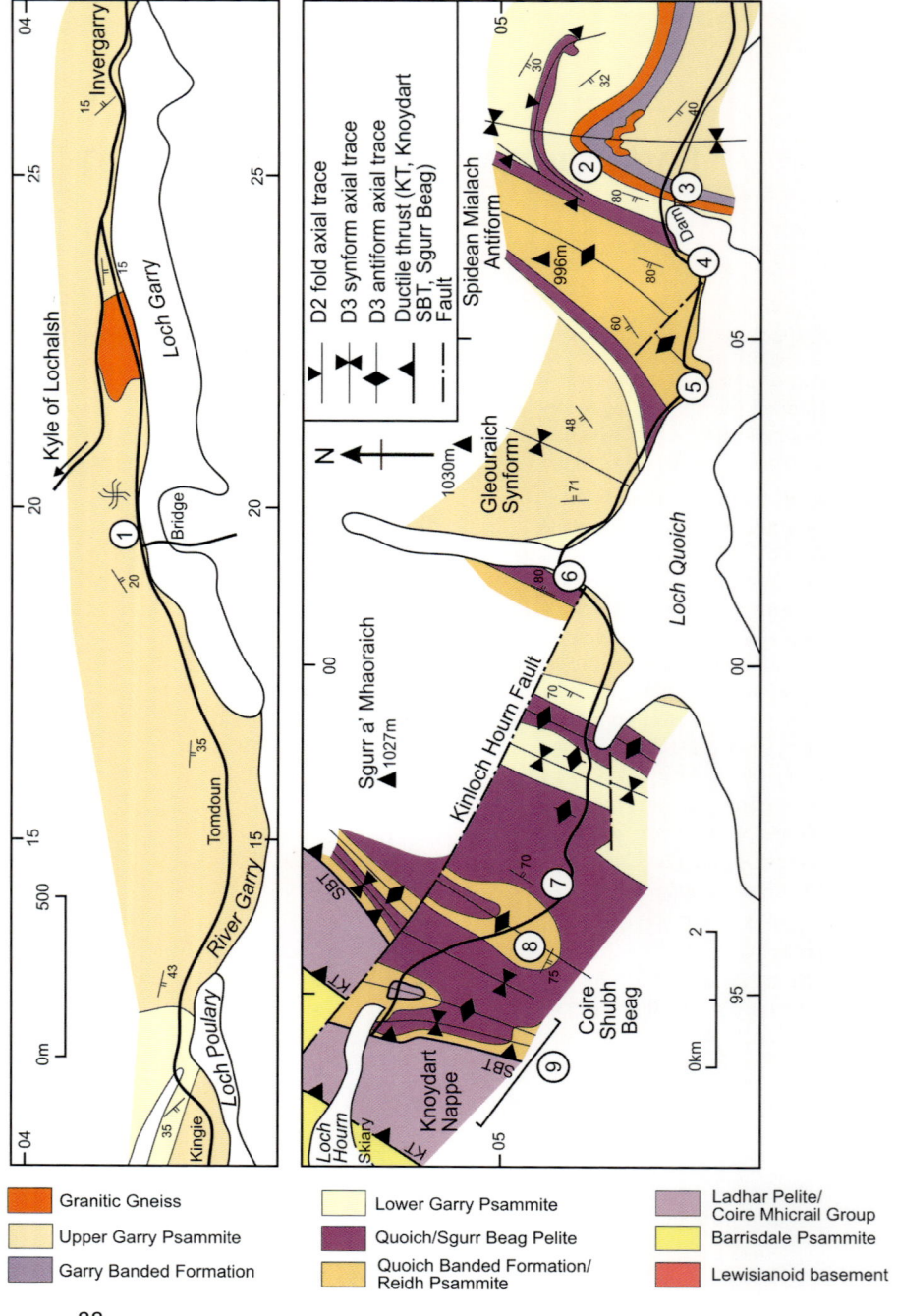

Locality 4.1 [NH 196 023]

Garry Quarry (Fig. 4.1). Curvilinear, recumbent folds in Loch Eil Group psammite.

The first locality is reached by driving west from Invergarry on the A87 until the turnoff for Kinloch Hourn and Tomdoun is reached [NH 243 029]. Turn left onto the Kinloch Hourn road (single track with passing places) and continue along it for 5 km until you reach a small quarry adjacent to the road, just short of the Greenfield bridge across Loch Garry. Parking is available opposite the quarry for no more than two cars. The quarry exposes Loch Eil Group psammites within the regional flat belt and allows an unequalled opportunity to examine the geometry of structures within the psammites in three dimensions. A detailed structural analysis of this quarry, including a map, 3D diagram and stereographic plot of structural elements can be found in Holdsworth & Roberts (1984).

The psammites are folded by tight, recumbent folds that can be traced laterally for some distance. Holdsworth & Roberts (1984) drew attention to the fact that the hinges of individual folds at this locality are curved rather than linear. The curvature of the hinges of the folds in the quarry is symmetrical about a very weak, north-south extension lineation. Such an extension direction is distinctly different in orientation from the usual WNW-directed extension lineations associated with overthrusting during the Scandian phase of the Caledonian orogeny, both in the Moine and the Moine Thrust Belt. It has therefore been suggested that the folds seen in the quarry (D_2 in the local deformation sequence) formed during an earlier tectonic event. Titanites aligned within the north-south extension lineation near Invergarry have yielded a U-Pb zircon age of 470 ± 2 Ma, perhaps indicating that the lineation and associated curvilinear folds formed during the Grampian phase of the Caledonian orogeny (Rogers *et al.*, 2001). Garry Quarry also exposes post-tectonic granitic veins belonging to the Glen Garry Vein Complex (Fettes & MacDonald, 1978), together with mineralised (pyrite, haematite, calcite) faults and joints.

Opposite page:
Fig. 4.1 Location map and general geology of Excursion 4.

Locality 4.2 [NH 076 035] to [NH 082 0390]

Coir' an t-Seasgaich (Figs 4.1 and 4.2). Sedimentary structures and cross-cutting amphibolite sheets preserved in a 'low-strain window' around the Loch Quoich Line.

From Locality 4.1 drive west for 12 km until the road crosses the Allt Coir' an t-Seasgaich at [NH 078 023] (Fig. 4.2). Parking is available for several cars or a small coach on the southwest side of the road. About 1.5 km north of here, on the east flank of Spidean Mialach, lies Coir' an t-Seasgaich, a corrie that contains many large, glacially-scoured exposures visible from the road. The corrie is best reached by walking up beside the Allt Coir' an t-Seasgaich, aiming eventually for the lower exposures in the corrie 2A (Fig. 4.2). A practised and fit hillwalker will reach these slabs in just over half an hour; others should allow longer and all should wear stout footwear. In total, a visit to this locality involves about 4 km of walking and 550 m of ascent.

The walk up the Allt Coir' an t-Seasgaich takes you from the Upper Garry Psammite of the Loch Eil Group (Fig. 4.2) through granitic gneiss, the Garry Banded Formation, another sheet of granitic gneiss, and finally into the Lower Garry Psammite that forms the slabs in Coir' an t-Seasgaich. This section is part of the Glenfinnan/Loch Eil group transition zone described by Roberts & Harris (1983).

On reaching the slabs, 2A (Fig. 4.2), the most obvious difference between the psammites here exposed and those seen in the flat belt at Locality 4.1 is the common preservation of sedimentary structures. A slow traverse across the corrie from this initial location to a point on the east ridge of Spidean Mialach at 2C (Fig. 4.2) should reveal abundant cross-lamination (younging towards the SE), graded beds, slump folds, contorted bedding and pebbly beds. Such excellent preservation of sedimentary structures is unknown elsewhere in this part of the Sgurr Beag Nappe, and illustrates the low state of strain of these rocks. Also indicative of low strain are the amphibolite sheets and pods. As will be seen at Locality 4.3, amphibolites normally occur as concordant sheets; however, the low strain state in Coir' an t-Seasgaich has allowed their original cross-cutting intrusive relationships to be preserved. An excellent example of a cross-cutting amphibolite is exposed in the slabs at 2B (Fig. 4.2). Such preservation of strongly cross-cutting relationships is rare within the Sgurr Beag Nappe.

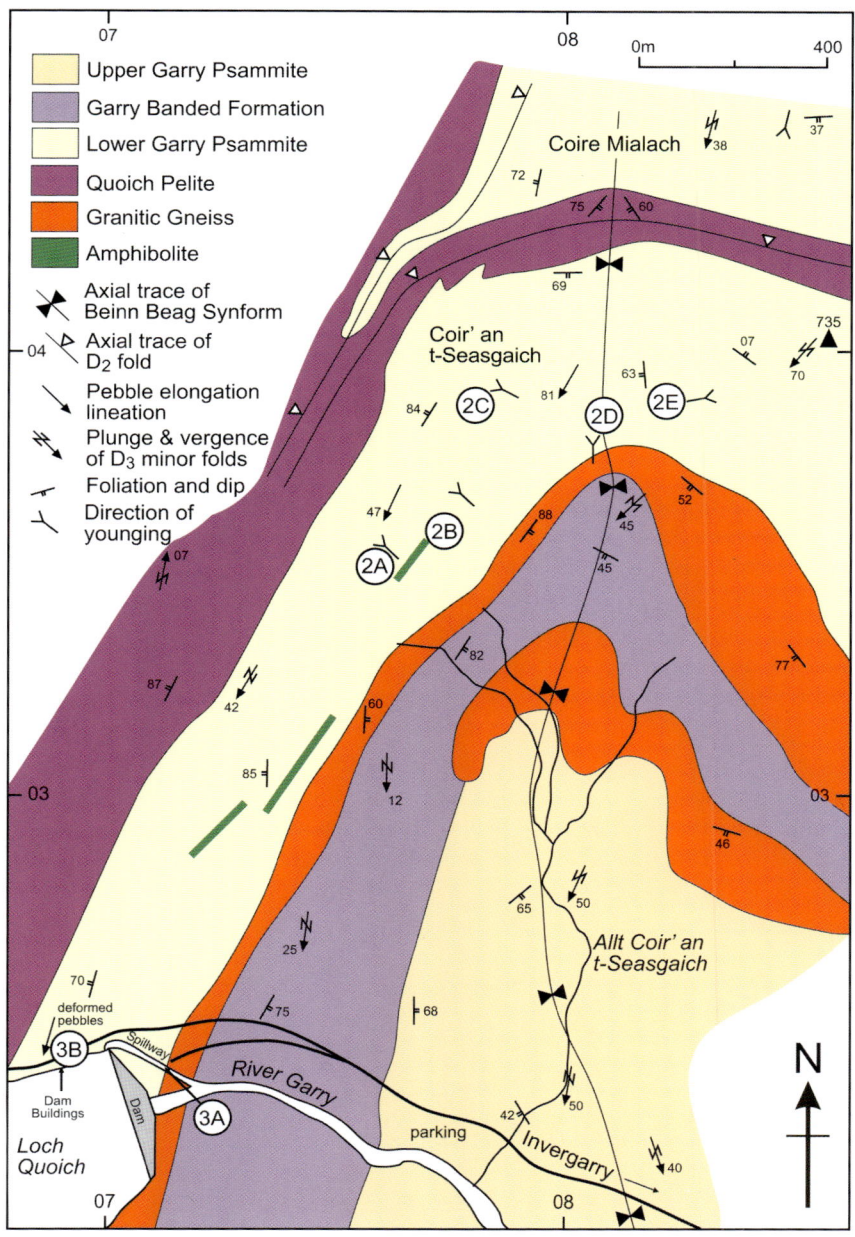

Fig. 4.2 Geological map of Locality 4.2, also showing location of Locality 4.3.

Although the psammites in the corrie are generally at a low state of strain, local areas of high strain, into which sedimentary structures are progressively deformed, can also be seen. Within these zones of higher strain, pebbles in conglomeratic horizons have been deformed to an elliptical shape.

On reaching the ridge 2C a rest should be taken to enjoy the view. Only 1 km to the west is the peak of Spidean Mialach, one of many Munros (3000-foot-high mountains) in this area, to the north the view extends to the high peaks above Loch Cluanie, to the east can be seen Ben Tee adjacent to the Great Glen, and the Monadhliaths beyond, and to the south (weather permitting!) can be seen Ben Nevis and the surrounding mountains.

The psammites observed at Locality 4.1 had a nearly flat sheet-dip, while those seen in Coir' an t-Seasgaich are steeply inclined or vertical. This is because the Loch Quoich Line, the boundary between the regional flat and steep belts (Leedal, 1952; Clifford, 1957; Roberts & Harris, 1983) has been crossed (see 'Summary of Moine Geology' for regional significance of the Loch Quoich Line). The synform marking the course of the Loch Quoich Line in this area can be examined by traversing the ridge on which you are now standing. At stop 2B, and elsewhere throughout the slabs examined so far, the psammites strike approximately NE-SW and young towards the SE. At 2E, 400 m to the east the psammites strike NW-SE and young towards the SW; the Beinn Beag Synform (= the Loch Quoich Line *sensu* Roberts & Harris, 1983) has been crossed. The axial trace of the synform, where the psammites young due south, can be seen at 2D (Fig. 4.2). The rocks immediately east of the Loch Quoich Line do not become flat immediately, but rather their sheet-dip decreases gradually across several kilometres. Walk back down the hill and rejoin the vehicles.

Locality 4.3 [NH 071 023]

Quoich dam spillway (Figs 4.1 and 4.3). Quoich Granite Gneiss, metasedimentary country rock within the Glenfinnan/Loch Eil group transition zone, amphibolites, microdiorite.

From the parking place for Locality 4.2, drive west for about 1 km and park in one of the large lay-bys beside the Quoich dam (Fig. 4.2). Descend SE from the lay-by, alongside the spillway, keeping to the north of the fence. At 3A

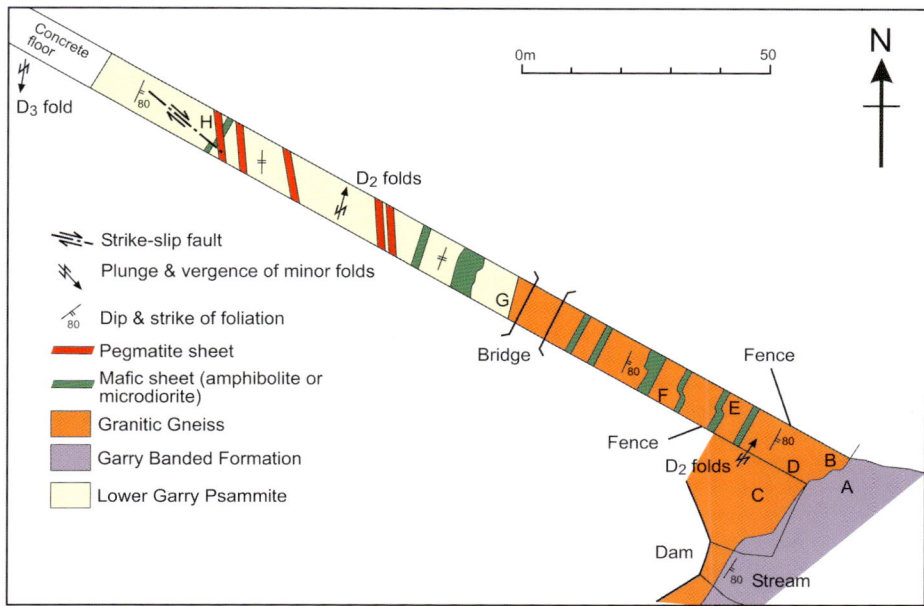

Fig. 4.3 Geological map of Locality 4.3, the Quoich dam spillway.

(Fig. 4.3), the contact is exposed between the Quoich Granitic Gneiss and the structurally overlying Garry Banded Formation. The granitic gneiss lying west of the contact is pale pink, medium-grained and consists mainly of K-feldspar, plagioclase (An_{15-20}) and quartz. The pervasive D_1/D_2 tectonic fabric is defined by discontinuous biotite-rich laminae and by concordant migmatitic *lits*. These have mafic selvedges and probably formed by *in situ* metamorphic segregation. The Quoich Granite Gneiss (part of the West Highland Granitic Gneiss) is regarded as a deformed and metamorphosed granite that was intruded at *c*. 870 Ma (Friend *et al.*, 1997; Rogers *et al.*, 2001). Whether migmatization also occurred at *c*. 870 Ma (Friend *et al.*, 1997) or much later during the Caledonian orogeny (Dalziel & Soper, 2001) is uncertain. The gneiss-metasedimentary rock contact is sharp but lacks the platy fabrics associated with tectonically emplaced basement bodies (see Locality 4.9). No transitional rocks indicative of a metasomatic origin are developed. The migmatitic pelite adjacent to the granite gneiss grades over about 5 m into interbedded psammite and pelitic gneiss. Leucosomes within the pelites are deformed by intrafolial isoclines and an unusual, antiperthitic-bearing variety was probably formed by partial melting (Barr, 1985).

At 3B, the core of a large Z-profile fold is occupied by a patch of remobilised granite gneiss with an undeformed, granitoid appearance. An early amphibolite traced through this locally pegmatitic body is progressively deformed, and altered to biotite and feldspar. Textures within this remobilised rock suggests an origin by partial melting but its chemistry favours subsolidus segregation or re-equilibration after consolidation (Barr, 1985), perhaps related to the intense retrogression evident in its vicinity.

At 3C, a metasedimentary enclave crops out in the spillway wall. It defines an S-profile, reclined D_2 fold pair, and consists of a thin but extensive band of psammite with concordant hornblende schist. At 3D, a second enclave comprising micaceous psammite and quartzite intruded by hornblende schist is also deformed by D_2 folds. Barr et al. (1985) interpret these and other enclaves as deformed xenoliths within an original magmatic granite. Also at 3C, a coarse granitoid patch disrupts the D_1/D_2 foliation and agmatizes an amphibolite sheet (Barr, 1985, Fig. 7.7b). It appears to be associated with a sinistral, extensional shear zone, one of several present in this area.

Return to the spillway at 3D and proceed upslope observing a number of sinistral, NW-SE- trending shear zones within the granitic gneiss. Pegmatitic segregations have developed within some but not all of these shear zones, destroying the foliation in the gneiss. Barr (1985) concluded that partial melting had been localized in the shear zones. This part of the section also exposes numerous isoclinal D_2 folds that deform pegmatites and an earlier D_1 gneissic fabric. In the SW wall of the spillway, an upright D_3 fold pair indicates an antiform to the west, the Spidean Mialach Antiform of Roberts & Harris (1983).

Over the next 30m of section, numerous amphibolite sheets crop out within the granitic gneiss. The margins of these amphibolites are generally concordant with the foliation in the gneiss, but the intrusive origin of one body at 3E is confirmed by the presence of several c.10cm diameter xenoliths of granitic gneiss. In this central portion of the gneiss, D_2 isoclinal folds of the earlier D_1 fabric are common, transposing the foliation everywhere but in the relic fold cores. Similar structures can also be seen in the Ardgour Granite Gneiss (Excursion 2, Localities 2.7 & 2.8; Barr et al., 1985, figure 4b). Where these D_2 folds deform amphibolite sheets, an axial planar hornblende fabric is produced.

At 3F (Fig. 4.3), a 2m-thick SE-dipping microdiorite sheet cuts the spillway. This sheet bifurcates and includes xenoliths of granitic gneiss. Unlike the older hornblende schists and amphibolites, the microdiorite cross-cuts the D_2 fabric in the gneiss. It also cuts across an earlier concordant amphibolite and is largely post-tectonic. It retains a coarse-grained centre and a fine-grained, amygdaloidal margin, but has recrystallized to an amphibolite-facies mineral assemblage. Several smaller microdiorites are present upslope.

At 3G (Fig. 4.3), a sharp contact is exposed between granitic gneiss and a 1m-thick, antiperthite-bearing migmatitic pelite. Upslope from this pelite band, the ground to the top of the spillway is occupied by psammites and quartzites of the Lower Garry Psammite (Roberts & Harris, 1983). These psammites are along strike from those seen in Coir' an t-Seasgaich (Locality 4.2), but lack their abundant sedimentary structures. Only a few highly deformed cross-beds are preserved, and the uniform, finely banded appearance of the psammites, which contain reclined, isoclinal folds, is thought to indicate that they have been highly strained. This high strain has obliterated any original angular discordances, and Roberts & Harris (1983) attribute it to severe upright Caledonian reworking, during the D_3 deformation that produced the regional steep belt west of the Loch Quoich Line.

A 4m-thick amphibolite sheet lies 10m into the psammites. Its northern margin has been interfolded with the psammites by reclined D_2 structures, and the amphibolite carries an axial planar D_2 fabric. Upslope, several north-south-trending pegmatites cut D_2 structures within the psammites and are deformed by late, semi-brittle kink bands. They are little deformed internally, and probably late Caledonian in age.

At 3H (Fig. 4.3), a 1.5m-thick, NE-SW-trending microdiorite sheet is displaced $c.2$m by a NW-SE-trending fault. This fault passes along its length into a brittle, dextral kink band. Adjacent to this fault, in the NE wall of the spillway, the psammites are intruded by a fine-grained, intermediate igneous rock that may be a member of the minette suite of Smith (1979). Complete this section by walking up to the concrete-floored part of the spillway, and look up at the SW wall to observe a mesoscopic, gently-plunging D_3 fold pair verging westwards to the Spidean Mialach Antiform.

Return to the bridge and climb back up to the road. Whilst walking past the dam buildings, note the presence, in psammites of the road section, of several south-plunging D_3 fold pairs that verge westwards towards the Spidean Mialach Antiform. These are cut by SE-dipping microdiorite sheets but fold an earlier amphibolite sheet. At [NH 069 025], opposite the dam, a stream passes beneath the road. On the north side of the road, on the left side of the stream, a deformed pebbly unit is exposed within the psammites. The pebbles of quartz and feldspar are much more strongly deformed than those seen in Coir' an t-Seasgaich. The quartz pebbles define a steep extension lineation and have axial ratios of approximately 3:1:0.2. They help to quantify the amount of ductile strain recorded by the enclosing platy psammites.

Locality 4.4 [NH 062 018]

Quoich Quarry (Fig. 4.1). Highly deformed Glenfinnan Group psammites; microdiorite and felsic porphyrite sheets.

Return to the vehicles and drive *c.* 1 km west to [NH 062 021], where a steep track descends to the lochside quarry. En route, check that the nearer of the two islands in Loch Quoich is connected to the shore; if not, the quarry floor may be flooded and access difficult. Coaches and minibuses should discharge their passengers on the main road and park in the large lay-by 250 m further SW. Once in the quarry, examine the steeply-dipping, planar-banded psammites exposed in the south and west walls. These psammites form part of the Quoich Banded Formation, lie on the eastern limb of the D_3 Spidean Mialach Antiform (Fig. 4.1: Roberts & Harris, 1983) and are very similar in rock type to the Reidh Psammite seen later at Locality 4.9. The psammites are quite extensively migmatized and the regular, planar banding, largely unaffected by minor folding, is typical of a structural setting on the limb of a major, upright Caledonian fold. Foliation surfaces within the psammites carry a steeply-plunging mineral and intersection lineation, that is related to a cleavage lying in places at a low angle to the main lithological/metamorphic banding. The cleavage strikes clockwise of banding and is related to the north-plunging Spidean Mialach Antiform. Between here and the Quoich dam, D_3 fold axes have rotated through the horizontal such that the adjacent Spidean Mialach Antiform and Beinn Beag Synform both open southwards (figure 4.1: Roberts & Harris 1983, figure 2).

Two sets of Caledonian pegmatites cut the psammites in the quarry: an early biotite-bearing, deformed set, typically dipping at 60° towards 270°, and a later, less-deformed, muscovite-rich set, typically dipping very steeply towards 250°, and comparable to those in the spillway.

At the east end of the south face of the quarry, a microdiorite sheet dipping at 45° towards 108° cuts across the psammites. It has a chilled margin and a coarse-grained core, and cuts both sets of pegmatites. It also forms a prominent feature in the north face of the quarry.

If time is pressing, return now to the vehicles. If not, the eastern side of the 'island' lying 200m SE of the quarry can be visited. The level of the loch is commonly low enough for the island to be reached on foot. The island consists of typical Glenfinnan Group migmatitic pelitic gneiss (the Quoich Pelite; Roberts & Harris, 1983), and the eastern side exposes complexly folded pelitic gneiss containing numerous pods and sheets of metabasic garnetiferous amphibolite. Having examined the pelitic gneiss and amphibolites, best exposed at the SE corner of the island, return to the vehicles.

Locality 4.5 [NH 042 019] to [NH 046 016]

Loch Quoich shore section (Figs 4.1, 4.4). D_2 sheath folds, fold interference patterns, boudinage, members of the microdiorite and appinite suites.

From Locality 4.4, drive west for about 2km and park in the large lay-by at [NH 045 018], below the radio mast. This parking space lies above a remarkable stretch of exposures along the north shore of Loch Quoich between [NH 042 019] and [NH 046 016]. The lower the level of the loch the more rock is exposed, but even if the water level is high there is much of interest to be seen.

The rocks of the shore section lie within the Quoich Banded Formation of the Glenfinnan Group (Roberts & Harris, 1983), but unlike the Quoich Banded Formation at Quoich Quarry (Locality 4.4), these rocks lie not on a major fold limb affected by high D_3 strains, but rather within the hinge zone of the D_3 Spidean Mialach Antiform (Fig. 4.1). The low D_3 strain within the hinge zone provides a 'window' back to the pre-D_3 history of the area. A similar low-strain hinge zone was examined in Coir' an t-Seasgaich (Locality 4.2).

The structural history of the shore section is extremely complex, and for a full structural analysis accompanied by detailed maps the reader is referred to Holdsworth & Roberts (1984) and Roberts (1984). In this account, only the salient features of the shore section, of interest to the general reader, are described.

Folds produced by all the major deformation events recognized in the area, D_1-D_4 *sensu* Roberts & Harris (1983) and Holdsworth & Roberts

EXCURSION 4

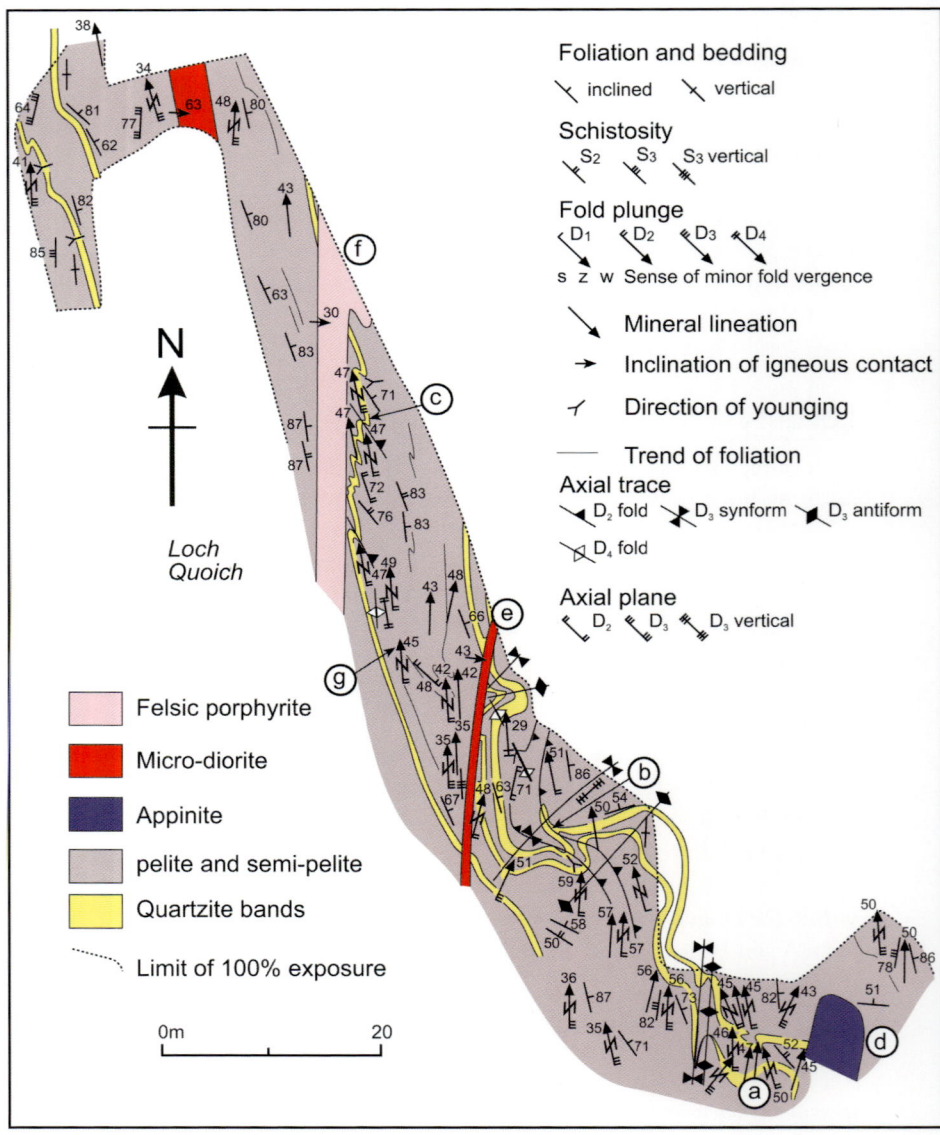

Fig. 4.4 Geological map of the central part of the Loch Quoich shore section (Locality 4.5) (from Holdsworth & Roberts, 1984).

(1984), can be seen in the shore section. D_1 folds are generally small-scale (1m) and most easily recognized where they are folded around D_2 folds. Quartzofeldspathic migmatitic segregations lying within S_1 occur in some pelitic units, but later deformation has resulted in this fabric being almost entirely transposed. Evidence for the early age of this migmatization is clear only in D_2 fold hinges, where S_1 passes around the fold. D_2 structures are ubiquitous in the southern two-thirds of the section. The folds are tight to isoclinal, with an axial planar crenulation fabric that is commonly so strongly developed that it appears penetrative to the naked eye. Open to close D_3 folds have nearly upright NE-SW-trending axial surfaces and an axial planar crenulation cleavage (S_3). These folds commonly reorientate the earlier, much tighter D_2 structures. A solitary, NW-SE-trending D_4 antiform, with no related axial planar fabric, occurs at [NH 043 018].

The shore section is best traversed from SE to NW, and in this guide will be discussed in three sections: (1) [NH 046 016 to NH 044 017]; (2) [NH 044 017 to NH 043 018] (Fig. 4.5); (3) [NH 043 018 to NH 042 019].

(1) [NH 046 016 to NH 044 017]

Begin this traverse at the most southerly set of exposures on the small headland. At this end of the section, tight to isoclinal D_2 folds are exposed, unaffected by later refolding. The folds, both here and throughout the section, are best defined by *c.* 1m-thick bands of quartzite that occur within a striped unit of pelitic gneiss, semipelite, psammite and quartzite. The competence of the quartzite bands in relation to the surrounding metasediments has commonly resulted in tight D_2 fold hinges being cut through by small thrust faults as the folds developed. This phenomenon is well developed in the southeastern part of the section. Examples of D_1 folds and an S_1 fabric can also be seen folded around the D_2 folds in the first 50m of the traverse.

About 60m into the traverse, the first effects of upright, NE-SW-trending D_3 folding can be seen, in the form of an antiform-synform pair with an associated axial planar crenulation cleavage. These D_3 folds are superimposed on earlier, tight D_2 folds. The D_2 hinges are in places cut through by small thrusts and spectacular boudinage occurs within quartzite bands on the long limbs of D_2 folds.

Continuing northwestwards, the slabs are dominated by interference between reclined D_2 folds and upright D_3 folds. Where the quartzite bands were boudinaged during D_2 the boudins have been refolded during D_3, adding to the complexity of the deformation. D_3 refolding of D_2 folds and boudins continues to the end of exposure within section 1 at [NH 045 017], some 140m NW of the start of the traverse.

Several examples of easterly-inclined felsic porphyrites, cutting across all structures within the host rocks, are exposed in this section, at 80m and 100m into the traverse, and at the last exposures 140m into the traverse. This final example is unusual in that the felsic porphyrite has been intruded into an already-present microdiorite sheet. Both sheets cut across minor D_3 folds and fabrics but the microdiorite has been extensively foliated whereas the porphyrite is unfoliated, indicating that the microdiorite is late-tectonic but the porphyrite is probably post-tectonic. Nevertheless, both microdiorites and felsic porphyrites have recrystallized to amphibolite facies mineral assemblages in this area.

From this multiple intrusion walk across $c.115$m of unexposed ground to the start of section 2 at [NH 044 017].

(2) [NH 044 017 to NH 043 018]

A published map (reproduced here, Fig. 4.4) of this part of the section can be found in Holdsworth & Roberts (1984, figure 4). An unpublished map of the whole shore section can be found in Roberts (1984).

Section 2 is again dominated by D_3 refolding of D_2 structures, although D_2 boudinage and fold hinges cut by thrust faults are less common than in section 1. Occasional examples of D_1 folds and fabrics can be seen in D_2 fold hinge zones, and a solitary, upright, NW-SE-trending D_4 antiform has been recognized.

The distinguishing feature of section 2 is the curved nature of the hinges of the D_2 folds. As at Garry Quarry (Locality 4.1), D_2 folds are curved about a N-S extension lineation; however, the curvature of individual folds is here much greater. In some cases the hinges of individual folds curve through nearly 180° in about 1m, giving them a conical or sheath-fold geometry. When such folds are viewed along their extension direction, closed outcrop patterns are seen (similar to those produced by dome-and-basin interference) as well as double vergence (S and Z) within a single fold pair (e.g. Holdsworth & Roberts, 1984, figure 6). The best examples of such closed outcrop or 'eye' structures are exposed in quartzite and psammite bands at stop 5A (Figs 4.4, 4.5). Note the strong extension lineation parallel to the hinges of the folds. Eye structures can also be seen elsewhere in section 2, and less spectacular examples occur in section 1.

Further NW, a number of quartzite bands trace out several upright, NE-SW-trending D_3 folds. However, at 5B (Fig. 4.4), two quartzite bands trace out a very obvious, reclined D_2 pair (openly folded during D_3), showing S vergence. The vergence of this fold pair should be compared with that of a train of D_2 folds exposed in a quartzite band 40m further northwest, 5C (Fig. 4.4), where the D_2 folds consistently show Z vergence. Examination of the completely exposed section between the S and Z folds shows no evidence for a major D_2 fold core. The absence of a major D_2 fold is also indicated by sparsely preserved cross lamination within the quartzites, that shows overall younging to the NE. The change in fold vergence from S to Z is attributed to the curved nature of the D_2 folds producing 'double vergence', the hinges of the D_2 folds having swung through $c.180°$ between the S and Z folds. This phenomenon represents a larger scale version of the eye structures seen earlier in the traverse.

From 5C to the end of section 2, approximately 70m, the main structural features seen are a number of upright, S-profile D_3 fold pairs.

A number of late Caledonian igneous rocks are also exposed in section 2. At 5D (Fig. 4.4), a $c.5$m-wide plug of typical hornblendic appinite is well exposed, cutting across the stratigraphically highest quartzite band. Two easterly-inclined microdiorite sheets are exposed in section 2. At 5C, a $c.1$m-thick microdiorite has been intensely sheared by late, localized movements within the sheet. The margins of the sheet, however, truncate D_2, D_3 and D_4 structures in the surrounding rocks. A larger, non-foliated microdiorite is exposed 60m further NW (Fig. 4.4). A single, 3m-wide felsic porphyrite is exposed at stop 5F immediately west of the quartzite band at 5C. The porphyrite sheet is inclined 30° to the east and is foliated internally. It clearly truncates the D_2 folds, although its relationship to later structures cannot be demonstrated. At 5G a number of structurally early pegmatites are exposed. These pegmatites are folded by D_2 structures and carry the strong north-plunging D_2 lineation present throughout section 2. The age of these pegmatites is unknown.

Fig. 4.5 (above) D_2 sheath fold viewed looking NW. Note the intense L_2 mineral lineation above coin (15 mm diameter) (Locality 4.5 [NH 0438 0174]).

Fig. 4.6 (right) West-vergent D_3 folds directly below Quoich Bridge (viewed north at Locality 4.6 [NH 0148 0406]).

(3) [NH 043 018 to NH 042 019]

There is no exposure gap between section 2 and 3. Section 3 starts at a 20 m-wide, south-facing cliff and is approximately 100 m long.

Only D_3 folds are exposed in section 3, accompanied by an axial planar S_3 crenulation fabric. The overall structure of section 3 is that of an open, NNE-plunging D_3 antiform, around which several quartzite bands can be traced. Sparse but consistent cross-lamination within the quartzites shows the antiform to be upward facing. There is therefore no major structural inversion between sections 2 and 3.

Three foliated microdiorite sheets, all inclined at $c.60°$ to the east, crop out in section 3, at 25 m, 30 m and 60 m into the traverse. These sheets cut across minor D_3 structures, and across the major antiform. The largest and most easterly microdiorite contains xenoliths of country rock quartzite, and also interbanded psammite and pelite containing a pre-existing S_3 fabric. This sheet shows a classic example of the sigmoidal internal fabric described by Smith (1979). The margins of the sheet are intensely foliated at a low angle to the contact, but as the centre of the sheet is approached the fabric weakens and its angle to the contact increases. The centre of the sheet is non-foliated.

Having completed section 1 to 3 of the shore section, most people will now wish to return to their vehicles. However, if time permits, those with more than a passing interest in the Moine will find the complete shore section between here and Quoich Bridge, $c.3$ km west, extremely interesting. It affords a well-exposed traverse across the Loch Eil Group outlier, and its Glenfinnan Group envelope, cropping out in the core of the Gleouraich Synform (Roberts & Harris, 1983), the northward continuation of the Glen Dessarry Synform (Roberts et al., 1984).

Locality 4.6 [NH 015 041]

Quoich Bridge (Fig. 4.1). Short stop to examine spectacular D_2 eye structures refolded by upright D_3 folds.

From Locality 4.5 drive west for $c.3.5$ km and park (space available for about five cars) on the west side of the bridge (weight limit 10 tons) over the north arm of Loch Quoich. Look over the SW corner of the bridge, and if the rocks are exposed approximately 6m or more above the level of the loch, descend to them. If not, continue to Locality 4.7 as the level of the loch is too high.

This locality exposes the Garry Banded Formation (Roberts & Harris, 1983) on the west limb of the Gleouraich Synform. Numerous upright, NNE-plunging D_3 folds, verging west away from the Gleouraich Synform to the east, can be seen (Fig. 4.6). These tight folds refold earlier, isoclinal D_2 folds. The D_2 folds have intensely curved hinges, and numerous three dimensional eye structures, with an accompanying axial extension lineation, are exposed. Some of the eye structures are folded by D_3 folds. When well exposed this locality is extremely photogenic. Return to the vehicles and proceed to Locality 4.7. Coaches, however, should proceed no further than the sheepfolds 1.5km to the SW [NH 004 033], where there is space to turn around.

Locality 4.7 [NG 9668 0407]

Coire Shubh road section (Fig. 4.1). Short stop to examine the envelope of the Coire Shubh Pegmatite Complex, D_3 folds in Glenfinnan Group semi-pelites, both cut by late-D_3 pegmatites.

From Locality 4.6, drive on past the western end of Loch Quoich and across the watershed (5.5km). Drive down the first steep hill (with a stone embankment on the left) and park on the firm ground to the right of the road at [NG 9668 0407], 70m past a stone bridge at the foot of the hill. A *roche moutonnée* beside the passing place consists of migmatitic Glenfinnan Group semipelite and contains several upright, NNE-SSW-trending, reclined refolds of D_3 (steep belt) age. Early leucosomes are folded and/or transposed into the strong axial-planar crenulation cleavage. Several NW-SE-trending white pegmatite veins are ptygmatically folded about D_3 axial planes. These belong to the suite of late-D_3, Caledonian pegmatites that will be visited at the next locality. Several large, white pegmatites are visible in surrounding exposures. The semi-pelites and pelites form the envelope to the Coire Shubh Pegmatite Complex, which is developed in a psammite-cored D_3 antiform. Proceed to Locality 4.8.

Locality 4.8 [NG 960 045]

Coire Shubh Beag (Figs 4.1, 4.7). Late Caledonian Pegmatite Complex in core of D_3 steep-belt fold.

From Locality 4.7, drive on *c.*900m until the road crosses a major stream, the Allt Coire Shubh Beag, at [NG 9610 0468], 200 m after cutting through a rocky spur. One car can be parked 100m east of the stream without blocking the passing place, and there are also two large passing places *c.*800m to the west, on either side of the abandoned building at Coire Shubh [NG 958 053]. The pegmatite complex is developed in the core of a major, SW-plunging D_3 antiform (Fig. 4.7). The psammites and semi-pelites that occupy the core of the fold are probably equivalent to the Reidh Psammite or the Quoich Banded Formation (Tanner, 1971; Roberts *et al.*, 1987). They define a major fold interference pattern, occupying the core of a (sheath-like?) D_2 recumbent fold that has been refolded by D_3. Climb the hillside to the top of the rocky spur, keeping *c.*100m SE of the Allt Coire Shubh Beag, observing the white pegmatite dykes visible in slabs across the stream. About 120m from the road, at 8A [NG 9612 0449], bedding strikes almost east-west on the SE limb of the D_3 antiform close to its core. Open to close, Z-profile D_3 folds have a weak NNE-SSW-trending axial planar fabric. Compare these with the tight D_3 folds at Locality 4.7 on the fold limb. Intrafolial, isoclinal D_2 folds of bedding and of early migmatitic leucosomes are locally preserved. These folds are present throughout the centre of the psammite, i.e. in the core of the major D_3 fold. Move WSW towards the stream and observe late pegmatites of the complex forming ESE-WSW-trending dykes. They define open D_3 folds and carry a weak axial-planar fabric.

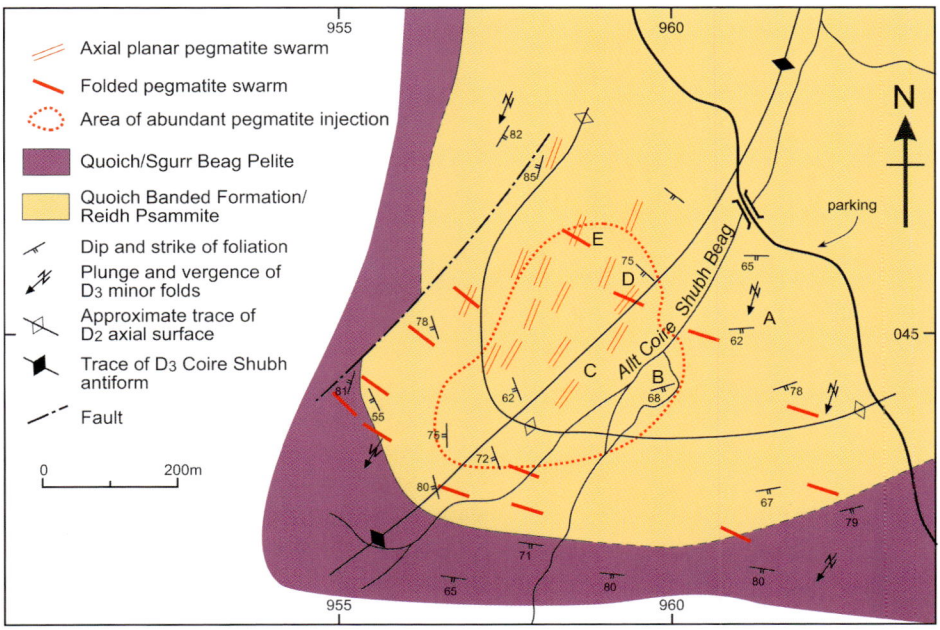

Fig. 4.7 Geological map of Locality 4.8, Allt Coire Shubh Beag.

Proceed upstream to the confluence, noting the small dams that divert water via a conduit to Loch Quoich. Between the two streams, 8B, bedding strikes WNW-ESE in the core of the D_3 fold and is disrupted by numerous pegmatites carrying a weak, NNE-SSW-trending D_3 fabric. At 8C, a number of sloping exposures consist almost entirely of pegmatite. Several generations of pegmatite are present, along with patches and streaks of micaceous restite and some little-modified psammite blocks. Some of the larger psammite rafts in the centre of the exposure contain tight to isoclinal, E-W-trending D_2 folds. The core of the D_3 antiform runs up the Allt Coire Shubh Beag to the ridge, and to the north it runs through the tree-covered crags below the pylon line. Walk downhill on a bearing of 025° to 8D, a large sloping exposure with abundant white pegmatites. Bedding trends NW-SE and the psammites are broken up by concordant, weakly foliated pegmatites to form a migmatite. This is cut by NE-SW-trending, muscovite-rich pegmatite dykes up to 4m thick, that in turn are cut by weakly foliated, muscovite-poor dykes striking at 120°. These dykes belong to two major swarms, one approximately axial planar to the major D_3 fold of bedding, and one that defines an open fold having the same axial plane as the major fold of bedding but a smaller interlimb angle. Apophyses of the axial planar set define open D_3 folds. Sheets, patches and swarms of micaceous restite are common.

Cross the gully and proceed to the large rocky spur at 8E. This locality contains the strongest evidence for the origin of the micaceous restite. The psammites strike at 100° and are cut by 025°-trending pegmatites that have micaceous reaction rims against the host psammite. These zones probably represent restite after extraction of quartzofeldspathic components from the psammite, possibly by partial melting (Barr, 1985). The pegmatites contain streaked-out relics of psammite and restite and, locally, zones of restite are developed with little or no pegmatite. The varying ratios of pegmatite to psammite and restite indicate that quartzofeldspathic material has moved some distance, perhaps as a melt, so that all stages are seen from restite with virtually no pegmatite to pegmatite with virtually no restite.

Walk NNW across the exposure, noting the presence of S-profile D_3 folds indicating that we are now on the NW limb of the major antiform. The D_3 fabric also becomes stronger to the NW. Both on the exposure scale and on the scale of the NW-SE dyke swarm (Fig. 4.5), the pegmatites show evidence for emplacement after some D_3 shortening, i.e. they cut tight D_3 folds but themselves carry a weak D_3 fabric or have been gently folded. Walk down the hillside to the road and return to the vehicles.

Invergarry to Kinloch Hourn

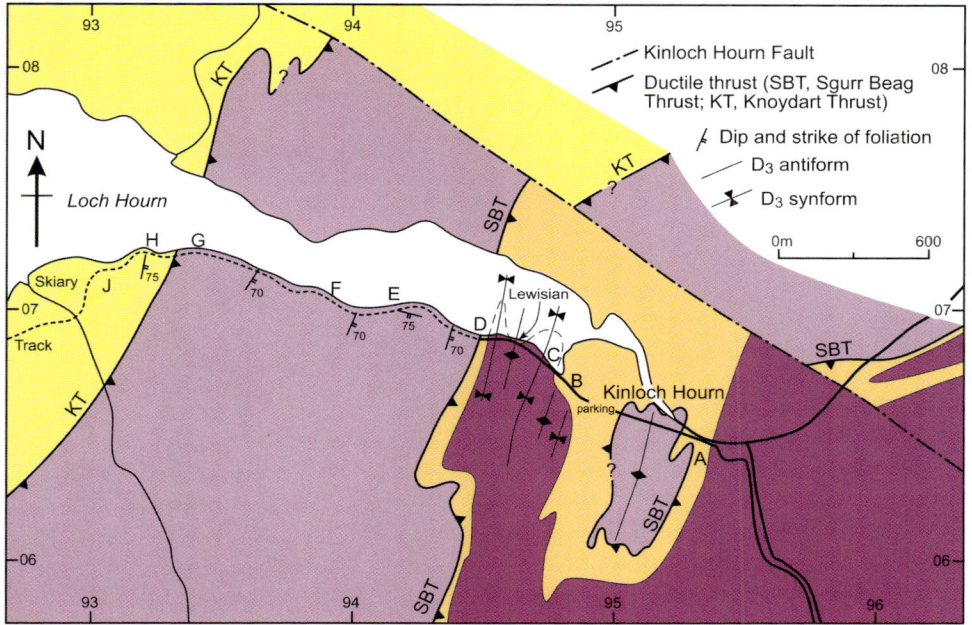

Fig. 4.8 Geological map of Locality 4.9, Kinloch Hourn to Skiary (lithostratigraphic symbols as Fig. 4.1).

Locality 4.9 [NG 953 064 to NG 933 072]

Kinloch Hourn to Skiary (Figs 4.1, 4.8). Knoydartian pegmatite, steep-belt folds, Sgurr Beag Thrust, Lewisianoid inlier, Caledonian pegmatites, Knoydart Thrust.

From Coire Shubh, drive NW along the road for *c.*2km to Kinloch Hourn Farm where parking is available for a nominal sum. Overnight parking is also available, as well as a toilet, tea room and accommodation. Details can be obtained by contacting Martin Riley (Tel: 01809 511253). Return on foot to the road cutting 70m east of the junction with the private road to Kinloch Hourn Lodge, 9A, [NG 954 064]. The Kinloch Hourn fault forms a marked topographic feature north of here and is followed closely by

107

the pylon line. A transitional contact is exposed between the Sgurr Beag (= Quoich or Lochailort) pelite to the east and the Reidh Psammite to the west (Tanner, 1971). This locality lies on the eastern limb of the D_3 Kinloch Hourn Antiform and so a D_3 synform is inferred to lie within the Sgurr Beag Pelite, between here and the Coire Shubh Antiform (Fig. 4.1). The rocks are steeply ESE-dipping and moderately platy, with early migmatitic leucosomes and garnets that form augen ('eyes') within the dominant fabric. Feldspar porphyroclasts are wrapped by a quartz ribbon fabric that is present in the pegmatites and in the more psammitic metasedimentary rocks. A number of deformed pegmatite and quartz-muscovite veins and pods are also present. If these are interpreted as syn-metamorphic segregations, the Rb-Sr muscovite age of 755 ± 19 Ma obtained from one of these veins by Piasecki & van Breemen (1983) suggests that the early high-grade metamorphism of these rocks occurred during the Neoproterozoic Knoydartian event.

The exposure of Reidh Psammite behind the fence on the south side of the road contains several reclined, tight-to-isoclinal S-profile fold pairs that occupy low strain augen. These have a strong axial planar quartz fabric but fold a fine lamination and, in more micaceous bands, an earlier fabric. Follow the exposure southwards for $c.$ 50m, to a point on a bearing of 152° (from grid north) from the south end of the bridge. A tight, upright fold pair deforms a planar fabric in which the early leucosomes are streaked out and form augen. The fold pair has an associated axial planar and an open-folded pegmatite, similar to those of the Coire Shubh Complex, supporting assignation of at least some of the upright folds to D_3 (steep belt), but much of the platiness to an earlier event.

Walk 70m west to the foot of the next spur. This platy semipelite is relatively non-migmatitic and fine-grained. It lies structurally below the Reidh Psammite in the core of the Kinloch Hourn Antiform, and may represent an upfold of the Morar Group lying below the Sgurr Beag Thrust (Barr, 1983, 1985). The semipelite is thoroughly recrystallized, but quartz and feldspar grains define a shape fabric, and quartzofeldspathic segregations, garnets and mica form augen in a manner typical of rocks from the vicinity of the Sgurr Beag Thrust. The platy fabric is crenulated by upright, steeply-plunging folds with variable vergence. Associated quartz rodding plunges at 75° to 148°. The Kinloch Hourn Antiform is inferred to be sheath-like, with variable plunges in an east-west zone near its centre and steep plunges at its northern and southern closures. The northern closure of the semipelite/psammite contact is not exposed, and is inferred from fold and foliation trends within the semipelite to be a steeply south-plunging *synform*. This is because fold axes have rotated from north-plunging, through vertical, to south-plunging in approaching the steeply south-plunging extension direction.

Return to the parking place and walk along the crags south of the tidal pool towards 9B [NG 9475 0675]. These crags expose the transitional contact between Reidh Psammite (east) and Sgurr Beag Pelite (west). Quartzose psammites and highly migmatitic micaceous psammites and semi-pelites all carry an intense planar fabric whose strike varies from north-south to east-west on open, south-plunging D_3 fold pairs verging towards the Kinloch Hourn Antiform to the east. Early leucosomes, quartz veins and large pegmatite pods are intensely deformed (Barr, 1985, Fig. 7.4), feldspar augen being wrapped by quartz ribbons. In non-migmatitic psammites, quartz ribbons parallel this fabric and intrafolial, S-profile isoclines are preserved. The platy fabric is present throughout the Reidh Psammite, and is probably related to the presence of the Sgurr Beag Thrust at its base.

Pelites exposed by the roadside strike north-south, and swing anticlockwise to east-west in the core of a steeply south-plunging D_3 synform. A down-dip mineral lineation is observed in the synform core, but the platy fabric clearly passes around this fold. Immediately before the jetty, at the top of the exposure south of the road, intrafolial D_2 folds can be seen in low strain augen within the platy fabric. Descend west of the jetty to the point at 9C [NG 9470 0683], where a series of south-plunging D_3 folds carry a north-south trending axial-planar crenulation cleavage. The earlier, D_2 platy fabric contains augen of leucosome and garnet, and in thin section it too is observed to be a crenulation cleavage. Walk west to the next bay, where a body of feldspathic augen gneiss occupies the core of a complex south-plunging antiform. It lies close to the boundary between the Reidh Psammite and the Sgurr Beag Pelite and contains isoclinally folded pegmatites to which the platy fabric is axial-planar. Also present are some late, cross-cutting pegmatites, and towards the western end of the exposure, several concordant pods and sheets of somewhat biotitized hornblendite. These hornblendites comprise one of the Lewisianoid basement bodies recognized by Tanner (1965, 1971) and interpreted as a tectonically-emplaced inlier along the course of the Sgurr Beag Thrust. The augen gneiss could also be part of the basement sheet, but it has a metasedimentary bulk chemistry (Barr, 1983) and could represent a basal Moine arkose.

Proceed to the west of the jetties, crossing a synform cored by Sgurr Beag Pelite that contains several tight, north-south trending D_3 folds. The exposure to the west of the boathouses comprises platy, migmatitic Reidh Psammite, intruded by late Caledonian pegmatites. The final 10m is more quartzitic, and the Sgurr Beag Thrust is inferred to lie in the rubble-choked gully at the end of this exposure (Tanner, 1965, 1971; Barr, 1983). The platy fabric within the Reidh Psammite and the lower part of the Sgurr Beag Pelite is considered to result from intense ductile strain (simple shear) in the hanging-wall of the Sgurr Beag Thrust, and the fact that this fabric is folded by the upright north-south D_3 structures indicates that thrusting occurred prior to formation of the steep belt.

The sequence of events observed within the Sgurr Beag Nappe at Kinloch Hourn is: D_1 – early fabric, high-grade metamorphism, migmatization, garnet growth; D_2 – ductile thrusting, development of platy fabrics; D_3 – upright refolding on NNE-SSW axial planes with curved hinges, pegmatite emplacement. It is not possible at this low level within the nappe to identify unambiguously the north-south curvilinear folds seen at Localities 4.1 and 4.5.

The Morar Group west of the Sgurr Beag Thrust carries an intense planar fabric with a down-dip mineral lineation, contains ribboned quartz veins and ?syn-thrusting pegmatites, and becomes less micaceous westwards. All planar discordances (e.g. cross-bedding, cross-cutting quartz veins and cleavages) have been eliminated (see also Rathbone & Harris, 1979; Rathbone et al., 1983). Towards the end of the road, micaceous psammite bands become common. The concordant quartz veins are joined by long-limbed isoclinal folds of quartz veins, pegmatites and calc-silicate ribs, indicating a westward reduction in ductile strain.

The calc-silicate rocks, seen as pale coloured pods and ribs a few centimetres thick, were described in detail by Tanner (1976). They contain garnet, hornblende, calcic plagioclase and occasional pyroxene, and indicate middle to upper amphibolite facies conditions (Tanner, 1976). The enclosing psammites were assigned to the Coire Mhicrail 'Group' by Tanner (1971), who suggested that they represent a local psammitic variant of the Morar Schist, rather than Lower Morar (= Barrisdale) Psammite.

Beyond the jetty at 9D, semi-pelites carry a semi-penetrative mica fabric with a strong down-dip lineation, presumably related to the Caledonian Sgurr Beag Thrust, and an earlier, slightly oblique schistosity lying anticlockwise of the dominant foliation. Together with garnet augen within the platy fabric, this observation demonstrates that the Morar Group had been deformed and metamorphosed prior to Caledonian ductile thrusting.

Continue westwards around the promontory to 9E where the psammites trend east-west and lack a strong fabric but are folded on north-south axial planes. In the next bay, open-folded zones with ESE-WNW-trending bedding and discordant pegmatites and quartz veins alternate with NNE-SSW-trending zones where a strong fabric dips at 70° towards 110° and is axial planar to tight folds of pegmatites. Pegmatites in the east-west domains both cut and are folded by north-south structures or are axial planar with folded east-west apophyses; they are probably syn- to late-tectonic. Also present are earlier, intrafolial folds and somewhat-deformed sedimentary structures. By 9F the east-west domains preserve sedimentary structures, younging southwards. Some cross-beds are oversteepened, either as a result of east-west shortening or as an original sedimentary feature. Round the promontory at the western end of the bay, isoclinal folds are seen in a north-south zone and boudins of Caledonian pegmatites become common. The age of these north-south folds is unclear. They could be related to the D_2 Sgurr Beag Thrust, the north-south trending platy zones representing minor ductile shear zones within the Morar Group, or they could be of D_3 'steep belt' age. The latter interpretation is favoured by the presence of syn-tectonic pegmatites, but it is also possible that these pink-weathering pegmatites are not the same as the white-weathering ones at Kinloch Hourn and Coire Shubh.

Continue to walk westwards along the straight section of coast. Note the general steep ESE dip of the psammites, that are flaggy but contain discordant quartz veins and occasional steeply-plunging upright folds. At 9G a Z-profile fold pair is succeeded westwards by a 50m-wide platy zone with abundant concordant quartz and pegmatite veins as well as some that are slightly cross-cutting. About 20m east of the small promontory in the centre of the bay, the micaceous and calc-silicate-bearing psammites of the Coire Mhicrail 'Group' give way abruptly to the grey and featureless Barrisdale Psammite. Both units are finely laminated and carry a strong platy fabric.

Walk westwards along the track and descend to the foreshore by the cairn, 9H, observing the variably platy psammites that include a pegmatite-rich zone of late, east-west folding. Towards the western end of this exposure, intrafolial folds and possible deformed cross-beds begin to appear. The next exposure, 50m further along the shore, contains quite large (0.5m wavelength) isoclinal folds. Return along the fence to the track, to a zone of tight to isoclinal folds that deform bedding and ribbon-like quartz veins, 9J. Some of these appear to define sheath folds with steeply-plunging axes. The strain gradient is similar to that observed at the Sgurr Beag Thrust, and, if anything, the high-strain zone in the footwall is wider (200m). As at the Sgurr Beag Thrust (Rathbone *et al.*, 1983), the footwall platy zone is much wider than the hanging-wall platy zone. The sharp boundary between the two psammitic units is considered by Barr *et al.* (1986) to represent the Knoydart Thrust, emplacing the Coire Mhicrail 'Group' onto the Barrisdale Psammite. The thrust cuts up-section to the west so that in Glen Barrisdale it lies at the base of the Knoydart Pelite (line 'C' of Tanner, 1971, figure 2) and in western Knoydart it lies at a low level within the Knoydart Pelite. Return along the track to Kinloch Hourn, but if the weather is clear it is worth diverting to the high point on the track, 800 m SW of Skiary, from where excellent views may be had of the surrounding hills. To the WSW lies Meall nan Eun, and in the background, Ladhar Bheinn and Beinn na Caillich. To the north lies Carn nan Caorach, with the Saddle ridge in the background, to the NE lie Sgurr na Sgine and Sgurr a' Bhac Chaolias, and to the ENE, Buidhe Bheinn.

Excursion 5
Glen Moriston and Glen Shiel

Tony Harris and Rob Strachan

Purpose: A general excursion across the Sgurr Beag Nappe.

Aspects covered: Metasedimentary lithologies typical of the Loch Eil and Glenfinnan groups, West Highland Granitic Gneiss, Caledonian (Cluanie) Granite, deformed cross-bedding, Sgurr Beag Thrust Zone, Lewisianoid slices.

Maps: OS: 1:25,000 sheets 414 Glen Shiel & Kintail Forest and 415 Glen Affric & Glen Moriston; BGS: 1:50,000 sheets 72W Kintail, 72E Glen Affric and 73W Invermoriston.

Type of terrain: Road-side outcrops, a quarry, and some rough hill walking.

Distance and time: 30km driving. 6 hours.

Short itinerary: Localities 5.2, 5.4A, 5.5A and 5.6A.

The excursion can be commenced by travelling from either Invermoriston or Invergarry. However, a start from Invermoriston is the more convenient if travelling by coach, as it is not then necessary to turn the vehicle at or near to Locality 5.1. To the west of Invergarry and Invermoriston, for many kilometres, the solid geology is dominated by flaggy, generally flat-lying psammitic rocks of the Loch Eil Group. These carry common to abundant thin ribs and lenses of pale-weathering calc-silicate rocks and often contain concordant stripes and bands of meta-igneous hornblende schist. Dykes, sheets and veins of microdiorite, granite and pegmatite represent Caledonian igneous activity.

If travelling from Invergarry, it is worth stopping on the watershed at the scenic viewpoint on the south side of the A87 at [NH 195 035], about 8km west of Invergarry, where there is parking for coaches. There are splendid views from here of the mountains to the west, and individual topographic features are identified on the metal surface of the viewpoint. Looking west, the rounded hilly topography (e.g. Glas Bheinn [NN 134 975]) has been carved from the generally flat-lying psammites of the Loch Eil Group, with their granitic vein complexes. In the distance to the west rise spectacular, craggy and generally higher mountains, such as Gairich, 919m (3015ft) [NH 026 995] and Spidean Mialach, 996m (3268ft) [NH 066 043], marking the steeply inclined, strongly folded, diverse pelitic and psammitic formations that comprise the Glenfinnan Group. These rocks occur to the west of the Loch Quoich Line (Clifford, 1957; Roberts & Harris, 1983) and are described later in this Excursion (Locality 5.5) and also in Excursion 4. The floor of Glen Garry is occupied by Lochs Garry and Poulary.

To reach Locality 5.1 from the scenic viewpoint, continue northwards to the foot of Glen Loyne where the A87 is joined from the east by the A887, and turn right. Drive eastwards along Glen Moriston for approximately 7.5 km. Locality 5.1 is a large cutting on the south side of the road just to the east of an entrance to the forest at [NH 2854 1219]. Parking for a coach or several smaller vehicles is available in the entrance, provided it is not in use at the time. If travelling from Invergarry, it is necessary at this point to turn the coach to face west. This can be accomplished here with care (*note: cars travel fast along this particular section of the A887*), but turning points can also be found further east.

Locality 5.1 [NH 2864 1219]

A887 road cutting (Fig. 5.1). Loch Eil Group psammites and veins of late Caledonian granite and pegmatite.

This locality exposes flat-lying, Loch Eil Group micaceous psammites that are typical of the 'flat-belt' in this part of Inverness-shire. The absence of sedimentary structures in combination with the very flaggy aspect of the psammites probably results from high tectonic strains imposed on these rocks during tight-to-isoclinal D_1 and D_2 fold phases (see also Excursion 4, Locality 4.1; Holdsworth & Roberts, 1984). The alignment of biotite and muscovite sub-parallel to compositional layering is thought to represent a composite S_1/S_2 fabric. Several layers of hornblende schist are present and are interpreted as early metabasic intrusions that record all the deformation events that have affected their host Moine rocks. The most prominent layer is $c.2$ m thick and extends for some distance along the lower level of the cutting. These metamorphic rocks are intruded by numerous coarse-

Glen Moriston and Glen Shiel

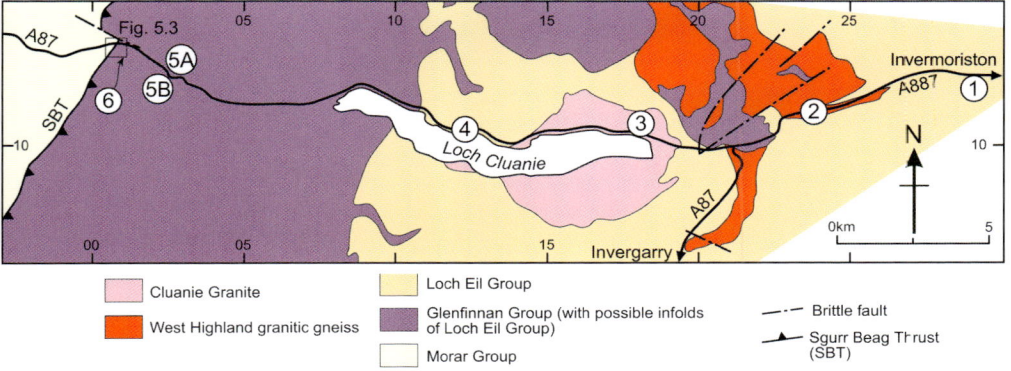

Fig. 5.1 Map showing the regional geology, route and approximate locations of the localities.

grained, granitic and pegmatitic veins and sheets, varying in width from a few centimetres to 10m at the west end of the locality (Fig. 5.2). Some are concordant with the flaggy banding in the Moine rocks, but most are steeply-dipping and strongly discordant. There is no consistent set of cross-cutting relationships between these intrusions which appear to have been emplaced more or less contemporaneously. Contacts with the Moine rocks are invariably sharp. There is no sign that any of the intrusive sheets or veins have been affected by any ductile deformation, and they therefore appear to have been intruded post-D_2. These intrusions represent part of the late Caledonian Glen Garry Vein Complex (Fettes & MacDonald, 1979).

En route westwards to Locality 5.2, various road cuttings expose sub-horizontal to gently-dipping Loch Eil Group psammites that are similar to those of Locality 5.1. There are few opportunities for parking a coach close

Fig. 5.2 Steep, ramifying pegmatites of the Glen Garry Vein Complex intruding flat-lying Moine psammites at Locality 5.1. Note the sheet of concordant amphibolite at the base of the exposure.

115

to any of these, but small groups in cars or minibuses could park with care on the side of the road if they should wish to inspect any of these exposures. Locality 5.2 lies approximately 5km west of Locality 5.1, on the south side of the A887. Parking is available in a lay-by for a coach or several smaller vehicles on the south side of the road at [NH 2364 1121] adjacent to the memorial to Roderick McKenzie.

Locality 5.2 [NH 2348 1115]

A887 road cutting (Fig. 5.1). Glen Doe Granitic Gneiss and hornblende schist bodies.

Walk west from the parking place along the extensive road cutting that exposes the Glen Doe Granitic Gneiss, a unit of the regionally extensive West Highland Granitic Gneiss (see also Excursions 2 and 4). The granitic gneiss is grey, medium-grained, and texturally homogenous, and carries a penetrative deformation fabric defined by aligned micas and quartzo-feldspathic aggregates. The planar component is interpreted as a composite S_1/S_2 fabric and is sub-horizontal to gently-dipping. Careful inspection of foliation planes reveals a N-S trending L_2 mineral and extension lineation. The granitic gneiss is intruded by several sheets of hornblende schist that vary in thickness from a few centimetres to one metre. These intrusions carry a penetrative S_1/S_2 fabric defined by aligned hornblende grains, and their margins are sub-concordant with the banding in the host granitic gneiss. Plausible examples occur of mesoscopic D_2 folds of hornblende schists, accompanied by axial-planar S_2 crenulations of S_1. However, in places the original intrusive geometry gives rise to 'pseudo-folds' where hornblende schist sheets bifurcate.

This locality is significant in discussion of the origin of the protolith of the West Highland Granitic Gneiss that was intruded at $c.870$ Ma (Friend *et al.*, 1997; Rogers *et al.*, 2001). Millar (1999) and Dalziel & Soper (2001) point out that both the Glen Doe Granitic Gneiss and the hornblende schists are structurally indistinguishable in that they have both been deformed by D_1. Barr *et al.* (1985) proposed a syn-D_1/MS_1 age of intrusion of the protolith of the granitic gneiss because of the absence of a distinct thermal aureole within adjacent Moine host rocks. However, in the view of Dalziel & Soper (2001) there is nothing to preclude an entirely pre-D_1 age.

Accordingly, the hornblende schists and the granitic gneisses have been reinterpreted by Millar (1999) and Dalziel & Soper (2001) as a pre-tectonic, bimodal igneous suite. The tholeiitic chemistry of the hornblende schists is consistent with emplacement into thinned continental crust during crustal extension and development of the Moine sedimentary basin (Millar, 1999). Melting of Moine sediments and/or underlying basement as a result of the advection of heat by emplacement of basaltic magma at deeper crustal levels is thought to have produced the protolith of the granitic gneiss (Dalziel & Soper, 2001; Ryan & Soper, 2001). Interested parties may wish to obtain permission from Ceannacroc Lodge to examine further the outcrops of granite gneiss and associated amphibolites and hornblende schists exposed in the River Doe. Publications by Peacock (1977), Millar (1999) and Dalziel & Soper (2001) provide sufficient information and grid references of critical localities.

Locality 5.3 is reached by driving westwards past the junction with the A87. The road then starts to climb towards the dam at the eastern end of Loch Cluanie, a hydroelectric reservoir. The nature of the glacial erratics at the roadside clearly indicates where the route has crossed the eastern margin of the Cluanie Granite; many natural exposures and quarries opened for dam and road construction are available near the road. The steep rocky mountain to the south is made entirely of Cluanie Granite. The largest of the quarries is recommended as Locality 5.3. Parking is available at the north end of the Cluanie dam for a coach or several smaller vehicles, and access is by a rough quarry track about 300m long. Care should be taken when approaching quarry faces as these may be unstable. Hard hats should be mandatory.

Locality 5.3 [NH 185 103]

Large quarry to the north of the Cluanie dam (Fig. 5.1). Cluanie Granite.

The Cluanie Granite (Leedal, 1952) is a handsome pink, megacrystic granite comprising quartz + oligoclase + microcline + microperthite + hornblende + titanite. Microcline-microperthite occurs as small megacrysts, while the oligoclase shows spectacular zoning that can readily be seen in hand specimen. Locally, the granite contains hornblende-rich mafic enclaves. There is no sign of any systematic orientation of either the megacrysts or the enclaves. The pluton has yielded a Rb-Sr whole rock isochron of 425 ± 4 Ma

(M. Brook, quoted in Pankhurst, 1982), that may approximate to the time of intrusion, but cannot be viewed as definitive. On the north face of the quarry, the granite is cut by a microdiorite sheet that dips southwards at about 60°. Detailed investigation reveals that the margins of the microdiorite sheet are lobate in places, perhaps indicating that it was emplaced into granite that was only partially crystalline at the time of intrusion.

Locality 5.4 lies 6.7km to the west of the Cluanie dam at [NH 1238 1048] where parking is available in a large lay-by on the south side of the A87.

Locality 5.4 [NH 1230 1034]

Loch Cluanie shoreline (Fig. 5.1). Sedimentary structures and polyphase folds within Loch Eil Group metasediments.

If water levels in the loch are low, as is commonly the case at the time of writing, walk SE from the lay-by to an obvious promontory on the loch shoreline at Locality 5.4A [NH 1230 1034]. Excellent washed surfaces expose banded psammites and semi-pelites that are probably assignable to the Loch Eil Group. The psammites contain abundant examples of small-scale cross-bedding, and stripes and laminae of micaceous material also assist in defining the original bedding surfaces. Tectonic strain is heterogeneous at this locality and its effect on the original sedimentary structures, therefore, varies. In places, original angles between foreset and topset have been increased by deformation, while elsewhere angles have been reduced such that way-up cannot be reliably read. Numerous folds are present, and the locality is therefore useful not only for demonstrating the effects of strain on sedimentary structures, but also for discussion of facing directions in the axial planes of folds. The difficulty of distinguishing truncated fold limbs from original cross-bedding, a common problem in the Moine, as in other moderately deformed terrains, can also be introduced.

Careful inspection of the surfaces exposed by low water levels indicates that the rocks preserve a polyphase deformation history. Tight-to-isoclinal D_2 folds that carry an axial-planar composite S_1/S_2 fabric are deformed by tight-to-open, steeply-plunging D_3 folds with an axial-planar crenulation fabric (S_3). These folds are of the same age as those that define the Loch Quoich Line (see Excursion 4). The metasediments are intruded by numerous leucocratic segregations and veins; some are parallel to lithological banding whereas others are strongly discordant. The earliest intrusions are rare quartz veins that are folded by the D_2 folds. Most common are quartzo-feldspathic veins that are coarse-grained and may be internally layered. These are clearly deformed by the D_3 folds and carry an internal fabric that is axial-planar to these structures. This locality provides numerous instructive examples of the influence of layer thickness on fold wavelength.

Further west along the shoreline at 4B [NH 1242 1038], a prominent knoll exposes a large pod (5m x 4m) of garnet-amphibolite within banded psammitic and semi-pelitic gneisses. The schistose and highly strained margins of the amphibolite pod lack garnet and are dominated by hornblende and biotite. The contact between the metasediments and the amphibolite is deformed by a D_3 fold. Within the host metasediments, abundant coarse-grained quartzo-feldspathic veins are deformed by D_3 folds. Discordant quartz veins appear to be late to post-D_3.

Return to the lay-by. If the water level in the loch prevents examination of the localities described above, the road cutting opposite the lay-by provides good examples of variably deformed cross-bedding and D_3 folds as described above. From Locality 5.4, the excursion passes westwards for about 9.9km into Glen Shiel, passing through rocks of the Glenfinnan Group, exposed in many road cuttings. A general locality in the hillside to the south of the road is recommended as Locality 5.5. A large lay-by, popular with summer visitors, which is part of the old road down Glen Shiel, is large enough for coaches or several cars; it lies on the south side of the main road.

Locality 5.5   [NH 0266 1234]

Glen Shiel (Fig. 5.1). Glenfinnan Group striped rocks and interference structures.

The purpose of this locality is to demonstrate characteristic rock types and deformation style within the Glenfinnan Group. For the faint-hearted, the road cutting on the north side of the A87 opposite forms Locality 5.5A and lies in striped and banded coarse siliceous psammite and coarse pelitic schists, the psammite carries occasional calc-silicate lenses. Folds are clearly seen to plunge steeply and are probably the same age as the D_3 folds at Locality 5.4. Caution should be exercised at this locality because of fast traffic and the instability of the high rock face.

EXCURSION 5

The road cutting, although of interest, is considerably less spectacular than Locality 5.5B – the rocks exposed on the hillside some 500m to the south, in the general vicinity of [NH 022 122] that can only be reached with some exertion and difficulty. The exertion derives from the $c.$ 500m of rough and soggy, moderately inclined moorland leading to the steep hillside whose crags expose the geological features described below. The difficulty lies in crossing two substantial streams that are not bridged in the vicinity but which can be crossed in wellington boots or by nimble footwork when water level is moderate-to-low. In the past, leaders have carried planks to aid the crossing of these streams.

The weathered rock faces and glaciated pavements on the hillside display spectacular folds of both the Loch Quoich Line (D_3) generation and earlier (D_2). The rocks involved are coarse siliceous psammites, disposed as stripes and bands, separated by coarse-grained pelitic and semi-pelitic schists that carry a coarse crenulation and numerous garnet and muscovite porphyroblasts. Locally, lenticles of calc-silicate, up to a metre long and 0.1m across, stand out from the surrounding metasedimentary rock and are readily recognized as being studded with reddish brown garnets up to 5mm across. They consist of garnet, amphibole, calcic plagioclase and quartz. A large range of interference structures can be demonstrated, from simple eye structures to hooked interference patterns. The present distribution of interference structures probably defines the hinge zone of a major fold that predated the imposition of the Loch Quoich Line (D_3) generation of structures. It is inferred that, whereas the D_3 folds are universally developed, the D_2 major folds were long-limbed structures in which minor- and intermediate-scale folds were confined to the hinge zones. Hence the areas in which D_3 folds *only* occur correspond to the D_2 major fold limbs, while the zones of interference correspond to the D_2 hinge zones. The hinges of the D_3 folds are spectacularly curvilinear in style with the majority being almost reclined, plunging steeply in a general easterly direction with axial traces trending NNE.

About 2km to the west of the lay-by used for Locality 5.5, lies Locality 5.6 where the Sgurr Beag Thrust Zone may be examined. Locality 5.6 can be easily located in the area next to a well-defined gap some 200m wide between forestry plantations on the right-hand side (north) of the road. A conventional lay-by, part of the old road below the trees, occurs on the north side of the road, but a metal barrier prevents its usage by any vehicles higher than 6 foot 9 inches (~2m).

Locality 5.6   [NH 006 135, NH 008 137]
Glen Shiel (Figs 5.1, 5.3). Sgurr Beag Thrust Zone, Lewisianoid sheet.

This general locality comprises several exposures within the Sgurr Beag Thrust Zone. Two stops will be visited. From the lay-by, cross the road and proceed to the south side of a small knoll at Locality 5.6A [NH 0073 1348]. Here lenticular garnetiferous amphibolite and biotite-hornblende schist are exposed. The foliation within these rock types anastomoses between a series of steeply-dipping, irregular shear zones along which platy quartzo-feldspathic veins occur. Minor crenulations detach along these zones and between them occur rootless fold hinges. These hornblendic rocks form part of a thin, but remarkably continuous strip of Lewisian that coincides with the ductile thrust zone separating the Morar and Glenfinnan groups.

Proceed to the river bank [NH 0068 1343]. Next to a small confluence platy gneissose psammites are exposed. These psammites have strong affinities with the Reidh Psammite of Tanner (1971). Westwards toward the thrust, the platy or lenticular gneissose segregation bands are progressively sheared into weakly asymmetric augen. On the south bank of the river [NH 0066 1342] where the surface of *decollement* is clearly exposed, the resultant augen gneiss passes abruptly into platy semi-pelitic schist. The progressive development of feldspar augen from more or less continuous gneissose banding reflects the steep strain gradient above the thrust.

From the above observations it is apparent that the Lewisianoid hornblendic rocks do not lie immediately above the thrust. Detailed mapping has shown that the Lewisianoid rock types occur within an augen of low strain, perhaps within a fold core. The underlying limb of this fold has been cut out along the thrust. Cross back over the road to the gap between the forestry plantations. Looking up the hill, stop 6B is located within the left-hand stream beneath a small waterfall [NH 0082 1366].

Exposed in this stream section, 6B, are Lewisianoid hornblendic gneisses with associated bands of diopsidic marble. Above these occurs a zone of chloritic schists that lie beneath a late major low-angle fault. This thrust, exposed halfway up the waterfall, places vertically-oriented gneissose psammites over the Lewisianoid rock types. The fault zone consists of a series of anastomosing fractures between undeformed but rotated blocks of gneissose psammite. Within these fractures occurs a foliated microdiorite that was presumably intruded before the fault formed.

EXCURSION 5

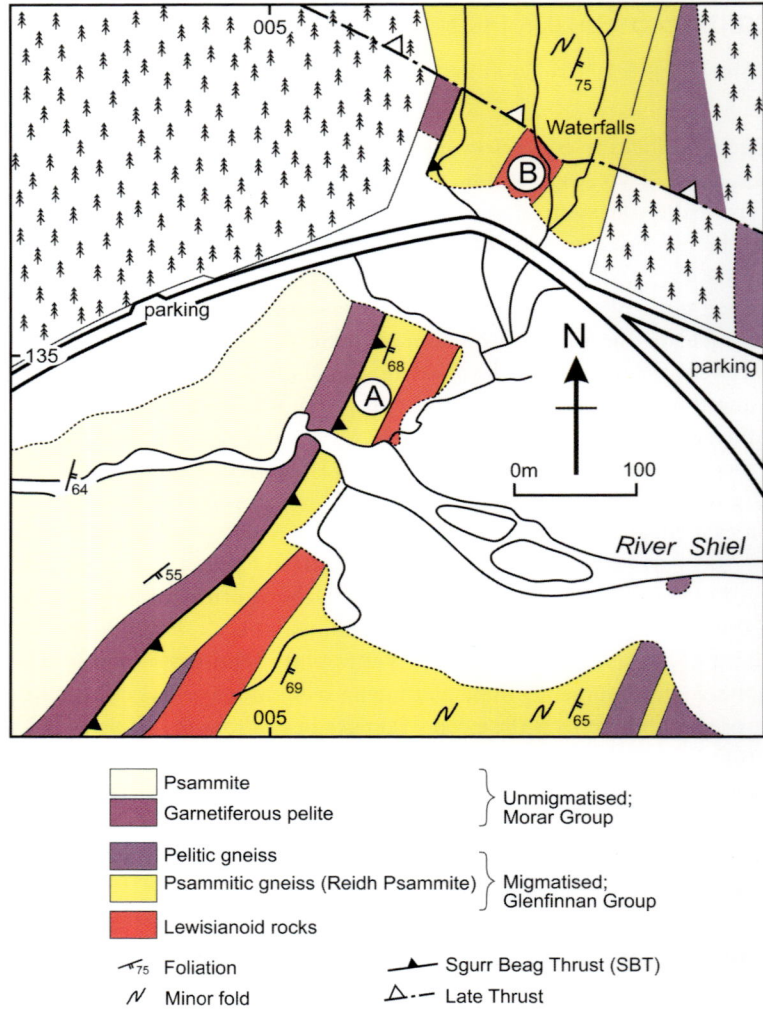

Fig. 5.3 Detailed sketch map of the Sgurr Beag Thrust zone at Locality 5.6. Refer to text for stops 5.6A and 5.6B. Dip and strike symbols refer to orientation of regional gneissosity.

Excursion 6

West Glenelg and Loch Hourn

John Ramsay

Purpose: To examine relationships between Moine metasediments and Lewisianoid basement rocks above the Moine Thrust, and the nature of the polyphase deformation that affects both units.

Aspects covered: Lewisian orthogneisses and associated mafic intrusions (locally eclogitic), Moine psammites and sedimentary structures, polyphase folds and fabrics.

Useful information: Hotel and B&B accommodation are available in and near Glenelg village.

Maps: OS: 1:25,000 sheet 413 Knoydart, Loch Hourn and Loch Duich; BGS: 1:50,000 sheet 71E Kyle of Lochalsh.

Type of terrain: Rocky coastline.

Distance and time: The excursion is best followed over 2 days; Localities 6.1-6.3 can be visited on the first day and 6.4-6.7 on the second.

Short itinerary: For visitors only interested in Moine-Lewisian relations, Localities 6.1, 6.3 and 6.6 can be completed in a single day.

This is a classic region for establishing the differences between Lewisian basement and Moine metasediments above the Moine Thrust. These are especially apparent in the west of the region but are progressively overprinted towards the east by the effects of superimposed (Caledonian?) deformation and metamorphism. Detailed descriptions, maps and structural analysis relevant to this excursion are provided by Peach *et al.* (1910), Ramsay (1958), Sutton & Watson (1959) and Ramsay & Spring (1962).

The Lewisian gneisses comprise 'Western' and 'Eastern' facies (Fig. 6.1). Both are dominated by acid and mafic orthogneisses, but the Eastern facies is particularly distinctive because of the presence of metasediments and eclogites (see also Excursion 7). The Lewisian rocks are found in three strips (Fig. 6.1) which are interpreted as the cores of D_1 anticlines. The two western fold cores are of Western facies aspect (Fig. 6.1) and the third is of Eastern

EXCURSION 6

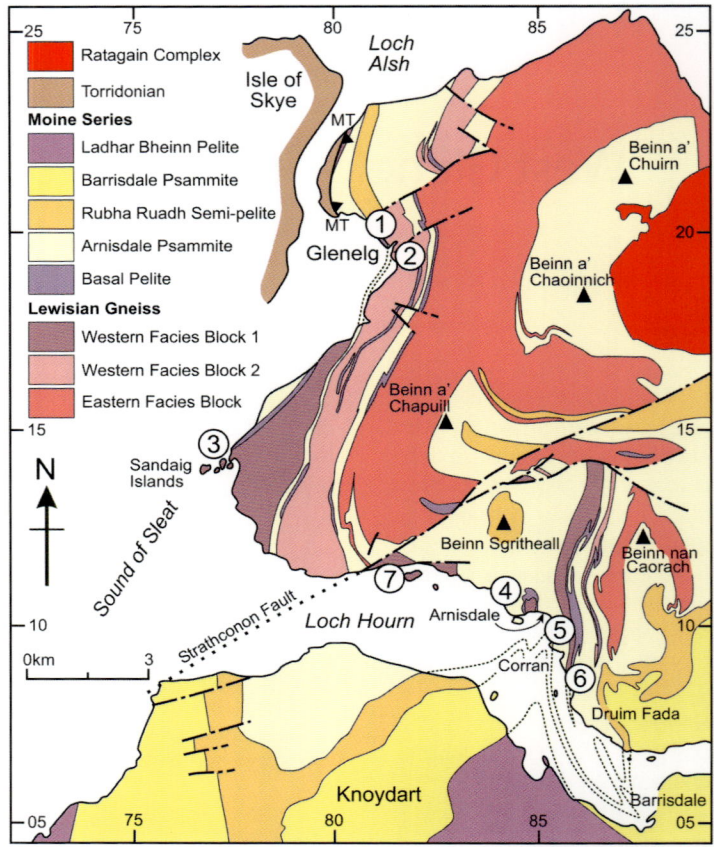

Fig. 6.1 General geological map of the Glenelg and Loch Hourn region.

facies aspect (Fig. 6.1). The Moine rocks comprise a well-defined stratigraphic sequence of psammites, semi-pelites and pelites that has been assigned to the Morar Group (Fig. 6.1). Towards the east, the Moine rocks become progressively migmatitic.

Three major deformation phases have been recognized in the Lewisian and Moine rocks (Fig. 6.2). The Lewisian and Moine were interfolded by D_1 tight to isoclinal folds (Fig. 6.2). Fold-axis parallel lineations are mainly quartz rods that probably formed at no higher than the low amphibolite facies. D_2 folding resulted in the Beinn a' Chapuill-Beinn nan Caorach Antiform and the Glen Beag Synform (Fig. 6.2, BC and GB respectively) as well as numerous associated minor folds. D_2 folding was accompanied by

West Glenelg and Loch Hourn

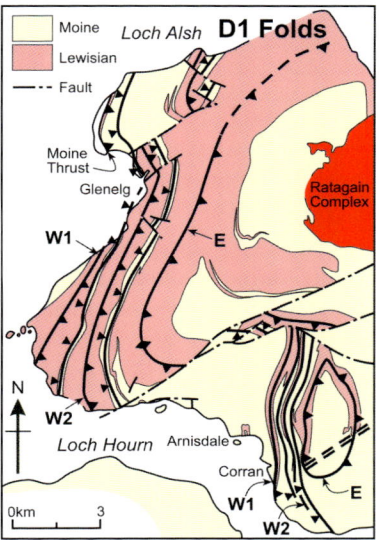

Fig. 6.2 Axial traces of the major D_1, D_2 and D_3 folds.

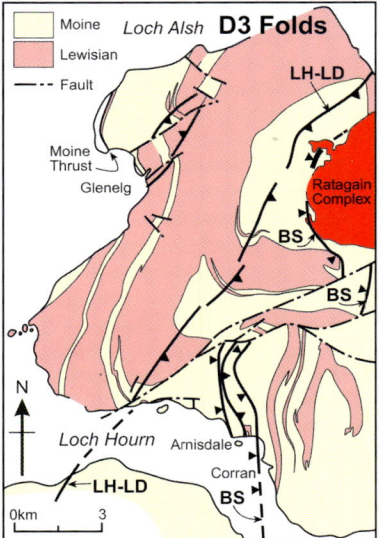

strong recrystallization and migmatization of both Lewisian and Moine. A strong axial-planar schistosity is present and a prominent D_2 lineation is parallel to local D_2 fold axes. D_3 folds trend NE-SW and have axial planes that dip SE. The most important regional folds are the Loch Hourn-Loch Duich anti-form and the Ben Sgriol Synform (Fig. 6.2, LH-LD and BS). D_3 folds crenulate earlier fabrics and were probably accompanied by low amphibolite facies metamorphism.

Locality 6.1 [NG 809 205]

North Glenelg Bay (Figs 6.1, 6.3). Tectonically modified unconformity between Western facies Lewisian gneisses and Moine psammites.

From the Y-road junction of the road northeast of Glenelg village, take the road towards the Kylerhea Ferry. After about 1.5 km [NG 8094 2049], a

125

small track on the west of the asphalt road and beneath an imposing old sea cliff with raised beach and sea caves leads down to the present beach where there is space to park. These outcrops will take around ½-1 hour of study.

Just to the north of a newly-built (2005) house 'Beach-Haven', a prominent red-brown lamprophyre dyke cuts grey Moine psammites (Fig. 6.3). South of this dyke, a contact between Moine psammites and Lewisian gneisses is exposed [NG 8080 2036]. The outcrops are best visited at low-tide. The best exposures are to be found just below high water mark. The mineralogical and textural contrasts between the grey Moine psammites and the orange-brown Lewisian gneisses are very marked, and at several localities the actual contact between the two groups of rocks may be investigated.

Most of the Western facies Lewisian gneiss consists of coarse-grained biotite-hornblende gneiss with a banding produced by variations in the biotite and hornblende content. The gneisses often show an agmatitic structure with clots and lenses of biotite and hornblende-rich material. There is no evidence of a sedimentary origin. The gneisses locally have boudinaged

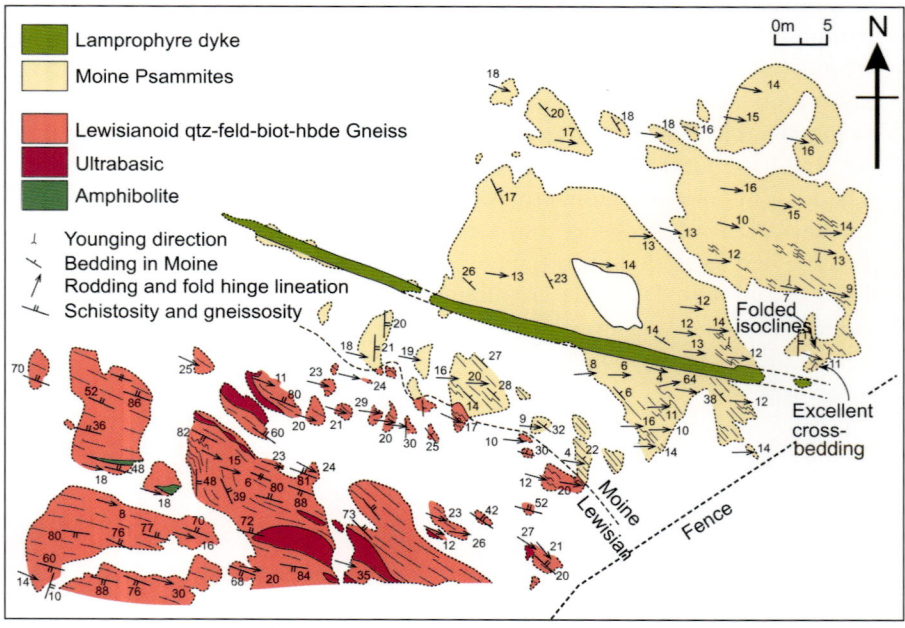

Fig. 6.3 A detailed outcrop map of the exposures in Glenelg Bay at Locality 6.1.

sheets of amphibolitic rock and, although these are here parallel to the banding in the enclosing biotite gneisses, their contacts are generally sharp. They probably represent amphibolitized basic dykes which at some other localities in the Glenelg region can be seen to be discordant to the gneissic banding. Lenticular masses of ultrabasic material are also common, consisting of green actinolite and diopside, dark brown biotite and black hornblende. At some localities, S-shaped minor folds (D_2) are present with their hinge lines parallel to an intense rodding lineation.

The Moine psammites are well-bedded, muscovite-biotite-quartz-feldspar rocks. Gneissic material is completely absent, although some quartz veins are present. Certain beds show their original clastic grain structure and are marked with small pock-marked weathering holes. At several localities, cross-bedding is preserved and truncated foreset bedding is often well preserved especially in the hinge zones of minor folds. These sedimentary structures consistently indicate that the Moine rocks young in a direction away from the Lewisian-Moine contact. There are abundant tight D_2 folds with regionally consistent S-forms. These folds generally show well-developed axial-plane schistosity, together with an E-plunging and intense schistosity-bedding intersection parallel to a rodding-lineation and to the fold hinge lines. At one locality (Fig. 6.3) small isoclinal folds are refolded by the dominant D_2 structures, but it is uncertain if these structures are tectonic (D_1) or sedimentary in origin.

The contrast of rock type between the Lewisian and Moine units is very marked and the contact is sharp with stratigraphic relationships that suggest that the Moine sediments were deposited on a basement that had already undergone upper amphibolite facies metamorphism and had been intruded by basic and ultrabasic sheets. At the actual contact here is no clear evidence of an angular unconformity and some of the finer grained bands of Lewisian gneiss superficially resemble the adjacent Moine psammite. This similarity in appearance is a good example of metamorphic and textural convergence, whereby two groups of quartz-feldspar-mica rocks with initially distinct textures and fabrics converge in grain size and overall appearance as a result of undergoing identical deformational and metamorphic processes. Both Lewisian and Moine have experienced intense common ductile folding and such deformation is well known to generally reduce angles of discordances (Ramsay, 1967; Ramsay & Lisle, 2000). However, certain structural features do suggest that the two groups were originally structurally discordant. Although the orientations of the main folds which affect the Lewisian and Moine are similar, there are slight but consistent differences between the common geometric features. For example, the orientations of the fold hinges and rodding structures in the Lewisian and Moine are not quite the same: those in the Lewisian plunge to the ESE with a generally more varied and steeper plunge than the more constant and easterly plunging structures of the Moines. These differences are probably best interpreted as being the result of differences in the orientations of the Lewisian banding (steeper and more variable) from those of the Moinian bedding surfaces (less steep and more constant).

If the visitor has extra time (10 minutes walk) some nearby coastal outcrops a little further to the south and to the SE of the headland of Creag Mhor at [NG 8086 2013] of Lewisian agmatitic gneisses can be highly recommended, and are especially photogenic (see Clough in Peach *et al.*, 1910, plate VI).

Locality 6.2 [NG 8092 1921]

Glenelg village (Fig. 6.1). Western facies Lewisian gneisses and amphibolite sheets.

At Rudha Mhic Cuinn, on the foreshore just west of the Glenelg War Memorial, banded hornblende-biotite gneisses show hornblende agmatites injected by acid veins [NG 8092 1921]. The gneisses are cut by massive sheets of amphibolite with sharp contacts and with little or no gneissic material. These sheets are interpreted as basic intrusions injected into the gneissic host and subsequently recrystallised in amphibolite facies metamorphic conditions during D_2. All the rocks show an intense D_2 rodding lineation plunging 15°-25° to the ESE. Just south of the monument are exceptionally clean outcrops of banded gneiss cross-cut by a massive amphibolite with a knife-sharp contact. The gneiss shows an easterly dipping schistosity oblique to the banding and with schistosity/banding intersection parallel to the prominent rodding which is exactly parallel to the hinge line of nearby D_2 folds.

Locality 6.3 [NG 7680 1463 to NG 7706 1520]

Sandaig (Fig. 6.1). Western facies Lewisian gneisses and amphibolites, some of which show eclogitic cores; tectonically modified unconformity with cross-bedded Moine psammites.

Drive along the asphalt surfaced road south from Glenelg village, and after 1.5 km turn right to cross the Gleann Beag river at Eilanreach. Continue along this road for about 4 km where a gravel track branches left into a forestry plantation just north of the reed-filled Loch Drabhaig [NG 7836 1517]. Park by the roadside and proceed through a locked forestry gate into the forestry plantation following the main track (there are various branching tracks) to cross the Allt Mor Shantaig by a bridge. About 600m from the bridge a complex of tracks meets at a cross road: take the track on the right which descends towards the sea. After several bends the track finishes and is replaced by a narrow footpath descending towards the Bay of Sandaig. This descent from the park place will take about 30 minutes. Sandaig [NG 7724 1472] is now a ruined croft, once inhabited by Gavin Maxwell who wrote a well known book describing his stay there (*The Ring of Bright Water*, Penguin Books, 2001). Pass through a gate and follow a footpath to two small monuments, one with a small bronze plaque in memory of Maxwell, the other to his otter Edal (site marked on the 1:25,000 map 'meml.'). If the season has been dry it is possible to cross the river (Allt Mor Shantaig) where it enters the sea. More normal Scottish conditions will require wading or by using the shaky double-rope bridge, one rope for the feet the other for the hands.

From the rope bridge a footpath leads westwards to the Sandaig Islands where there are excellent wave-washed outcrops [NG 7680 1463] of quartz-feldspar-hornblende gneiss with a strongly developed, easterly plunging rodding lineation (D_2) (Locality 6.3A). The gneissic banding is often cross-cut by dark-green hornblende-garbenschiefer spears. Some of these hornblendes are up to 4cm long, while others occur as stellate aggregates following hornblende-rich layers. Because they cross-cut the rodding structures it is clear that the metamorphism which led to their development must be post D_2. If tide and time allow it is well worth visiting the more westerly of the islands (although that in the far west where the lighthouse is located is generally inaccessible except by boat). Here the quartz-feldspar rich gneisses contain amphibolite sheets free of gneissic material and are sometimes boudinaged. Their contacts are invariably sharp and locally discordant to the gneissic banding, and they are best interpreted as representing basic intrusions into the gneisses.

From the Sandaig Islands, proceed northward past the headland of An Gurraban. In a small bay (Locality 6.3B, [NG 7696 1506]) a contact between Lewisian gneisses and Moine metasediments is exposed. At the actual contact, the SE-dipping Moinian sediments contain elongated blocks of Lewisian gneiss material which has been interpreted as representing a true basal conglomerate of the Moine Series (Clough in Peach *et al.*, 1910, p. 50). Unfortunately the best and most convincing exposures of this conglomerate are now practically inaccessible, overgrown by moss and situated in a steep forest-covered hillside further to the northeast [NG 7868 1649]. The basal Moine sediments form alternating bands of garnetiferous pelitic micaschist and semipelitic material (the so-called 'basal Moine pelite', Ramsay & Spring, 1962). In contrast to the immediately adjacent Lewisian rocks, the Moine metasediments contain no gneissic material, only a few crosscutting quartz veins. The pelitic Moine sediments contain perfectly idiomorphic dark red-brown garnets up to 8mm in diameter and which, under the microscope, are seen to contain spiral inclusion trails suggesting that they grew during and after the D_2 deformation. The pelitic bands nearest to the Lewisian gneisses contain some small hornblende-garbenschiefer spears, and these are believed to have formed at the same period of metamorphism as those seen in the Lewisian gneisses described earlier (post D_2). Generally the Moine sediments of the Glenelg region lack amphibole. The origin of the hornblendic material is uncertain: it could be have been derived from the erosion of Lewisian material, or it could be the result of migration of the mineral constituents from the adjacent gneisses during metamorphism.

A broken wire fence comes down to the coast just north of the Moine-Lewisian contact. Just north of this fence the more psammitic parts of the Moine are well exposed at Locality 6.3C [NG 7706 1520] and show good cross-bedding structures indicating that the SE-dipping metasediments are inverted and young away from the Lewisian. The psammitic Moines show beds with well preserved (although tectonically stretched and recrystallised) clastic feldspars. The cross-bedding features are especially well developed as the distance from the Lewisian-Moine contact increases and are especially well preserved in the sea cliffs on the headland of Rubha na h-Airde Beithe ([NG 7785 1640]; and see Peach *et al.*, 1910, p. 53), but a visit to this headland will take at least 20 minutes (low tide conditions necessary).

If the tide is low and the visitor has time (½-1 hour) it is worth visiting the coastal outcrops of the Lewisian south of Sandaig. The Lewisian along this stretch of coast, in contrast to that seen in the Sandaig Islands, has been very highly deformed, and much of the coarse-grained fabrics have been replaced by laminated, very fine grained quartz-feldspar-hornblende-biotite rocks. Because of their fine- and rather even-grain size, these rocks were once described as 'granulites', this description applied because of their granular nature and not because of their metamorphic state. The rocks are intensely and finely banded, and the sheets of hornblende schist have been boudinaged into pod-like pieces. At Locality 6.3D [NG 7675 1404] the centres of the largest pods sometimes have an altered eclogitic mineralogy (the eclogite transforming to garnet amphibolite and amphibolite), a feature which is more characteristic of the Eastern Lewisian facies and generally absent in the Western facies. In fact, these outcrops are the only ones seen in the Western facies and the eclogitic remnants here have been dated at ~1700 Ma (Storey *et al.*, 2010), markedly different from the ~1100 Ma eclogites in the Eastern Lewisian facies (Excursion 7; Sanders *et al.*, 1984; Sanders, 1988). The strongly banded Lewisian gneisses show lenticulate blebs of feldspar, while the quartz crystals are found as sinuous streaks around the feldspar porphyroclasts. The timing of this granulation of the rocks is uncertain at this locality. Elsewhere in the Glenelg region, such highly deformed fabrics are also found in Moinian rocks, appear to predate the strong recrystallization that is attributed to D_2 and post D_2, and are attributed to the first of the deformation in common to both Lewisian and Moine (D_1) (Ramsay & Spring, 1962). From south Sandaig return to the road.

Locality 6.4 [NG 8320 1118]

Road section west of Arnisdale (Fig. 6.1). Relationships between D_2 and D_3 structures within Moine psammites.

From Upper Sandaig drive along the road in the direction of Arnisdale. The road climbs to a hilltop [NG 7802 1302] where it is worth stopping to view to the SW of the Tertiary Volcanic islands of Rhum and Eigg. Continue downhill, passing into a forestry plantation (cattle grid) and through roadside outcrops situated in the complex core zone of the Loch Hourn-Loch Duich

D_3 antiform. On emerging from this forest (cattle grid), the road passes close to the shore line passing the cottage at Rarsaidh. After about 0.5 km, park by roadside exposures of gently inclined Moine psammites which are Locality 6.4 [NG 8320 1118] (Fig. 6.1). Proceed by foot along the road towards Arnisdale.

The Moine psammites progressively change dip from SE to west as the trough line of a D_3 syncline is crossed. Excellent roadside exposures on the north of the road show sections of typical Moine psammites with thin bands of pelitic material with well developed D_3 crenulation cleavage. As one proceeds along the road to the SSE, the bedding surfaces of the Moines are exposed and one can see the development of two distinct crossing lineations: a D_2 quartz feldspar rodding is cut and refolded by strongly developed D_3 crenulation folds. The dip of the bedding gradually increases as one passes through the hinge of the D_3 synformal fold (the Arnisdale Synform, Fig. 6.4). The dip gradually increases to vertical: the D_3 linear features remain constantly oriented, with sub-horizontal axes, whereas the D_2 lineations become progressively steeper, passing through the vertical to plunge towards the south and SE at the top of the hill [NG 8370 1054]. The angular relations between the constant D_3 direction and the variable D_2 lineation seen here indicates that the D_3 fold geometry cannot be that of a simple flexural slip fold. In fact, this geometry is directly comparable with the small scale lineation refoldings seen at Loch Monar (Excursion 8), only here the deformed lineations are on a regional scale. The geometrical distribution of the D_2 lineations is best explained by the flow kinematics described at Loch Monar, with the principal fold-forming flow ('a'-direction) calculated from the intersection of the mean D_2 great circle and the axial planes of the D_3 folds (Ramsay, 1960). The significance of this locality in understanding the kinematics of superimposed folding is of fundamental importance here and elsewhere in the Scottish Highlands (Ramsay, 1960).

Return to the vehicle, and drive along this same road descending into the village of Arnisdale. Continue through the village and where the coast road bends abruptly to the left [NG 8468 1006] find a park place by a gate. Walk 400 m south along the beach to the rocks at Crudh 'Ard [NG 8464 0963].

Locality 6.5 [NG 8464 0963]

Crudh 'Ard (Fig. 6.4). Relationships between D_2 and D_3 structures within Moine psammites.

This locality is situated close to the hinge of the D_3 Ben Sgriol Synform. On the east side of the outcrop, Moine psammites dip steeply to the east on the overturned limb of the synform and these psammites (ii) become folded in the hinge (M-shaped D_3 folds) of the main fold which is occupied by semi-pelitic rocks (Fig. 6.4). This is an excellent outcrop to study the deformation of the D_2 linear structures around the hinges of small scale D_3 folds. The D_2 quartz-feldspar rods undulate over the D_3 hinges. In the more pelitic parts the D_3 folds show excellent coarse crenulation structures. Before leaving this locality it is good idea to view the hillside on the south face of Ben Sgriol above Arnisdale village. The Ben Sgriol Synform is extremely well displayed in this face. To the left (on Creag an Fhithich), ribs of Moine psammite dip eastwards (to the right) at ~20°. The actual curve of the fold hinge is seen just to the right of a prominent stream (Allt an Fhuarain) and further to the right the psammites turn through the vertical into an overturned position.

Return to the vehicle and drive along the road to a large parking place just before the village of Corran.

Locality 6.6 [NG 85 08]

Corran to Rudha Camas na Caillin (Fig. 6.4). D_1 interfolding of Lewisian gneisses and Moine psammites.

Cross the bridge over the River Arnisdale and proceed southward along the foreshore of a raised beach for about 300m to the first prominent coastal outcrops at Locality 6.6A [NG 8500 0886]. These outcrops show the contact relations between strongly banded Lewisian gneisses forming the western part of the anticlinal strip of Western facies Lewisian (Fig. 6.4). The Moine rocks are a mixture of psammitic and muscovite-biotite pelitic bands with bedding planes parallel to the Lewisian banding. A strong D_2 lineation is developed plunging down the dip of the layering to the ESE. In the Lewisian rocks, the distinctions between the original deformed gneissic bands and fresh augen and veins of new D_2 quartz-feldspar are clear. Continue southwards. If the tide is low it is possible to skirt the outcrops along

EXCURSION 6

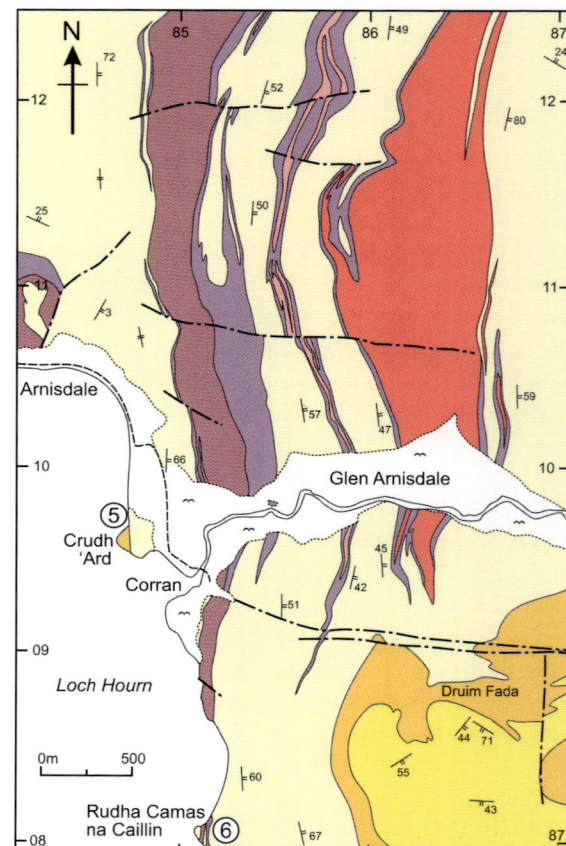

Fig. 6.4 Map of the Moine and Lewisian rocks in the Arnisdale-Corran region. The Lewisian is found in three fold hinges, two of western facies (A and B) and one of eastern facies (C), an arrangement exactly comparable with that seen on the NW side of the Strathconon Fault. The Moine strip between B and C opens southward into a major D_1 syncline on Druim Fada.

the shore; if not, it is best to follow a small footpath in the woods above to coastal exposures into a sandy bay. There are very clean outcrops of Moine psammites with strongly developed D_1 isoclines plunging towards the south and SSE, with D_2 migmatite veining. Leave the footpath and aim for the low rocky headland of Rudha Camas na Caillin which is Locality 6.6B [NG 8500 0800] (Figs 6.4, 6.5). The western Lewisian D_1 anticlinal core, which has been narrowing to the south on account of its southern plunge, splits into three narrow southward plunging isoclinal fold cores at this locality and perfect exposures of all the Moine-Lewisian contact relations and three sets of superposed folds are to be seen (Fig. 6.5). The lithological differences between Lewisian and Moine occur at knife sharp contacts, yet there are no anomalous high strain zones at these contacts. Clear D_1 southward-opening synformal folds are seen in the Moine psammites, and D_1 southward closing antiformal folds in the Lewisian and the Moine-Lewisian relationships appear to be those characteristic of sets of isoclinal D_1 folds. All these D_1 folds have been overprinted by Z-shaped D_2 folds which plunge to the ESE accompanied by a strong rodding, especially well developed in the hinge zones of the D_2 folds. The refolding of small D_1 isoclines in the cores of the D_2 folds is exceptionally clear. Locally within the banded Lewisian gneisses, small, angular, almost kink-like D_3 folds deform the D_2 linear fabric. These are not present in the Moine psammites presumably because the massive recrystallised nature of the psammites proved too competent for such folds to form. These outcrops are exceptionally clear in the view that they present of the differences between Moine and Lewisian and the successively overprinted folds of the three deformation events: they should on no account be missed. All the geological relationships seen in this region and shown in Figs 6.4 and 6.5 confirm Clough's interpretation that the primary relationship between Lewisian and Moine is that of early isoclinal folding.

If the visitor has time it would be instructive to spend a further half hour proceeding eastwards along the Loch Hourn coast where there are many examples of SSE-plunging isoclinal D_1 folds (this represents the synclinal zone between the two western facies anticlines [Fig. 6.4] superposed by D_2 folds and intense SE-plunging linear fabrics associated with the development of migmatite veins).

Return to Corran where a welcome cup of tea can be obtained in 'Sheena's Tea Hut'. Return along the road through the village of Arnisdale and just after entering a forestry plantation west of Rarsaidh one crosses a cattle grid. Park in a large site on the left of the road at [NG 8094 1193].

Locality 6.7 [NG 8193 1181]

Loch Hourn (Fig. 6.1). Western facies Lewisian gneisses; Strathconon Fault Zone.

The outcrops to be visited are coastal exposures which lie directly below the park site. One can descend directly through the forest, but the recent state of the newly-cut trees with a jungle of small branches knit together with thorny brambles is not encouraging. The best solution is to re-cross the cattle grid and descend to the foreshore at some convenient place at Leac Glas, then walk towards the west along the shore. At Locality 6.7 (Fig. 6.1,

EXCURSION 6

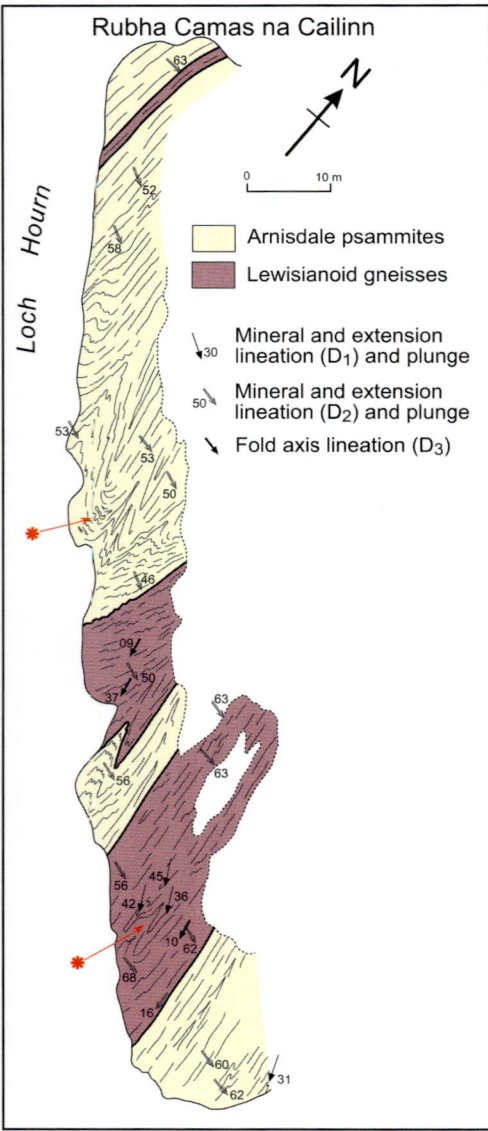

[NG 8193 1181]) banded coarse-grained Lewisian hornblende-gneisses of western facies with lenticular masses of ultrabasic rock are cut by discordant amphibolite sheets which are clearly intrusive basic dykes. These dykes completely lack any gneissic material, but are cut by en-echelon quartz veins.

If time allows, it is instructive to proceed westward from these outcrops to visit the zone of faulting associated with the main Strathconon fault. Lewisian gneisses and Moine psammites are traversed by NE-SW-trending zones of shattered rock with subsidiary faults, and both become intensely crushed towards the main fault plane seen at Sgeir a'Chuirn-uisge [NG 8014 1181]. Some mono-clinal and conjugate folds, probably related to the fault movements, are found with their kinked sectors fractured and brecciated. Occasional dykes of brick-red-coloured late Caledonian lamprophyre show partial brecciation

Fig. 6.5 Detailed map of the coastal exposures on Rudha Camas na Caillin (Locality 6.6B). The Lewisian gneisses are found in the cores of SE-plunging D_1 anticlines and the Moine rocks in the cores of D_1 synclines. The red stars refer to outcrops of special interest.

Excursion 7
East Glenelg and Loch Duich

Craig Storey

Purpose: To investigate the distinctive Grenvillian eclogite facies rocks of the eastern Glenelg basement inlier, the exhumation history of the inlier and its relationship with adjacent Moine units.

Aspects covered: Lewisianoid ortho- and paragneisses with eclogite facies mineral assemblages; mylonites; Moine metasediments; polyphase folds and fabrics; granite and pegmatite intrusions.

Useful information: Hotel, B&B, hostel and campsite facilities are available in the Shiel Bridge-Dornie area; accommodation is also available in Glenelg village.

Maps: OS: 1:25,000 sheet 413 Knoydart, Loch Hourn and Loch Duich; BGS: 1:50,000 sheets 71E Kyle of Lochalsh and 72W Kintail.

Type of terrain: Roadside, rocky coastline and moorland exposures.

Distance and time: The excursion could be followed from either Glenelg or the Dornie-Shiel Bridge area, taking 2 days. See each locality for suggested times.

Short itinerary: Locality 7.2 for a traverse across the shear zone contact between the western and eastern Glenelg inliers, incorporating intervening Moine rocks; Locality 7.3 for a similar roadside traverse across the sheared contact between the eastern and western Glenelg inliers, without intervening Moine rocks.

The eastern Glenelg basement inlier (Fig. 7.1) contains eclogites formed during the $c.$ 1.1-1.0 Ga Grenvillian orogeny, enclosed within orthogneisses that resemble those found within the Lewisian Gneiss Complex of the Caledonian foreland. The inlier also contains abundant metasediments and there are some important lithological and metamorphic differences between the eastern inlier and the Lewisian, and also the adjacent western Glenelg inlier described in Excursion 6. This area is unique in the Scottish Caledonides in preserving evidence for high-grade metamorphism during the Grenvillian orogeny. The aims of this excursion are to: (1) investigate the nature and

EXCURSION 7

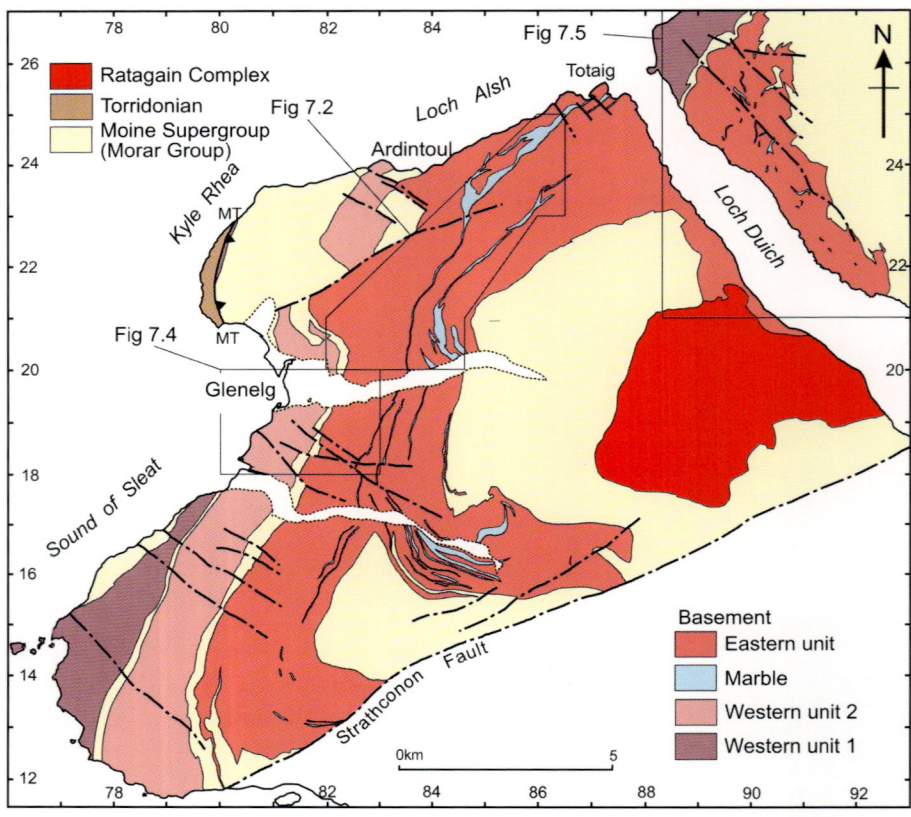

Fig. 7.1 Geological map of Glenelg and Loch Duich, showing the locations of Figs 7.2, 7.4 and 7.5.

evolution of the pre-eclogite crust; (2) observe a variety of lithologies and structures that formed at eclogite facies and therefore provide insights into deep crustal processes; (3) examine the ductile shear zones that border the eastern Glenelg inlier and provide evidence for its exhumation history; and 4) understand the relationships of the inlier to structurally underlying and overlying Moine rocks.

Locality 7.1 [NG 8397 2039 to NG 860 234]

North of Glen More (Fig. 7.2). A range of characteristic lithologies within the eastern Glenelg inlier, including spectacularly preserved eclogite facies assemblages and deformation fabrics.

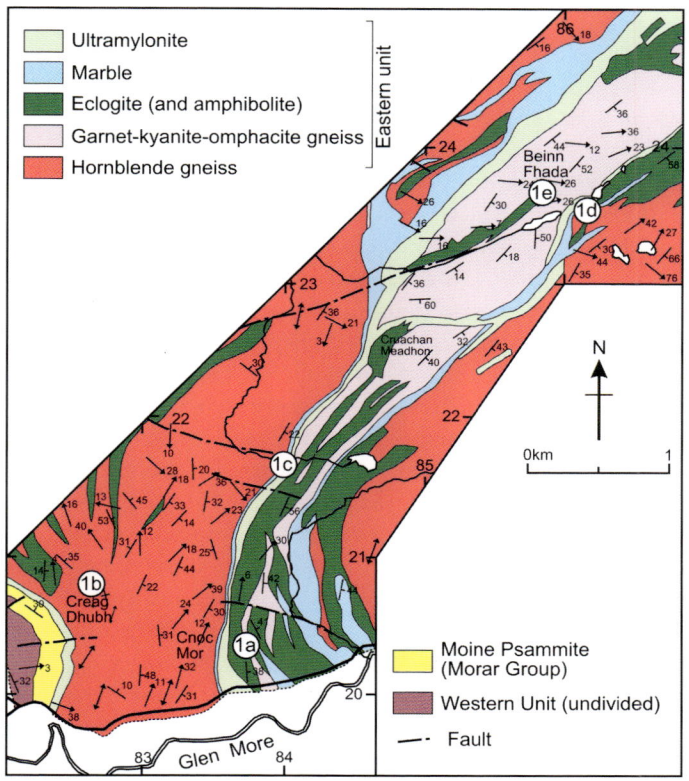

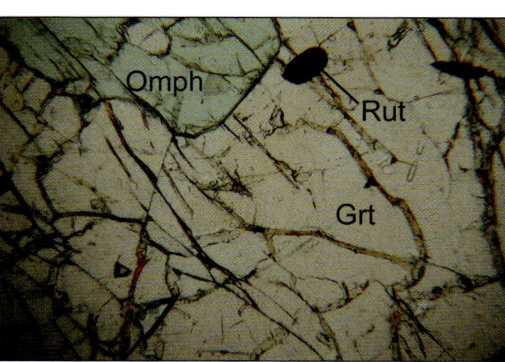

Fig. 7.2 (above) Geological map north of Glen More covering Localities 7.1A-E (modified from Sutton & Watson, 1959, plate 9).

Fig. 7.3 (left) Thin section of typical eclogite from Locality 7.1E.
Grt = garnet; Rut = rutile; Omph = omphacite. Field of view = 3 mm.

From Glenelg village, drive east up Glen More and after ~3km park opposite a small farm (Iomairaghradain) that is on the south side of the road close to the Lamont Holiday Homes at Creag Mhor. Parking here is limited to two cars or one minibus, but passing places are closely spaced along the road so further vehicles could be parked close by. Allow a whole day for this locality.

A path leads up on the north side of the road through a gate and skirts the east side of Cnoc Mór which is Locality 7.1A [NG 836 203]. To the right (east) of this path, on the hillside, black-weathering marble contains numerous disharmonic folds. Head westwards uphill towards the summit where eclogite and retrograde amphibolite are widely exposed. Coarse-grained dark green omphacite and dark red garnet are preserved in patches, often transected by mm-scale veinlets where the rock has been converted chiefly to pargasitic amphibole. The best preserved eclogite occurs where the amphibolite veins are less intense. Where the mafic rock has a strong fabric, wholesale retrogression has generally occurred and often the only remnant of the high-pressure paragenesis is in the form of kelyphitic (symplectic) amphibole and plagioclase rims partially replacing garnet.

Head WNW towards Creag Dubh [NG 826 207], approximately 1 km from Cnoc Mór. On the highest point of Creag Dubh, Locality 7.1B, are exposures of interlayered felsic and mafic rock. The mafic layers occasionally contain relict garnet surrounded by kelyphites (symplectites), but generally are retrogressed completely to amphibolite and epidote-amphibolite. In places, mafic layers are very quartz-rich and omphacite is sporadically preserved. This may indicate dehydration reactions that occurred during eclogite facies metamorphism, particularly in the surrounding felsic gneisses. Garnet is mostly only preserved in the felsic gneisses as relicts or pseudomorphic replacements by amphibole and/or biotite. However, SW of the summit, at [NG 8255 2055], a coarse-grained felsic layer contains omphacite, garnet and kyanite, recording eclogite facies conditions of 20 kbar and 750°C (Storey *et al.*, 2005). These rocks are strictly high-pressure granulites due to the presence of plagioclase, although their peak metamorphism occurred at eclogite facies. The basement here is moderately strained, but not mylonitised. However, the relative ages of the protoliths of the mafic and felsic gneisses are difficult to establish on the basis of field evidence alone. The author's unpublished work indicates that the protoliths of the felsic gneisses are Late Archaean in age, whereas geochemical evidence, from Hf isotopes in zircon from the eclogites, suggests that their protolith may have formed close to 2.0 Ga (Brewer *et al.*, 2003). It therefore seems likely that the protoliths of the eclogites intruded the pre-existing felsic basement.

Retrace the route back towards Locality 7.1A until you find the path that led up from the road. Continue uphill on the path which finishes as the hillside flattens out; exposures ahead around [NG 838 209] continue NW to the southern termination of a prominent ridge at [NG 8365 2105]. Follow the ridge in a NNE direction examining the excellent exposure along the way. This is Locality 7.1C. These rocks are a different type of eclogite, termed by Sanders (1988) 'streaky eclogite'. The rock is mafic with white quartzo-feldspathic streaks that vary from mm-scale isolated threads up to cm-dcm-scale networks of veins that form locally up to half of the rock mass. They typically are intensely rodded with a dominant L>S D_1 fabric. Intervening eclogite patches also have a tectonic fabric defined by aligned omphacite grains and tabular garnet. Kyanite, sometimes preserved within the streaks at the microscopic scale, is also aligned and occasionally forms asymmetric fish, indicating non-coaxial shearing. Sanders (1988) demonstrated that the streaks formed during eclogite facies metamorphism, and Storey et al. (2005) estimated peak conditions at around 20 kbar and 750°C.

Along the length of this ridge, up to around [NG 834 223] (area marked on OS maps as Cruachan Meadhon), the rodding lineation plunges shallowly towards between 010° and 040°, which has been attributed to possible transcurrent shearing during eclogite facies metamorphism (Sanders, 1988).

Continue NE from Cruachan Meadhon towards the east side of Lochan na Beinne Faide. At [NG 86251 23497], in low cliff exposures on the SW side of the loch, browny-grey weathered ultrabasic rocks occur at Locality 7.1D. Rawson et al. (2001) described both garnet-bearing olivine websterites and their garnet-absent equivalents, websterite. The olivine websterites contain two pyroxenes, olivine, garnet, amphibole and minor magnetite and spinel. Websterites are essentially identical in mineralogy but lack garnet and, generally, the olivine has been replaced by serpentine and the pyroxenes by amphibole as a retrograde reaction. Similar rocks are reported by Rawson et al. (2001) from the north side of a small loch approximately 0.5 km SE of Lochan na Beinne Faide [NG 866 233]. The garnet-bearing olivine websterite gave a pressure-temperature estimate of ~20 kbar and 730°C (Rawson et al., 2001). The relationship of the ultrabasic rocks to the other lithologies is unclear and it cannot be determined whether they represent tectonically emplaced alpine-type peridotites or whether they crystallized from a basaltic magma as a result of crystal accumulation. The agreement of the pressure-temperature estimate with that of the surrounding high-pressure granulites (felsic gneisses) and the streaky eclogites implies that they are cofacial, and thus achieved peak conditions at ~1082 Ma (during the

Grenvillian orogeny), but their earlier history is enigmatic. If they are related to the basic protoliths of the eclogites, then they may have formed at ~2.0 Ga.

On the north side of Lochan na Beinne Faide lies a large mass of exposure elongated in a NE-SW direction for approximately 2 km, known as Beinn Fhada, Locality 7.1E. This mainly comprises a distinctive type of garnet and pyroxene-bearing felsic gneiss, which is trondhjemitic in bulk-rock composition. In fresh outcrops, omphacite and garnet co-exist with plagioclase. These rocks are strictly high-pressure granulites, although the presence of omphacite indicates eclogite facies conditions (Sanders, 1979). These rocks are considered identical to the more limited exposure of felsic high-pressure granulite at Creag Dubh, Locality 7.1B. Concordant mm-dcm-scale eclogite layers are interbanded with the felsic gneisses. It is within this exposure, within a thick eclogite layer at ([NG 860 234]; Fig. 7.3), that Sanders *et al.* (1984) obtained a garnet-clinopyroxene-whole rock Sm-Nd age of 1082 ± 24 Ma, interpreted to be close to the peak of eclogite facies metamorphism.

Summary of the early history

It is thought that the majority of the basement gneisses formed from trondhjemitic and granitic protoliths in the Late Archaean, and thus there are similarities with gneisses of the Lewisian Gneiss Complex within the Caledonian foreland and also with the gneisses of the western Glenelg inlier (see also Friend *et al.*, 2008). The major difference is the preponderance of eclogite and paragneisses, which do not have a direct comparison, although possibly similar metasediments occur within the Gairloch region of the Lewisian outcrop. It is considered likely that the metasediments and the majority of the eclogite protoliths formed as part of a volcano-sedimentary sequence that accumulated upon pre-existing trondhjemitic gneisses possibly at around 2.0 Ga. The eastern Glenelg inlier was subsequently buried to depths of around 70 km and metamorphosed within the eclogite facies during the ~1.1-1.0 Ga Grenvillian orogeny, most likely as a result of the deep subduction of continental crust in a collision zone.

East Glenelg and Loch Duich

Locality 7.2 [NG 823 193 to NG 8223 1873]

Between Glen More and Glen Beag (Fig. 7.4). A traverse from the uppermost part of the western Glenelg inlier across the intervening 'Moine strip' and into the shear zone which defines the structural base of the eastern Glenelg inlier.

Either park in Glenelg village or it is possible to drive up a small partially made road that turns sharply to the left off the main road through the village as you head southwards; the junction is between the two entrances to the Glenelg Inn car park. The road follows the edge of the south side of Glen More towards a farm at Cósag [NG 823 193]. The farm is gated at the end of the road, but approximately 200m before the gate there is a disused croft and it is possible to park here on the side of the track. Space is limited to two cars or one minibus. Allow 3-4 hours for this locality.

Walk back towards Glenelg and, where the road turns a first right bend, cross the barbed wire fence and a small stream and walk uphill to the SW. After about 0.5km, around Locality 7.2A [NG 817 187], low hummocky outcrops expose highly strained gneisses of the western Glenelg inlier. It is recommended that some of the localities in Excursion 6 are visited first in order to be able to recognize these lithologies in their lower strain state. Here, these gneisses commonly contain a strong ductile D_2 L-S fabric, with a mineral stretching lineation plunging moderately down-dip to the east. In

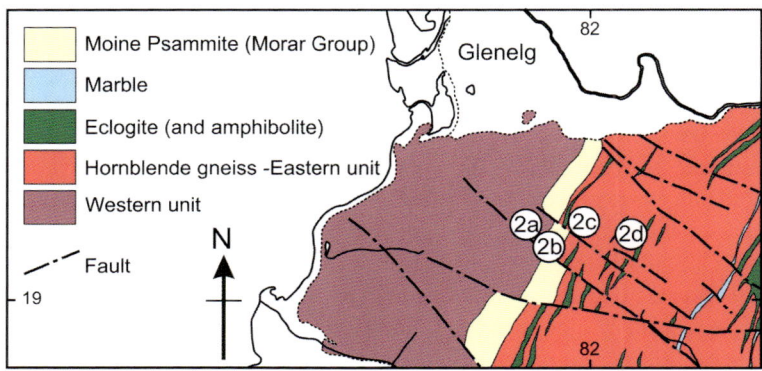

Fig. 7.4 Geological map south of Glenmore covering Locality 7.2A-C (modified from Ramsay, 1958, plates 37 & 38).

exposures where banded tectonised mafic and felsic layers occur, surfaces parallel to the lineation show low-angle C' shear bands that indicate a top-to-the-west sense of shear. Upright D_3 dcm-scale extensional shear bands disrupt the earlier fabric and are associated with a mineral elongation lineation that plunges fairly steeply towards the SE.

To the east, around Locality 7.2B [NG 818 187] low angular outcrops expose Moine psammites assigned to the Morar Group with variably deformed feldspathic sedimentary clasts and thin pegmatitic segregations. The fabric is mylonitic and contains a strong mineral stretching lineation formed mainly by muscovite. Low-angle truncations within the mylonitic foliation are thought by Ramsay (*pers. comm.*) to be relicts of cross-bedding and therefore that the Moine has a modified unconformable relationship with the western basement inlier. However, the high state of strain makes this conclusion difficult to accept and, with a critical eye and looking on variously oriented surfaces, it is more likely in the view of the present author that these low-angle truncations originated as rootless, detached isoclinal folds within the mylonitic foliation. There are two mineral stretching lineations within the Moine rocks: the earliest plunges moderately towards the east (L_2), whereas a later lineation, often associated with upright extensional shear bands, formed during D_3. Occasional kinematic evidence from D_2 fabrics can be found, particularly where pegmatitic layers are disrupted, and these indicate a top-to-the-west sense of shear.

To the east, around Locality 7.2C [NG 819 188] are exposed strongly banded ultramylonites of the eastern Glenelg inlier. Felsic and mafic components are interlayered on the mm scale, the latter commonly containing highly rounded porphyroclasts of garnet and dark green-brown amphibole. The ultramylonitic foliation is often disrupted by disharmonic folds with curvilinear hinges and sometimes eye structures can be observed. A strong mineral stretching lineation plunges moderately towards the east and the fold hinges are generally sub-parallel to this lineation. These are sheath folds formed during shearing. The state of strain is so high that it is impossible to gain kinematic information. The rocks remain mylonitic towards the east for several hundred metres, forming part of the Barnhill Shear Zone which separates the Moine rocks and the eastern Glenelg inlier (Storey, 2002). At the summit of Sgiath Bheinn, around Locality 7.2D [NG 8223 1873], the rocks are locally in a lower state of strain and it is possible to see disrupted migmatitic textures within the trondhjemitic gneisses.

Locality 7.3 [NG 885 254 to NG 8430 2445]

Eilean Donan Castle (Fig. 7.5). A traverse across the boundary between the western and eastern Glenelg inliers; this is similar to Locality 7.2 (although no intervening Moine strip is present), but road cuttings offer more complete exposure.

Allow 2-3 hours for this locality. Park either at the car park at Eilean Donan Castle or at a more convenient location approximately 0.5 km south along the main A87 road where there is a conspicuous crag of green-black rock on the loch side of the road and an unofficial parking area for several vehicles (Locality 7.3A, [NG 885 254]). This is within the western Glenelg inlier. The outcrop is high-pressure mafic granulite of Late Archaean age (Storey, 2002; Friend et al., 2008) and clinopyroxene, garnet, plagioclase, hornblende and minor quartz form a coarse granoblastic polygonal texture (Fig. 7.6). Patches and veinlets of trondhjemitic leucosome result from partial melting. On the opposite side of the road, melt veins are more widespread and have locally coalesced to form dcm-scale sheets. A serpentinite body is typical of the rare ultrabasic bodies that are ubiquitously altered to low-grade hydrous assemblages. Walk southwards along the road examining the roadside exposures; a strong, composite steep SE-dipping fabric characterizes the banded gneisses. After ~300m (Locality 7.3B, [NG 887 252]) there is a large sub-vertical road cutting on a shallow left hand bend covered by wire netting. Behind the netting is exposed the contact between the western and eastern Glenelg inliers. It is cryptic as there are no intervening Moine sedimentary rocks to guide. At the southern end of the netting, dark strongly banded ultramylonites contain rounded porphyroclasts of garnet and amphibole and are thought to have been derived by shearing of the eastern basement inlier.

The strong mylonitic fabric persists for up to 1 km to the SE along the road. There are two dominant sets of fabrics. D_2 structures often comprise rootless isoclinal folds within the mylonitic foliation and have a strong associated mineral stretching lineation and axial planar mylonitic fabric. These are reworked by D_3 structures which comprise a strong L_3 mineral elongation steeply plunging towards the SE and F_3 curvilinear fold hinges generally close to parallel with the principal D_3 stretching orientation. F_2 isoclines and the mylonitic S_2 fabric are refolded around F_3 folds. Hence, D_2 was the main ductile shearing event, whilst D_3 involved both shearing and

EXCURSION 7

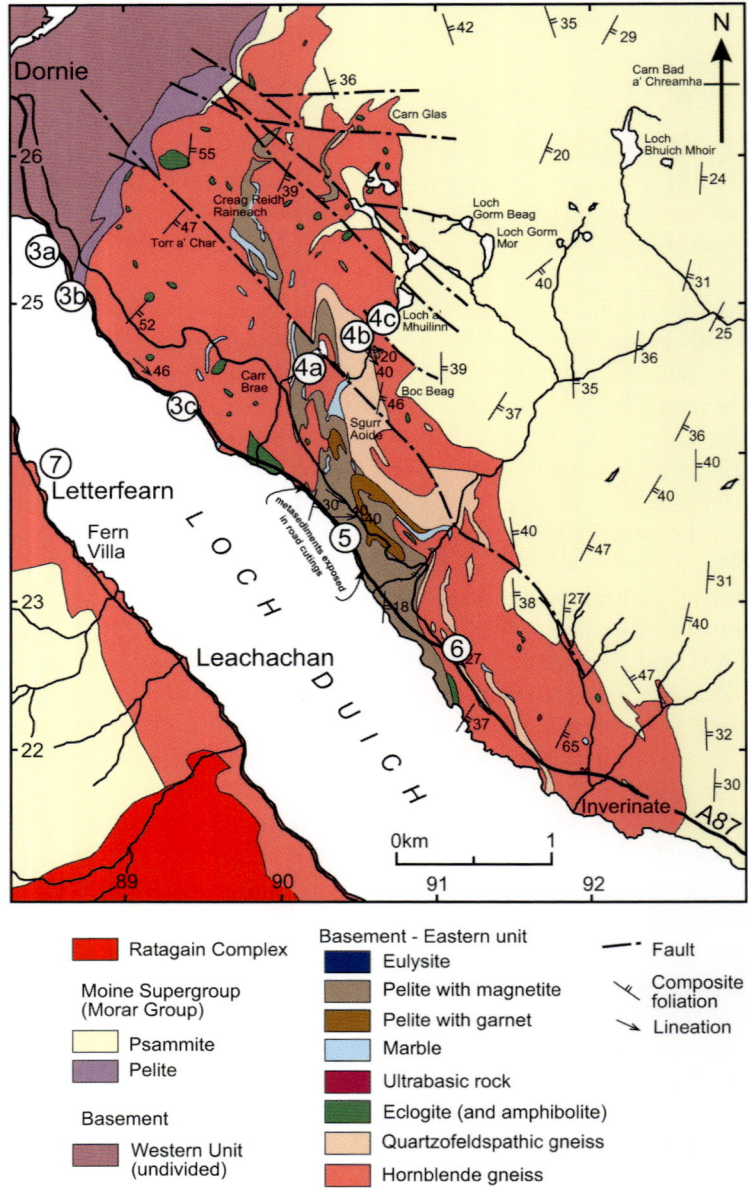

Fig. 7.5 Geological map north of Loch Duich covering Localities 7.3 to 7.7 (based on May et al., 1993, figure 5).

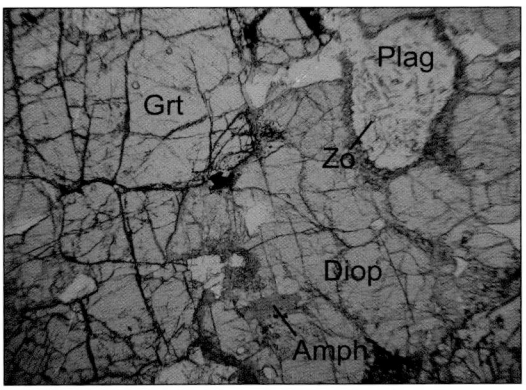

Fig. 7.6 Thin section of high-pressure mafic granulite from Locality 7.3A.

Grt = garnet
Plag = plagioclase
Zo = zoisite
Diop = diopside
Amph = amphibole
Field of view = 3 mm

folding but, in this area at least, generally not a new penetrative axial planar foliation. Temperley & Windley (1997) describe the same structures; although their interpretation that the D_2 fabric was extensional is questionable, their observations that the later D_3 deformation is accompanied by steep, extensional shear bands associated with the SE-plunging mineral stretching lineation appears sound. At Locality 7.3C [NG 8930 2445], low outcrops on the landward side of the road demonstrate that the strain here is lower because recognizable trondhjemitic rocks are exposed with migmatitic textures, enclosing partially melted xenoliths of basic rock. These xenoliths must presumably be older than the protoliths of the eclogites (see discussion at end of Locality 7.1 above). By the lochside here, banded trondhjemitic gneisses can be seen translated into high strain zones and, thus, afford an excellent glimpse into the protoliths at different states of strain.

Locality 7.4 [NG 8995 2445 to NG 9065 2480]

Carr Brae to Loch a' Mhuilinn (Fig. 7.5). A section across the highly tectonized contact between the upper boundary of the eastern Glenelg inlier and the overlying Moine rocks.

Allow 3-4 hours for this locality. Park at Carr Brae [NG 8995 2445] on the old road along the east side of Loch Duich where there is ample parking for a small coach/minibus and several cars. There is a small picnic area and a pleasant view, ideal for breaking up a day involving one of the other half-day excursions. Head directly uphill to the NE, following the Allt a'

Mhuilinn stream, to outcrops on a steep section of the stream approximately 200m from Carr Brae at Locality 7.4A [NG 901 245]. Small layers of eulysite within marble and garnet-biotite gneisses outcrop on the immediate northern side of the stream. Eulysites are metamorphosed manganiferous rocks, characteristically containing fayalite, but hedenbergite, Fe-hypersthene, garnet, magnetite and grunerite are also common. They were first described from Glenelg by Tilley (1936). The eulysites typically have a bluish-black weathered surface (hydrated Mn-oxide) and are very hard; hammering is required to search out these layers for fresh samples. The intimate association of the eulysites with other metasediments can be demonstrated here and they are probably sedimentary (exhalative?) in origin.

Walk ~500m upstream to the NE to Locality 7.4B [NG 906 247] to low-lying outcrops of mylonitised felsic and mafic gneisses. The latter contain rounded porphyroclasts of garnet and amphibole, very similar to the highly deformed units at the base of the eastern inlier, suggesting a comparable state of strain and metamorphic evolution. Garnet-biotite gneisses (metapelites) occur sporadically and contain conspicuous leucosome, which has been remobilized during shearing. It is common to see two lineations on the foliation surfaces throughout this zone, with the earlier L_2 plunging gently towards the east, and a later, steeper L_3 (around 40°) plunging towards the SE (130-140°). The mylonitic S_2 fabric and L_2 are folded around curvilinear F_3 hinges. Locally, a new penetrative S_3 fabric develops axial planar to the F_3 folds and this is in the form of a platy biotite-dominated foliation, which also contains the L_3 mineral stretching lineation. In contrast, S_2 is defined by amphibole rather than biotite, and thus indicates higher grade conditions (i.e. D_3 is retrograde).

The boundary with the overlying Morar Group psammites is exposed at Locality 7.4C [NG 9065 2480]. The contact is intensely sheared and all rocks are ultramylonitic. The contact is marked by a ~5cm layer of ultraphyllonite, which would correlate with the 'basal pelite' of Ramsay & Spring (1962). If this is truly sedimentary in origin, rather than a tectonic phyllonite derived from breakdown of mafic basement rocks, then its preservation is truly remarkable. Although it has been proposed that this contact marks a modified unconformity between the Moines and underlying basement rocks (Ramsay, 1958; Ramsay & Spring, 1962), this view is difficult to sustain given the lack of anything resembling a basal conglomerate and the uniformly high tectonic strain. About 10m above the contact, the Moine psammites contain cm-dcm layers of intensely sheared and friable coarse grained pegmatite. The common presence of such pegmatites within the Moine in the vicinity of contacts with the basement suggests that they may be syn-kinematic, but fixing their precise timing with respect to D_2 and D_3 remains elusive. The ultramylonitic contact can be followed to the SE around the southern side of Boc Beag, and the zone of highly sheared rocks (the Inverinate Shear Zone of Storey, 2002) is at least 300-400m thick within this part of the eastern Glenelg inlier.

Interpretation of the D2 history of ductile shearing

D_2 mylonitization of basement lithologies and Moine rocks occurred under upper amphibolite facies conditions (13 kbar and 650-700°C; Storey *et al.*, 2005) and has been correlated with static replacement of the eclogite paragenesis within low strain eclogite boudins and layers. Brewer *et al.* (2003) dated this retrogression at ~995 Ma, by the U-Pb method on zircon. However, this is older than the youngest detrital zircon age obtained so far within the Moine rocks of the Morar Group (980 ± 2 Ma, Peters, *pers. comm.* in Cawood *et al.*, 2007), and so it is likely that D_2 is composite, comprising an older phase associated with initial retrogression of the eclogites before the Moine rocks were deposited, and a younger phase common to both the basement and the Moines. Dating of syn-D_2 titanite within the shear zone at the base of the eastern Glenelg inlier yields an age of ~670 Ma (Storey *et al.*, 2004) and thus the basement and the Moines may have been juxtaposed by shearing within the middle crust at some stage in the late Neoproterozoic.

Locality 7.5 [NG 905 233 to NG 900 239]

Loch Duich roadside (Fig. 7.5). Metasedimentary rocks within the eastern Glenelg inlier.

A superb road cutting that exposes metasediments of the eastern Glenelg inlier occurs behind double crash barriers beside the main A87 road on the east side of Loch Duich at [NG 905 233] northward to [NG 900 239]. This is Locality 7.5. There is a lay-by just to the south of the crash barriers in which to park and there is plenty of room for a coach/minibus and several cars. Beware of common sheep and deer ticks at this locality! At the south end of the cutting, pelitic schists contain biotite, plagioclase, garnet, kyanite, quartz, muscovite and chlorite. Leucosome is ubiquitous, demonstrating partial melting and, as much of the muscovite is retrograde, indicates that the muscovite-out melt reaction has been crossed. Rawson (2004) recorded phengitic (= high pressure) mica in preserved microlithons within these schists, which along with kyanite is probably the only remaining evidence of the earlier eclogite facies history. Further northwards, the rocks grade into calc-pelitic lithologies, with ubiquitous epidote imparting a greeny colouration. At the furthest north part of the outcrop the succession is capped by a marble horizon, although at the time of writing this has been largely overgrown. The rocks here display M and W folds as they are in the hinge zone of a major F_3 fold. Rootless F_2 isoclines occur within the folded S_2 fabric and melt veins cut across S_2 but are folded around F_3 hinges and have undergone boudinage. Hence, partial melting occurred between D_2 and D_3; an attempt was made to date these melts, but they did not yield uranium-bearing accessory minerals.

Locality 7.6 [NG 9115 2260]

Loch Duich roadside (Fig. 7.5). Relationships between dated pegmatite and D_2 and D_3 structures.

Head south to the junction where the old road over to Dornie, via Carr Brae, meets the A87 at [NG 911 227]. It is possible to park in a lay-by on the A87 here and there is room for a coach/minibus and several cars. Walk approximately 100m south along the road and cross to low-lying outcrops on the landward side of the road [NG 9115 2260]. These expose typical banded mafic and felsic gneisses of the eastern Glenelg inlier with a dominant high-strain S_2 fabric. Conspicuous granitic pegmatites cut the dominant S_2 fabric in a number of places. F_3 folds, with curvilinear hinges, refold the S_2 fabric and a pegmatite. The pegmatite has yielded a U-Pb titanite age of 437 ± 6 Ma (Storey *et al.*, 2004); D_3 must therefore be younger than this and D_2 older.

Locality 7.7 [NG 8845 2380]

Loch Duich lochside (Fig. 7.5). Relationships between dated granite sheet and D_2 and D_3 structures.

Head southwards to Shiel Bridge on the A87 and take the turn towards Ratagan and Glenelg. After about 1 km take the sharp right turn downhill towards Ratagan on the small single lane lochside road that ends at Totaig. Continue past the Ratagan Youth Hostel to Letterfearn and use one of the passing places to park, leaving plenty of room for other cars to pass by. Space is restricted to a small minibus or two cars.

Exposures to be visited are on the loch shore at [NG 8845 2380] on the north side of a small bay at Letterfearn. Care is required as the tide imparts a treacherous slippery surface to the outcrops! A ~3m high south-facing exposure contains a granite sheet about 20cm wide coincident with a fracture in the rock. The margin of the granite cuts the dominant S_2 fabric. The S_2 fabric is within amphibolite and is defined chiefly by amphibole and plagioclase; on the east-dipping foliation surfaces a dip-slip L_2 mineral stretching lineation plunges moderately towards the east, but is variable as it is reworked by D_3 deformation. In the highest part of the outcrop, to the right (lochward) of the granite sheet, the amphibolite contains tabular garnet and elongated prisms of relict omphacite, replaced by symplectites of amphibole and plagioclase and neoblastic garnet. The retrogressed eclogite is of the streaky variety, containing quartzo-feldspathic threads, but no kyanite has survived. This is a relict of the D_1 eclogite fabric that has been statically overprinted during upper amphibolite facies retrogression. The fabric is coplanar and colinear with D_2 and implies that D_1 had a similar principal stretching axis to D_2. At the margins of the granite sheet the D_2 (and relict D_1) fabric is deflected into an upright attitude and a new S_3 foliation and L_3 lineation, defined by aligned amphibole and biotite, is developed that dips towards the SE with a dip-slip lineation. This D_3 fabric is also well developed within the granite sheet. Minor fold hinges throughout the host amphibolite are markedly curvilinear and vary in plunge by up to 80°. A metre behind this small exposure, on the beach, a prominent F_3 fold hinge plunges gently towards the SE. Note that the earlier L_2 lineation is folded around the hinge and reoriented into near horizontal with a N-S azimuth. The granite sheet was therefore intruded after D_2 and before D_3. Zircon fractions from the granite give a U-Pb discordia lower intercept age of 672

± 75 Ma, whereas euhedral titanite gives a concordant age of 520 ± 11 Ma (Storey *et al.*, 2004). D_2 is therefore at least late Neoproterozoic in age, and D_3 must be younger.

Interpretation of the D3 history of folding and shearing

The evidence obtained from Locality 7.6 suggests that D_3 structures are younger than 437 ± 6 Ma and hence must have developed during the Lower Palaeozoic Caledonian orogeny. The D_3 fabric is reworked by brittle-ductile chlorite grade (greenschist facies) shear zones that can be confidently correlated with movements on the Moine Thrust Zone at 437-430 Ma (Freeman *et al.*, 1998). Hence, it appears that D_3 is fairly tightly bracketed. Whilst it is clear that D_3 involved large-scale folding with axes trending NE-SW affecting the Glenelg inlier and surrounding Moine and giving rise to the spectacular fold interference patterns described by Ramsay (1958 and this volume), what has been understated is the amount of shearing associated with this deformation. Evidence of a locally penetrative lower amphibolite facies S_3 and associated L_3 mineral stretching lineation has been presented and is often associated with upright extensional top-to-the-SE shears. Temperley & Windley (1997) presented kinematic evidence for this episode of extensional shearing, but due to the lack of geochronology at the time interpreted this as being part of the extensional exhumation history of the eastern Glenelg inlier. However, since this is not the case, an explanation must be sought within the framework of the Caledonian orogeny. The Glenelg area is in the footwall of the Sgurr Beag Thrust, which has an earlier history than the Scandian (*c.*435-430 Ma) deformation that typifies much of the NW Highlands (Kelley & Powell, 1985; Kinny *et al.*, 2003b). There is evidence of earlier Grampian (*c.*470-460 Ma) crustal thickening within the NW Highlands (Kinny *et al.*, 1999; Friend *et al.*, 2000) and one possible explanation is that the Morar Group and underlying basement underwent crustal thickening during the Grampian that was followed by extensional reworking prior to Moine Thrust (Scandian) times.

Excursion 8
Glen Strathfarrar and Loch Monar

John Ramsay

Purpose: To study classic exposures of superposed fold systems, and Moine-Lewisian relationships in the central part of the Caledonian orogen.

Aspects covered: Lewisian orthogneisses; Moine psammites, including a probable basal conglomerate; polyphase deformation structures; Sgurr Beag Thrust.

Useful information: Hotel and B&B accommodation are available at Inverness, Beauly and Struy. Camp sites are situated at Lovat Bridge (2km south of Beauly), Cannich, and holiday chalets are available in Glen Strathfarrar. A car or small van is essential for this excursion as the road is not suited for coaches or trailer caravans.

Maps: OS: 1:25,000 sheet 430 Loch Monar, Glen Cannich & Glen Strathfarrar; BGS: 1:50,000 82E Scardroy. Ramsay (1957) contains a folded structural map at 1:21,120 scale of the Monar region, but made before the hydroelectric dams were built, so the shores of Loch Monar as depicted are not correct today.

Type of terrain: A mixture of lochside, roadside, quarry, river and hillside exposures, all of easy access.

Distance and time: The excursion can be accomplished in one day.

Short itinerary: If the main interest of the visit is to study the classic superimposed folding outcrops, Localities 8.1-8.4 can be completed in half a day.

Forward planning is advised to obtain access to Glen Strathfarrar. Follow the A831 towards Cannich and 800m south of Erchless Castle turn right along a narrow road into Glen Strathfarrar, which branches off the main road just before the bridge at Struy. At Inchmore the road is barred by a multi-locked gate. The road is now a private road and it is necessary to obtain a permit to proceed further at the gatekeeper's house (Tel: Struy 01463 761260) on the left of the road immediately in front of the gate. This permit is readily granted, but note that the Glen is only open from April until October. During this period, access is between the hours of 0900 to 1800-2000 depending on

EXCURSION 8

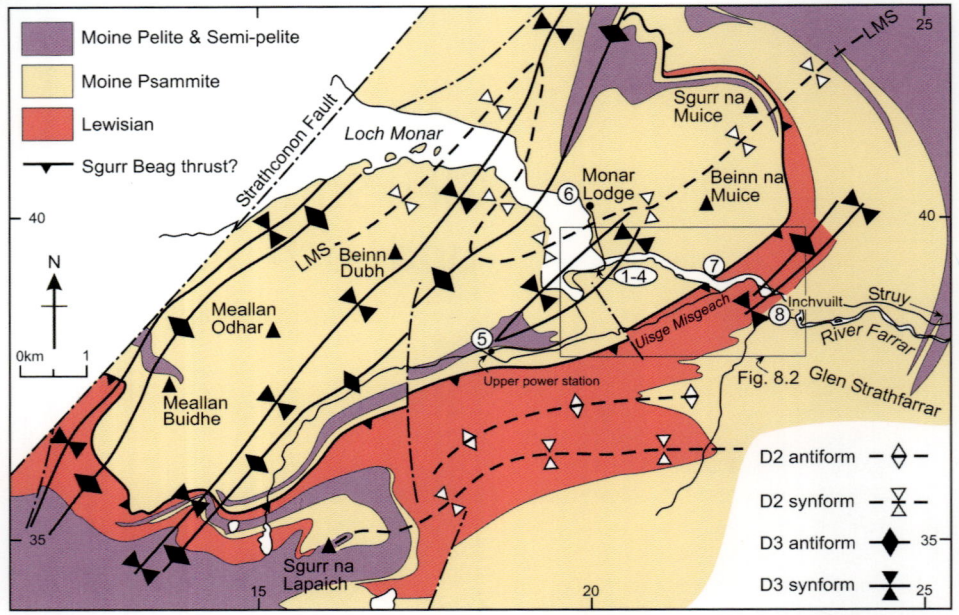

season, and it is closed on Tuesdays and Wednesdays until 1300. There are no places to overnight in the glen and camping is not allowed as it is a protected nature reserve (Scottish National Heritage). Proceed along Glen Strathfarrar for 20km as far as the Monar North dam site (Scottish and Southern Energy plc). The valley is especially beautiful, almost unspoilt and containing many stands of Scots Pine, a tree of slow growth characteristic of the Caledonian forest which once covered much of the Highlands.

The region consists of strongly deformed Moine metasediments and Lewisian basement. The Lewisian was emplaced as a sheet into the Moine rocks along a zone of intense D_1 deformation (the Sgurr Beag Thrust, Figs 8.1, 8.2) before D_2 and D_3 folding. This region has played a critical role in the general interpretation of fold interference structures. The geometry of the Monar region is dominated by the D_2 Loch Monar Synform (Fig. 8.1) which has a very steeply dipping southern limb, a more gently inclined northern limb and a hinge zone which generally plunges to the west. D_2 parasitic folds with S-shaped forms are apparent along the Moine-Lewisian contact along the southern part of the area (Figs 8.1, 8.2). These do not affect the D_1 Sgurr Beag Thrust on the northwest side of the Lewisian sheet, perhaps because the structure was also active during D_2 folding. In contrast to the D_2 folds, the D_3 folds which are superimposed on the limbs of the Loch Monar Synform generally have overall Z-shaped forms and have

Glen Strathfarrar and Loch Monar

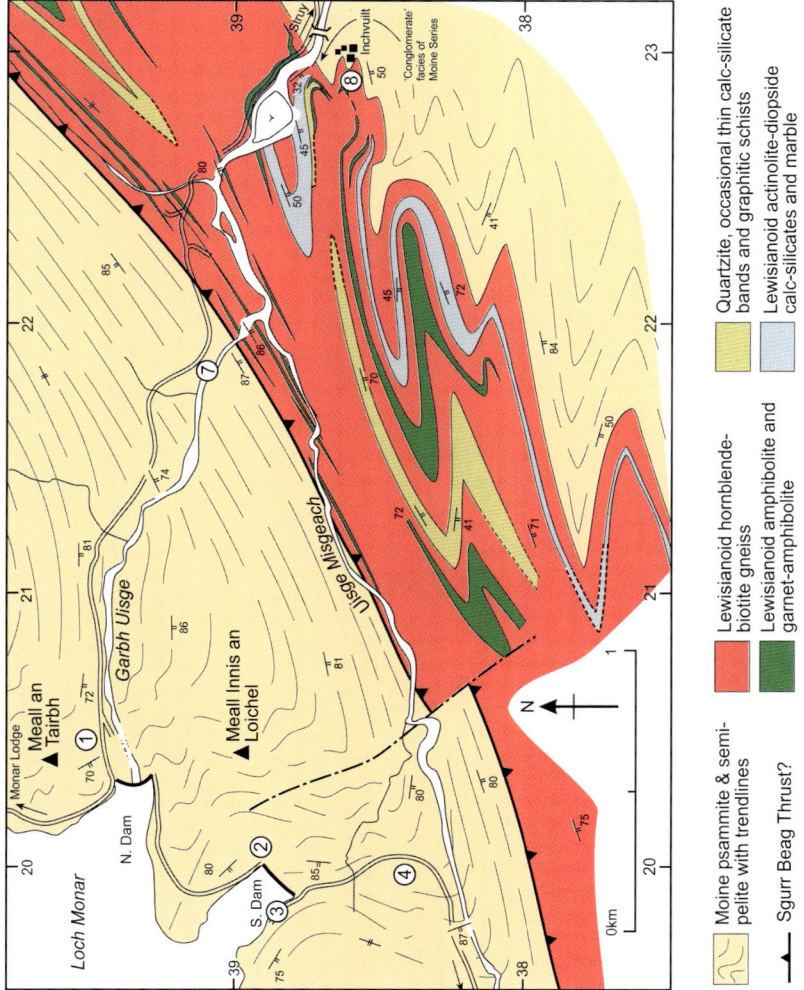

plunge directions inherited from the initial limb dips of the earlier fold (Ramsay, 1957). The significance of the special and systematic fold geometry seen at Monar has been analysed in some detail in Ramsay & Huber (1987, pp. 480-84, 499-502).

Opposite page:
Fig. 8.1 General geological map of the Monar region and upper parts of Glen Strathfarrar.

Fig. 8.2 (above) Geological map of the southern limb of the Loch Monar Synform showing localities around the eastern end of Loch Monar, together with details of the internal lithologies found in the Lewisian sheet.

Locality 8.1 [NH 2038 3938]

Roadside just NE of North dam. D_2 and D_3 folding within Moine psammites.

Park just below the North dam [NH 2035 3940]. The outcrops along the northern side of the road are Locality 8.1 and show two sets of fold structures. Earlier D_2 folds generally plunge to the west, but their hinge-lines and associated linear structures frequently undulate in a sinusoidal manner. The later D_3 folds frequently appear as corrugated forms with about the same wavelength as typical corrugated iron. The fold hinges generally plunge steeply or even vertically as is the norm throughout the southern limb of the D_2 Loch Monar Synform. Where the D_3 folds are especially strongly developed the recrystallisation associated with their development has locally, partially or even completely obliterated the linear fabrics associated with the D_2 folds and these linear fabrics are overprinted by new schistosity-bedding intersections parallel to the D_3 fold hinges.

Locality 8.2 [NH 2002 3895]

Hillside just NE of North dam. Deformation of D_2 structures by D_3 folds within Moine psammites.

Cross the North dam and park just before arriving at the South dam. On the hillside on the north side of the South dam is a large cement block used as an anchor during the construction of the dam. A flat-topped outcrop 15 m southeast of this block is Locality 8.2 at [NH 1998 3897] (Fig. 8.2). It shows a perfect miniature model of the fold intersection geometry with eye-like basin forms being located at the intersections of crossing D_2 and D_3 synforms. The fold hinge-lines of the late D_3 structures are variable and controlled by the dips of the limbs of the D_2 synform. On the roadside just below this outcrop are excellent examples of D_3 corrugations superimposed on a moderately inclined pre-existing surface of a D_2 fold. The limbs of these D_3 folds show traces of an early lineation parallel to the D_2 fold hinge lines which may be traced over the fold hinges. Each of these curving lineations is aligned in a plane (strike 095° dip 75° S). The significance of this geometry has been discussed by Ramsay (1960, 1967, pp. 470–82) and appears best explained by attributing the forms to flow in a direction which is not perpendicular to the fold axes, but controlled by flow quite oblique to the fold axis. This has been termed the 'a'-direction, and it can be found at the intersection of the lineation plane with the axial planes of the deforming folds (here with strike 037° dip 82° E). This calculation, easily carried out using a projection net, shows that the flow direction plunges 75° to 181° (see also a worked example and discussion in Ramsay & Huber, 1987, pp. 484 and 501). It can be deduced that the D_3 fold hinge lines associated with an intense rodding fabric are not simply related to the stretching direction, although they may lie close to this direction. Outcrops lower down this roadside section show almost vertically-plunging D_3 corrugations (superposed on initially steep limbs of D_2 folds) which also show D_2 lineations deformed by the folds, and calculations show that the 'a'-directions lies very close to that of the previous calculation. *Please do not hammer any of these most instructive outcrops.*

Locality 8.3 [NG 1990 3884]

Glacial pavement immediately west of South dam. D_2-D_3 interference patterns within Moine psammites.

Cross the South dam and park at the road junction. Descend to the lake side on the south side of Loch Monar to reach Locality 8.3 [NG 1990 3884] (Fig. 8.2). Here are almost unbelievably wonderful outcrops of folded Moine sediments with abundant very clearly preserved interference patterns on very clean glaciated surfaces. These are classic outcrops, well suited to photography (Fig. 8.3; and see Ramsay, 1962; Ramsay & Huber, 1987, p. 498; Ramsay & Lisle, 2000, p. 905). The Moine sediments are mostly psammite with bands of darker semi-pelitic and pelitic material and occasional bands rich in heavy minerals. The cuspate-lobate forms of D_3 folds developed at the interfaces of psammitic and pelitic bands indicative of buckling instabilities are extremely clear, and examples of competent layers with inward pointing cusps and incompetent pelitic bands with outward directed cusps abound. It is also clear that the basic fold model here is that of overall harmonic folding, related to the close packing of the initial layers, with some polyharmonic folding forms where especially competent or incompetent sheets are involved (Ramsay & Huber, 1987,

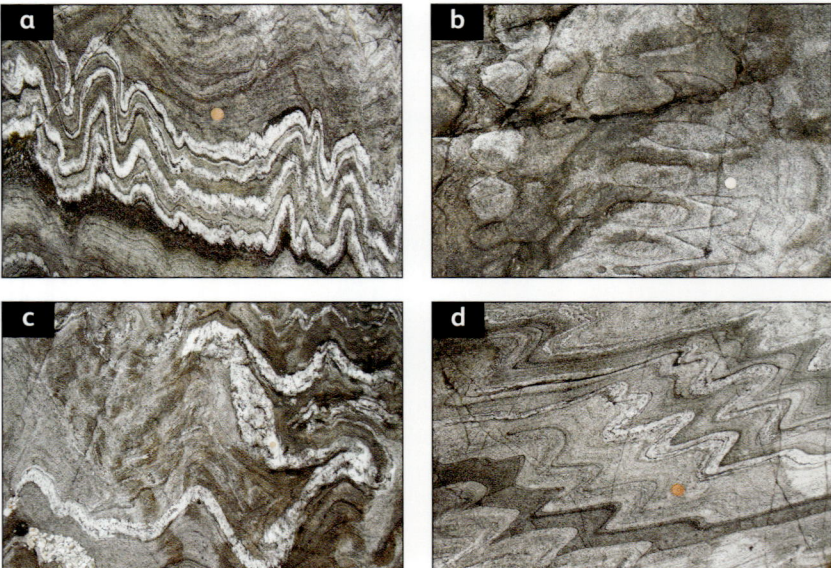

Fig. 8.3 Structures visible at Locality 8.3, all around [NH 1989 3882]:

(a) upright D_3 folds deforming migmatitic veins;

(b) 'dome and basin' (type 2) patterns produced by the interference of D_2 and D_3 fold sets;

(c) D_2 isoclinal fold of a migmatitic vein deformed by upright D_3 folds;

(d) D_3 fold set showing cuspate-lobate shapes developed at the interface between psammitic and pelitic lithologies.

p. 406). A ductility sequence can be established: quartz-feldspar D_2 pegmatite and gneissic veins > psammite > semi-pelite > pelite. Although these folds mostly show that buckling components are present, one should not forget the geometry of the folds seen just north of the South dam indicating that they are not just simple buckle folds but have strong flattening and shear components controlled by steeply inclined flow directions and might be best termed oblique buckle folds (Ramsay, 1967, p. 396; Ramsay & Huber, 1987, p. 489).

As well as having been subjected to amphibolite facies metamorphism during both folding phases, the rocks here have been subjected to melting and the development of migmatite veins. Those veins formed during the D_2 events form coarse feldspar-quartz-mica sheets sub-parallel to the bedding. These are plicated by well developed D_3 folds, often of ptygmatic habit, whereas those veins developed during the D_3 events form narrow pegmatite sheets located in the limbs of the D_3 folds and which cross-cut the earlier folded D_2 sheets. Note that these D_3 sheets rarely occur along the actual hinge zones of the D_3 folds, but are practically always developed in the fold limb which has been most strongly deflected from that of the original orientation of the banding after D_2 fold development. Other features of these outcrops are the presence of occasional late dykes of the schistose lamprophyre type and a few thin late cross-cutting pegmatite dykes which, although they cross-cut the D_3 folds, have been slightly affected by the last shortenings of the D_3 events. These pegmatites are the outermost part of an intense pegmatite swarm (Ramsay & Huber, 1987, p. 324) which may well be related to an unexposed Caledonian granite situated at depth. Return to the vehicles and proceed along the road south of the South dam.

Locality 8.4 [NH 1996 3848]

Roadside outcrops west of the road. D_3 folding of D_2 linear structures within Moine psammites.

On the west of the road at Locality 8.4 [NH 1996 3848] are very good examples of vertical or steeply plunging D_3 folds (imposed on the steep limb of the Loch Monar D_2 synform) folding D_2 linear structures. There are thick bands of pelitic material interbedded with the psammites. These show few of the D_3 corrugations seen in banded psammites, but contain a well-developed alignment of micas sub-parallel to the axial planes of the D_3 folds. Proceed further along this road towards the Upper Power station in the Uisge Misgeach river valley.

Locality 8.5 [NH 1840 3772]

South of the Upper Power Station. Lewisian-Moine relationships.

Park at [NH 1832 3810] (Fig. 8.1). Follow the pipeline south-southeast to a small quarry which is Locality 8.5. Here the contact between Lewisian gneisses and Moine psammites is well exposed. The Lewisian consists of orange-weathering quartz-feldspar gneiss with lenses of hornblende- and biotite-rich material. Although this gneiss has been reduced in grain-size it is quite unlike the pegmatitic gneissic bands seen in the Moine rocks. The adjacent Moines are psammites with pelitic bands clearly of sedimentary aspect. Both Lewisian and Moine rocks show steeply plunging D_3 folds. Return to the parked vehicles and drive to the main park place [NH 2035 3940].

Locality 8.6 [NH 1988 4050]

Lakeside outcrops west of Monar Lodge. D_3 folds cut by pegmatite dykes.

From the parking place, proceed by foot along the road towards Monar Lodge. From the north of the Lodge take a small footpath along the loch side to glaciated pavements exposed on the foreshore which are Locality 8.6 [NH 1988 4050] (Fig. 8.1). There are excellent examples of Z-shaped D_3 folds plunging 30° to 40° to the SW. We have crossed the axial surface of the D_2 Loch Monar Synform (it passes through Monar lodge) and are now situated on the northern limb of this fold, hence the plunge of the D_3 folds is significantly less than most of those seen in the Monar damsite outcrops. The D_3 folds are cross-cut by impressive pegmatite dykes. These are coarse-grained (crystals up to 10cm) quartz-feldspar-mica dykes with more finely crystalline margins. The dykes, although cross cutting the D_3 folds, are themselves quite strongly folded and shortened by overlapping thrust-like sectors. Return to the vehicles.

Locality 8.7 [NH 2180 3927]

River section of the Garbh Uisge. Highly strained Lewisian and Moine rocks within the D_1 Sgurr Beag Thrust Zone.

Return along the road descending into Glen Strathfarrar. After about 1.5km make a roadside stop [NH 2180 3919] and descend about 30m into the river section of the Garbh Uisge to Locality 8.7 at the northern contact of the Moine and Lewisian (Fig. 8.2). Here one becomes aware of the difficulties in distinguishing the two units. The banding in Moine and Lewisian is almost vertical, quite parallel and all the rocks show a well developed D_2 linear fabric plunging 30° towards 225° (the axial direction of the Loch Monar Synform). To the NW, the rocks are grey, well-banded psammites clearly part of the Moine. Descending the river to the SE, the rocks are banded orange-buff coloured gneisses containing sheets and boudins of amphibolite and clearly of Lewisian aspect. However, because of the convergent appearance of Moine psammites and Lewisian quartz-feldspar gneisses, the contact is difficult to locate with precision, even though the outcrops are extremely good. The well-banded and foliated nature of these outcrops may reflect the presence of a D_1 tectonic slide at this contact, probably the Sgurr Beag Thrust (Tanner *et al.*, 1970). Another point of interest at this locality is the presence of some dark, non-metamorphic camptonitic dykes (Permian) intruded parallel to the layering. Return to the vehicles.

Locality 8.8 [NH 2294 3874]

Moine-Lewisian contact north of Inchvuilt; probable basal Moine conglomerate on the east side of the Lewisian inlier.

Drive to the bridge which crosses the River Farrar at Inchvuilt (Fig. 8.2) [NH 2304 3874]. Park and cross the footbridge to outcrops at Locality 8.8 of Moinian rocks on the south side of the river by an electric pylon [NH 2294 3874]. Lewisian rocks of calc-silicate and amphibolitic types are exposed a few metres to the west. The Moine semi-pelites contain elongated and rodded lumps of quartz and quartz-feldspar-mica rock resembling pebbles and interpreted by E. M. Anderson as being the basal conglomerate of the Moine with the adjacent Lewisian gneiss (Anderson, in Peach *et al.*, 1913). Although this interpretation has been questioned (Ramsay, 1956) it is probably correct. Return down the valley to the gate where it will be necessary to log-out of the valley.

Excursion 9

Loch a' Bhraoin, Braemore and Loch Broom

Simon Kelley

Purpose: A traverse across the Moine from a major internal ductile thrust, the Sgurr Beag Thrust, to the margin of the orogen, the Moine Thrust Zone.

Aspects covered: The increase of deformation into the Sgurr Beag Thrust, post-thrusting deformation and formation of the Moine mylonites, shear fabrics in psammites and pelites, the Moine Thrust plane.

Useful addresses: The Tourist Information Office, Ullapool; Inverbroom Estate (Mr Cameron, Home Farm, Inverbroom [Tel: 01854 655252]).

Maps: OS: 1:25,000 sheets 435 An Teallach & Slioch, 436 Beinn Dearg & Loch Fannich; BGS: 1:63,360 sheets 92 Inverbroom, 101E Ullapool.

Types of terrain: The Sgurr Beag Thrust traverse has easy walking on paths and stream sides. The traverse up to the Moine Thrust has:

Option A – open country, climbing up to 500m; and

Option B – easy walking on seashore and roadside.

Distance and time:

Option A – car not used after Locality 9.1, on foot 19km. A full day (8 hours).

Option B – by car, 20km (1 hour), on foot 9.5km (4 hours).

Short intinerary: The Sgurr Beag Thrust traverse (Locality 9.2) is a complete excursion on its own and can be covered in about 3.5 hours. The Moine Thrust can be covered briefly by a quick visit to the Corrieshalloch Gorge (Locality 9.7) and the lochside exposures (Locality 9.9), all of which can be seen in 3-4 hours.

Loch a' Bhraoin, Braemore and Loch Broom

The first section of this excursion consists of a traverse from Morar Group psammites and pelites across the Sgurr Beag Thrust into Glenfinnan Group migmatites and Lewisianoid basement gneisses. The second part, a traverse into the Moine Thrust Zone, can be approached in two ways. *Option A* (Fig. 9.1 and Localities 9.1-9.7) is a walk along the ridge from Meall an t-Sithe to Creag Rainich, affording an excellent opportunity to observe the relationship between the Sgurr Beag Thrust and the Moine Thrust at their closest approach. However, this option involves a 500 m ascent and the area does not afford much shelter in bad weather. Further, the ground is used for deer stalking from August to February. *Option B* (Fig. 9.1 and Localities 9.8-9.10) illustrates the increasing breakdown of earlier peak metamorphic fabrics in psammites and pelites of the Morar Group as the Moine Thrust Zone is approached. Viewing of the low-level roadside and shoreline exposures is not restricted by weather or deer stalking.

The apparently simple stratigraphy of Moine rocks in the Loch a' Bhraoin, Braemore and Loch Broom area is misleading. At the base of the succession, the Inverbroom Psammite (Fig. 9.1) varies from 2.7km thick to less than 400m where it is cut out against the Moine Thrust. Common sedimentary structures indicate that it is 'right way up' throughout the area. The Sgurr Mor Pelite (Fig. 9.1) lies above the Inverbroom Psammite (it also exhibits sedimentary structures indicating 'right way up') and ranges from 950m to less than 20m thick at Meall Dubh where it interfingers with the Inverbroom Psammite. This pelite is important for correlations since it contains stratabound amphibolites (Winchester, 1976). The Sgurr Mor Pelite passes upward into the Meall a' Chrasgaidh Psammite through a sedimentary transition, and this psammite occupies a belt 100m-130m wide adjacent to the Sgurr Beag Thrust. No sedimentary structures have been preserved in this highly deformed psammite.

All the units below the Sgurr Beag Thrust are part of the Morar Group. However, direct correlations between these units and those of the type area in Morar, over 150km to the south, are tenuous because units such as the Sgurr Mor Pelite are clearly laterally discontinuous. All Morar Group rocks in the area contain calc-silicate pods that have been used to determine metamorphic grade in the absence of aluminosilicate indicator minerals (Winchester, 1974).

Lying above the Sgurr Beag Thrust, the Meall an t-Sithe Pelite is the only representative of the Glenfinnan Group (Fig. 9.1). This coarse-grained migmatitic pelite has clearly undergone a very different tectonometamorphic

EXCURSION 9

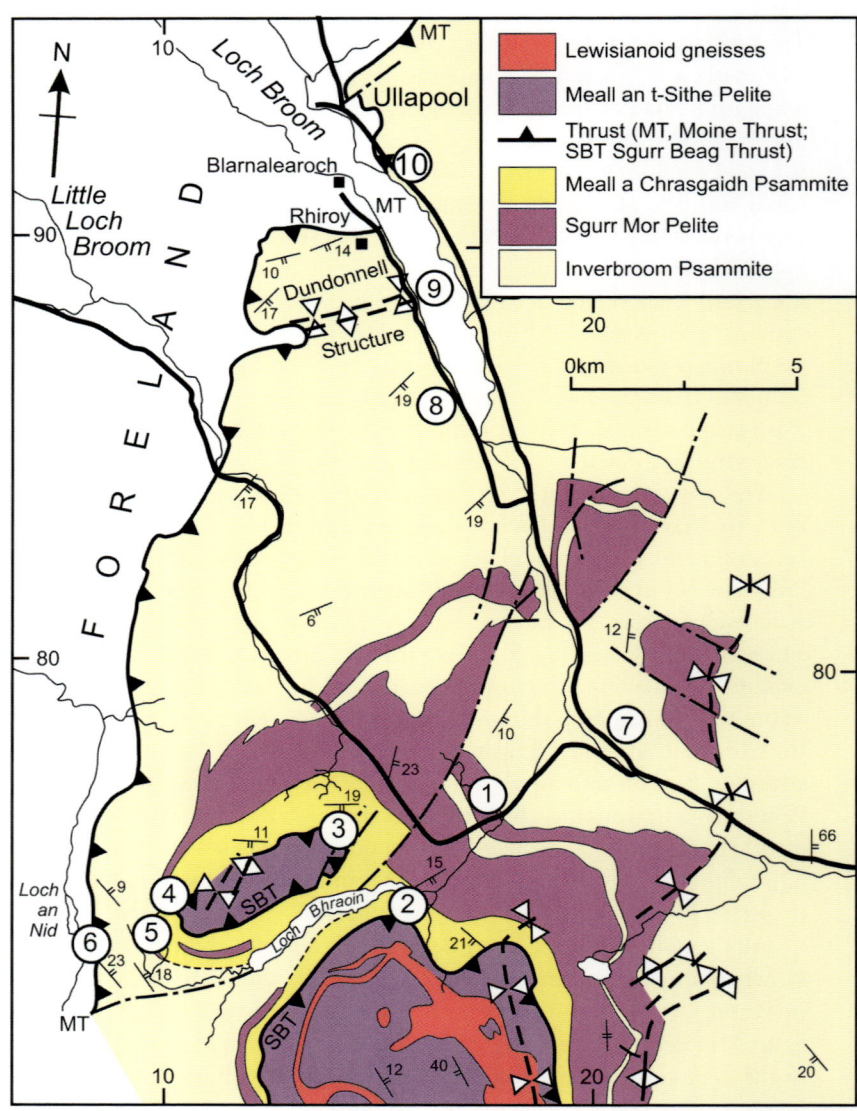

Fig. 9.1 Geology of the Moine rocks between the Fannich mountains and Ullapool, showing the localities to be visited.

history to rock types below the thrust. The Meall an t-Sithe Pelite contains amphibolites of tholeiitic origin (Winchester, 1976), but no calc-silicate pods nor sedimentary structures.

Lewisianoid gneisses lie between two lithologically and chemically identical units of the Meall an t-Sithe Pelite (Fig. 9.1). The basement gneisses are isoclinally folded with the Glenfinnan Group migmatites. Though only a 100m-150m-wide outcrop of the Lewisianoid gneisses is present, they are distinct from the Moine rocks, consisting of mainly siliceous acidic and hornblendic gneisses with minor calcareous, epidotic and pyroxenic horizons.

The Sgurr Beag Thrust section [NH 177 768 to NH 116 723]

The first locality (Fig. 9.1) of the section is on the A832 Gairloch road 4.5 km from Braemore junction [NH 209 777] with the A835 Ullapool road (not to be confused with the junction close to Garve) and about 16 km from Dundonnell. Roadside exposures between Braemore junction and the parking place are psammites of the Inverbroom Psammite, containing occasional interbedded pelites (Fig. 9.1). Parking for three cars is available by the bridge over Allt Leacach (Figs 9.1, 9.2). This section of the A832 road ('Destitution Road') to Dundonnell was first built in 1846 to assist years of poverty and famine in the area after several consecutive years of bad weather causing crop failures.

Locality 9.1   [NH 177 768]

Allt Leacach (Fig. 9.1). Start of the Sgurr Beag Thrust traverse (900-800m structurally below the Sgurr Beag Thrust).

The Inverbroom Psammite exposed in the waterfalls above and below the bridge, consists of banded psammites and semi-pelites with a flat-lying D_2 planar fabric crenulated by D_3. Occasional minor D_3 folds are exposed in the stream below the bridge. Note particularly the undulating style of the bedding planes with rare cross-bedding, the angular discordance between early quartz veins and the foliation/bedding plane, and the intensity of the stretching lineation in the psammites.

EXCURSION 9

About 1.5km further down the road towards Dundonnell, an untarred track leads from the road to the SW [NH 162 761]. Parking for five cars is available in a lay-by, 100m further along the road.

Locality 9.2 [NH 156 752 to NH 166 723]

Allt Breabaig (Fig. 9.2). A section from the Morar Group rocks across the Sgurr Beag Thrust into Glenfinnan Group migmatites and the Fannich Lewisianoid basement inlier (Fig. 9.1).

From the parking place, follow the track for about 1km to the bridge [NH 1585 7500]. Note the exposure of the Sgurr Mor Pelite in the stream by the lochside, 2A [NH 1565 7490] (350m below the Sgurr Beag Thrust, Fig. 9.2) which exhibits deformed quartz veining and a stretching lineation (trending at 135°) that is stronger than that at Locality 9.1. Rare sedimentary structures are flattened in the plane of the foliation and pelite-rich horizons have been crenulated by the post-thrusting D_3 deformation.

Follow the path leading from the bridge and up the glen between Druim Reidh and Meall a' Chrasgaidh. As the path drops down to follow the burn, exposures of the Meall a' Chrasgaidh Psammite are visible in its banks, 2B (250m below the Sgurr Beag Thrust, Fig. 9.2). Compare the near-parallelism

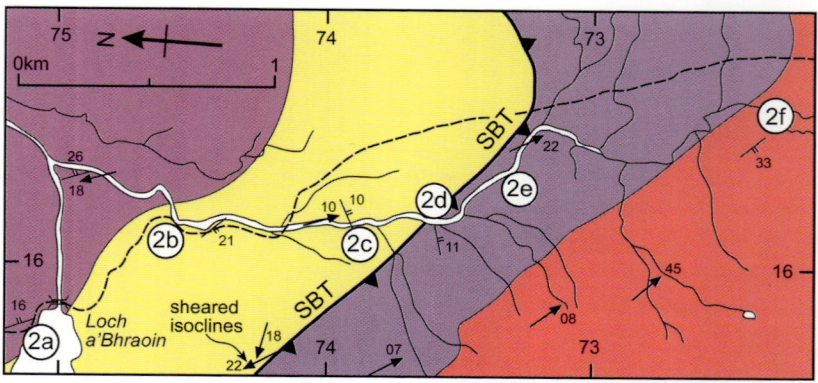

Fig. 9.2 Sketch map for Locality 9.2. See Fig. 9.1 for key.

of sedimentary structures, foliation and early quartz veins with the equivalent relationships in Locality 9.1 and stop 2A that were further beneath the thrust. Note the increased intensity of the stretching lineation in these psammites. A later minor curved D_3 fold hinge in one of these exposures folds the foliation and lineation related to the development of the D_2 Sgurr Beag Thrust. All localities in this excursion lie on the shallow-dipping limb of a major N-S-trending D_3 syncline (Fig. 9.1) overturned towards the west. Many minor folds of the D_3 generation exhibit curved hinges.

Large exposures of the planar-bedded psammites, 2C (75m-25m below the thrust, Fig. 9.2) have been created by waterfalls further up the stream. These exposures have a 'tramline' style foliation and increasingly intense stretching lineations. It is difficult, by this stage of the deformation, to distinguish any angular discordance between the foliation and early quartz veins.

Exposures of the planar-bedded psammites above the waterfalls exhibit occasional dark bands, less than 1cm thick, lying nearly parallel to the foliation. These bands are haematized cataclastic zones, almost certainly associated with late brittle movements on the Moine Thrust Zone that lies only 6km to the west or 1.5km vertically below these exposures.

The lower boundary of the Meall an t-Sithe Pelite is heralded by water-worn exposures of pelitic migmatites in the banks of the burn, 2D (Fig. 9.2). The actual contact is exposed on the Meall an t-Sithe to Creag Rainich ridge (*Option A*) where it is found in its original shallow lying position, in contrast to the steeply-dipping outcrops in the southwest Moine (see Excursions 4 and 5). Folds of thrust-generation age can be seen at [NH 155 741] by making a diversion at the end of the traverse on the way back to the farmhouse (or close to the summit of Meall an t-Sithe, *Option A*). These are a feature of the Sgurr Beag Thrust elsewhere (Rathbone & Harris, 1979), but are rare in this area.

In the first exposures above the Sgurr Beag Thrust, the *lit-par-lit* texture that is characteristic of these rocks has been destroyed by deformation associated with displacement along the thrust. Note that the shallow-lying foliation and stretching lineation (still trending 135°) are stronger than those further from the thrust.

Pelitic migmatites have retained their *lit-par-lit* texture further up the stream, 2E (50m above the thrust, Fig. 9.2). The migmatization is thought to have occurred during the Neoproterozoic Knoydartian event and the migmatitic fabric was subsequently deformed during Caledonian thrusting.

The composite fabric exhibits later crenulations, due to the post-thrust D_3 deformation.

Pale coloured acidic Lewisianoid basement gneisses are exposed in the stream bed and by the stream a few metres above the ruined croft, 2F (150m above the thrust, Fig. 9.2). These are characterized by stronger foliations and stretching lineations than those in the Moine rocks further downstream. The orientations of the foliation and lineation correspond closely with those seen in the Moine psammites. Refolded isoclines in some exposures testify to the complex polydeformational history of the rocks.

The Fannich outcrops of Lewisianoid basement gneiss have been the subject of controversy throughout the history of their study. The original Geological Survey workers disagreed over their outcrop pattern, some believing that it represented a Lewisian basement core to the Moine rocks in a 'fountain of nappes', and others that the outcrop represented a block of Lewisian translated to its present position along a thrust plane. No field evidence was produced for the latter view, but they believed that it lay at the boundary between the Meall an t-Sithe Pelite and the Meall a' Chrasgaidh Psammite (now generally accepted to be the position of the Sgurr Beag Thrust). Some years later, Sutton & Watson (1954) discounted the idea that the gneisses of Fannich were basement at all, and maintained that they were an integral part of the Moine succession. Subsequent work on the geochemistry of the gneisses (Winchester, 1971) re-established their basement affinities. Although the Lewisianoid gneisses and Moine migmatites (above and below the gneisses) are certainly interfolded, the precise structural relationship still remains uncertain.

Return to the bridge [NH 1585 7500]. Exposures of sheared isoclines in the Meall a' Chrasgaidh Psammite lie close to the 450m contour on the NE-trending spur at the head of the glen [NH 155 742]. From these exposures, simply walk down the spur and rejoin the path to the bridge.

Loch a' Bhraoin, Braemore and Loch Broom

Option A

From the bridge, proceed NW to the summit of Meall an t-Sithe (2 km distance, 350m climb).

Locality 9.3   [NH 141 765]

Meall an t-Sithe (Fig. 9.1). An exposure of the Sgurr Beag Thrust.

The contact between the Meall an t-Sithe Pelite (Glenfinnan Group) and the Meall a' Chrasgaidh Psammite (Morar Group) is exposed just below the summit to the north (Fig. 9.1). Many intrafolial folds of thrust-generation age (D_2) are exposed in the Meall a' Chrasgaidh Psammite below the thrust to the NE. A suite of pegmatites characterized by large K-feldspars (10-50mm) and post-dating the thrust movements, occur sporadically in all rock types, exhibiting low levels of post-intrusion deformation. This suite becomes highly deformed in the structurally underlying belt of mylonites associated with the Moine Thrust Zone.

Follow the ridge westward for about 3km to the summit of Meall Dubh at [NH 103 748] (Fig. 9.1). Between Meall an t-Sithe and Meall Dubh, psammites that were intensely deformed during displacements on the overlying ductile thrust are increasingly affected by cataclastic bands nearly parallel to the foliation, and the post-thrusting pegmatites become folded or exhibit boudinage. Close to tight D_3 folds also affect the thrust-related foliation in this area.

Locality 9.4   [NH 103 748]

Meall Dubh (Fig. 9.1). Closest approach of the Sgurr Beag Thrust and the Moine Thrust in the Northern Highlands.

The outcrop of the Sgurr Beag Thrust on the summit of Meall Dubh is 2 km east of the (brittle) Moine Thrust plane, is only 400m directly above it and within the zone of deformation associated with the thrust zone.

The Meall a' Chrasgaidh psammite in this area (Fig. 9.1) represents an intermediate stage between the coarse, annealed psammites forming the Sgurr Beag Thrust foliation and the fine-grained mylonites of the Moine Thrust Zone. The psammites exhibit kink folds, bands of cataclasite and zones of breccia up to 50cm thick. The foliation of the pelitic migmatites above the thrust is disrupted but not destroyed by the mylonitization. Shear bands indicating a top-to-the-WNW sense of overthrusting are found in both migmatites and pelitic horizons of the Meall a' Chrasgaidh Psammite.

A detour to the summit of Creag Rainich affords a superb view of the whole Moine Thrust Belt. Note the juxtaposition of the foreland Lewisian gneisses, Lewisian slices in the thrust zone, and the Lewisianoid basement inlier of Fannich. These were originally at least 150km apart and have been brought into juxtaposition having undergone very different histories. Descend SW, from the summit of Creag Rainich to the westernmost burn leading into Allt Teanga nan Caiseachan (Fig. 9.1).

Locality 9.5   [NH 090 746 to NH 093 736]

Allt Teanga nan Caiseachan (Fig. 9.1). Break-up of the Sgurr Beag Thrust fabric during mylonitization.

Exposures of the Meall a' Chrasgaidh Psammite are crossed in the upper part of the stream by shear bands up to 10cm across, forming lenses or pods of undeformed, pre-mylonitic psammite up to 3m long. The movement sense of the shear bands indicates displacement towards the WNW. In the centres of the shear bands, a fine, thinly-banded mylonitic foliation and stretching lineation are parallel to the same features in the Moine mylonites (i.e. shallowly-dipping foliation with the lineation trending towards 110°). In the unaffected rocks, the foliation still carries the characteristics of the coarse-grained Sgurr Beag Thrust fabric (stretching lineation trending towards 135°).

About 1km downstream, as another stream joins from the east, leave Allt Teanga nan Caiseachan and contour round to the west towards the Moine Thrust. The first exposures of Moine mylonites form a small 'quarry-like' area with a stream running down the centre [NH 085 733], as Loch an Nid comes into view.

Locality 9.6 [NH 085 733]

Loch an Nid (Fig. 9.1). The Moine Thrust.

The Moine Thrust plane passes between finely banded psammitic Moine mylonites, with breccia zones and kink bands and coarse-grained Lewisian amphibolites cut by late shear zones. The Moine mylonites in the slopes beneath the cliffs exhibit an intense mylonitic foliation, though low strain augen of less deformed Moine rocks up to 50m long remain, within 100m of the thrust plane. The thrust itself cuts through the Lewisian-Torridonian unconformity and thus the Lewisian gneisses below the thrust give way to Torridonian sandstones further north. Pods of Cambrian quartzite strung out along the thrust plane form prominent knolls on the hillside [NH 085 739 and NH 084 746]. These probably represent remnants of a horst block eroded from a ramp at a deeper level in the thrust zone.

Return to the farmhouse and the A832 by the path along the northern shore of Loch a' Bhraoin.

Option B

From the bridge [NH 1585 7500], return to the road and drive back along the A832 to the Braemore junction (Fig. 9.1). The first locality of this option lies about 1km along the A835 towards Ullapool. Parking is available for many cars and coaches in the Corrieshalloch Gorge car park on the north side of the road.

Locality 9.7 [NH 204 782[

The Corrieshalloch Gorge (Fig. 9.1). Initial deformation associated with the Moine Thrust in psammites and pelites.

Exposures in the immediate vicinity of the car park are not very informative. Walk westwards downhill along the A835 to a large roadside cutting at [NH 2007 7849]. *Care should be taken here as traffic can be fast-moving along this stretch of road.* The exposures here are interbedded psammites and pelites of the Inverbroom Psammite (Fig. 9.1). The psammites acted as rheologically competent layers and suffered relatively little deformation during the thrusting. Note that the quartz veins and pegmatites cross the psammites at high angles. Pelite rock types on the other hand acted as less competent layers and there is widespread evidence for simple shear across these layers, giving shear strain values as high as eight. Quartz veins and pegmatites are highly deformed within the pelites, indicating shear towards the WNW. A particularly important aspect of the metamorphic fabric within the pelites is the development of shear bands, creating lenses of undeformed, pre-mylonitic fabric between 0.25 mm and 10 cm long, and indicating overthrusting towards the WNW.

No stop is complete without seeing the Falls of Measach, formed by run-off water during the last glacial retreat. The view from the suspension bridge is truly spectacular, especially after heavy rain.

Return to the parking area and follow the road towards Ullapool. Approximately 6 km further on, turn left along the single track road signposted for Letters, Ardindrean and Rhiroy. (*This road is unsuitable for coaches; the turning place at Rhiroy is only large enough for cars or minibuses*). Turning right at the first crossroads, follow the road past the chapel and along the shore of Loch Broom. The road along the southern shore of Loch Broom originally reached only as far as the first houses and was extended to its present length in the mid 1930s. Prior to this time the only way in and out was by boat. The population which was at its peak around 1846 lived by crafting and fishing, but the number of people living in this community has now dwindled to less than 35% of its earlier peak, excepting weekend visitors and geologists.

Locality 9.8 [NH 171 858]

Loch Broom (Fig. 9.1). Increasing deformation associated with the Moine Thrust Zone mylonites.

As the level of the road rises above the shore of Loch Broom, about 1 km past the chapel, massively bedded Inverbroom Psammites and interbedded pelites are exposed just above and occasionally at the roadside. Early pegmatites within the psammitic rock types lie at low angles to the foliation and exhibit pinch and swell structures (incipient boudinage). The fabric in the pelitic rocks is dominated by shear bands, and the grain sizes of the pelites have been reduced from their peak metamorphic sizes (250-500 μm) to typical mylonitic sizes (10-50 μm), apart from resistant muscovites that are rotated toward parallelism causing a strong planar fabric.

The crags of Cnoc an Droighinn [NH 149 888] represent the end of an ENE-trending ridge reaching to the Moine Thrust in the west (Fig. 9.1) and continuing across Loch Broom to the east. The structure of the ridge is an anticline in the Moine rocks, caused by a ramp structure in the underlying thrust zone. The ramp structure is exposed at Dundonnell and as such can only be recommended as a separate excursion. Coarse-grained psammites and pebble bands in the Inverbroom Psammite can be traced from Cnoc an Droighinn into the Moine mylonites at [NH 118 874], demonstrating that the mylonite foliation is not parallel to the earlier bedding/foliation planes.

About 0.5 km towards Rhiroy, the wood between the road and the shore gives way to open pasture. Parking is available in two long passing places less than 100 m further along the road.

Locality 9.9 [NH 156 892 to NH 149 911]

Loch Broom shoreline (Fig. 9.3). Traverse across the Moine Thrust.

The traverse is best attempted at low tide, though this is not essential (consult the Ullapool Tourist Office for times of the tides).

Descend through the field to the shoreline where the Inverbroom Psammite is exposed 9A (Fig. 9.3). Massive psammite banding seen in the Inverbroom Psammite further from the thrust is broken up into irregular lenses of psammite separated by small shear zones or shear bands. The lenses

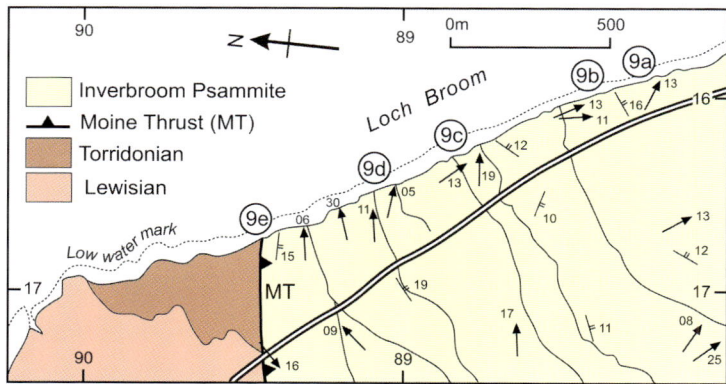

Fig. 9.3 Sketch map for Locality 9.9.

vary in size, reaching up to 3m in length. Lenses of remaining psammite are increasingly broken into smaller lenses or flattened into the new foliation as the process of shear band formation increases in intensity towards the thrust. The early foliation in all pelitic horizons is broken up by shear bands, all indicating thrusting towards the WNW (Kelley & Powell, 1985).

A few metres further along the shore, 9B, minor folds ranging from close to isoclinal in style deform the foliation. The folds exhibit a wide range of styles and axial trends, though their axial planes coincide to within 10°. The minor folds in these exposures are most probably of one generation apart from late kink zones. Curved hinge lines in some exposures indicate that the folds, which formed during the shearing event causing the mylonitization, nucleated perpendicular to the movement direction and were rotated towards it during progressive deformation (Kelley & Powell, 1985). This pattern of minor shear-related folds is characteristic of shear zones (Alsop & Holdsworth, 2004a and references therein).

The rocks develop a stronger planar foliation, 9C, with thinly interbanded psammitic and pelitic mylonites resembling slates in some exposures. However, the increase in the intensity of planar structures is disrupted by shears (see 9A) and minor folds (see 9B) (Fig. 9.3).

The already complicated pattern of the psammitic mylonites is further disrupted by kink zones that are related to brittle movement of the thrust, 9D. These exposures are less than 50m above the thrust plane, and the mylonites become extensively brecciated as the thrust is approached with thin zones of cataclasite occurring parallel to the earlier foliation.

The thrust plane is exposed as a sloping exposure, 9E, of Torridonian sandstone forming the footwall, as the Moine mylonites that formed the hanging-wall have been eroded. The sandstones do not have any apparent internal structures, but are heavily brecciated and cut extensively by quartz veins. A few metres further on, the sandstones exhibit sedimentary layering and cross-bedding that are undeformed. The Torridonian sandstones are not thick in this thrust sheet (the Kinlochewe Thrust sheet) as shown by the outcrop of coarse-grained Lewisian amphibolitic gneisses on the shoreline at Blarnalearoch [NH 148 912] (Fig. 9.3).

Return to the parking place via the shoreline or follow the stream at the Lewisian exposure to the track and return along the road.

The two stone forts marked as brochs on the 1:25,000 map, Dun Lagaidh [NH 142 913] and Dun an Ruigh Ruadh [NH 149 901], are two of the rare stone forts on the west coast of Ross-shire. Not much remains of their walls, but Dun an Ruigh Ruadh is easily accessible above the road at Rhiroy. They were excavated by Mackie (1975) who found evidence of occupation at Dun Lagaidh as early as the 7th century BC; thick deposits suggest a lengthy period of use. A second phase of occupation gave rise to a circular dry stone walled defensive structure, though it was not a broch. Mackie considers that the structure was an Iron Age galleried dun, which would have been built earlier than the classic brochs of Glenelg and the Western Isles. Later occupiers in the 12th and 13th centuries repaired the structures using mortar. For those continuing towards Ullapool, one further locality situated just above the Moine Thrust is worthy of a visit.

Locality 9.10 [NH 149 922]

The Moine mylonites (Fig. 9.1). Roadside exposures of the Moine mylonites on the A835.

Park in the large lay-by on the south side of the road at [NH 151 920]. The geology in the road cutting immediately opposite is now rather obscured by wire mesh, so walk carefully along the north side of the road for about 200m to the NW. A large road cutting exposes flaggy Moine psammitic and semi-pelitic mylonites that are only a few metres above the Moine Thrust (although this is not exposed on the road section). Quartz and pegmatite veins occur nearly parallel to the intense foliation and both carry the mylonitic stretching lineation (plunging towards 110°). In contrast to the previous locality, at least two sets of folds are present here. A series of small-scale intrafolial isoclines were probably formed during early stages of mylonitization. A more prominent set of mesoscopic folds have S-geometry and are open in style; axes plunge to the SE, down the dip of the fold axial planes. These folds deform the mylonitic foliation and carry a tight axial-planar crenulation fabric in pelitic bands. The two sets of folds are analogous to the 'F_2' and 'F_3' folds described within the Morar Group in Excursions 10 and 13, and are thought to have formed during a single phase of progressive ductile deformation associated with westerly-directed overthrusting of the Moine rocks onto the Caledonian foreland.

Excursion 10
South and Central Sutherland

Rob Strachan, Bob Holdsworth, Maarten Krabbendam, Graham Leslie and Jack Soper

Purpose: To examine various phenomena within the Moine, including basement-cover relationships, Caledonian ductile thrusts, migmatites, and minor syn-tectonic igneous intrusions.

Aspects covered: Lewisianoid basement inliers; metasedimentary rocks and structures of the Morar and Glenfinnan groups; Sgurr Beag and Naver thrusts; Naver Nappe migmatites; the Vagastie Bridge Granite.

Useful information: Hotel and B&B accommodation are available in Lairg and Bonar Bridge (camping is also available in Lairg). Vehicular access to Locality 10.2 requires permission from Alan Wyatt, Caplich Estate (Tel: 01549 441356). Permission to access the Airde of Shin should be sought from Mrs A. Parrot, The Croft House, West Shinness, Lairg, IV27 4DN (Tel: 01549 402095).

Maps: OS: 1:25,000 sheets 443 Ben Klibreck and Ben Armine, 440 Glen Cassley and Glen Oykell, 438 Dornoch and Tain; BGS: 1:50,000 sheets 93E Alness, 102E Loch Shin, 108E Loch Naver.

Type of terrain: Stream sections, moorland, lochside and coastal exposures.

Distance and time: The complete excursion involves driving some 180km, assuming overnight accommodation in either Lairg or Bonar Bridge. The total distance covered on foot is *c.*28km. 2½ days should be allowed for the whole excursion. See each locality for suggested times.

Short itinerary: A shorter excursion could include Locality 10.1 and then either Locality 10.3 or 10.4. Locality 10.5 could be visited en route to North Sutherland if travelling from Lairg.

South and Central Sutherland

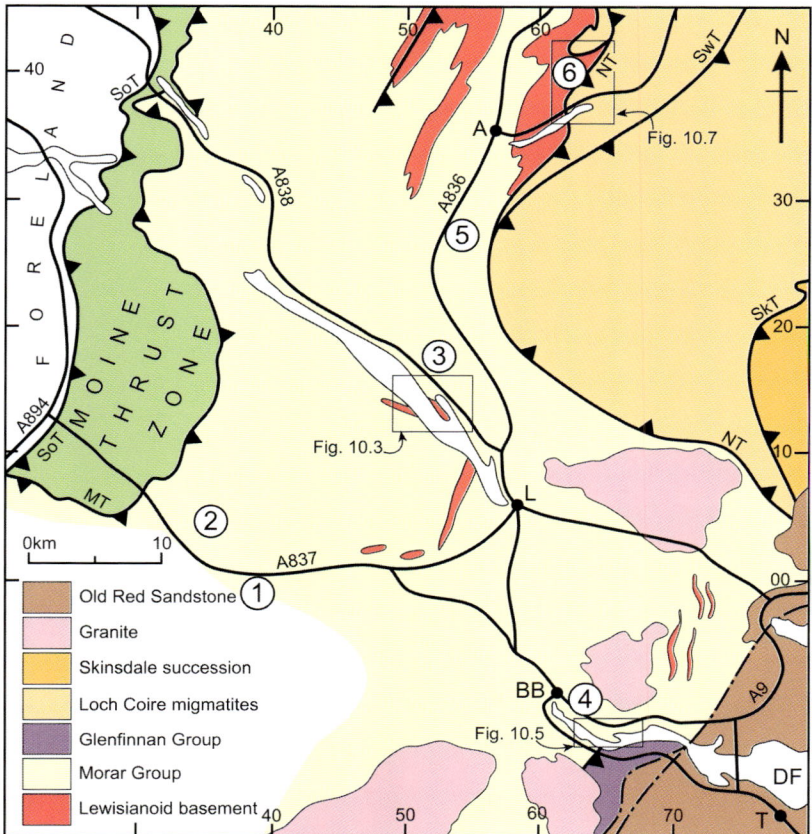

Fig. 10.1 Simplified geology of south and central Sutherland together with the localities for the excursion.
A = Altnaharra; BB = Bonar Bridge; DF = Dornoch Firth; L = Lairg; T = Tain; ST = Sole Thrust; MT = Moine Thrust; NT = Naver Thrust; ST = Swordly Thrust; SKT = Skinsdale Thrust

The large and often remote tract of ground that forms south and central Sutherland (Fig. 10.1) is generally poorly exposed, but forms a critical link between the hitherto better-known areas of northern Ross-shire and north Sutherland. Recent remapping of some of this ground, partly in association with the production of revised British Geological Survey maps, has drawn attention to this area which contains some key sections for demonstrating the nature of basement-cover relationships as well as the existence of regional-scale Caledonian ductile thrusts (Strachan & Holdsworth, 1988).

177

Locality 10.1 [NC 3859 0086]

Oykell Bridge (Fig. 10.1). Mullion structures within Moine psammites.

Parking is available (with permission) at the car park of the Oykell Bridge Hotel [NC 3843 0083]; allocate 1 hour for this locality. Walk east to the road bridge [NC 3859 0086]. Below the bridge, and easily accessible from the banks, are Morar Group psammites which are generally fine-grained and composed of variable proportions of quartz, feldspar, muscovite and biotite. Impressive mullion structures (Fig. 10.2; Wilson, 1953) plunge to the ESE parallel to the hinges of mainly reclined, tight to open, asymmetric folds. A well-developed mineral elongation alignment lies parallel to the axes of the mullions: this is the regional lineation, designated L_2 (e.g. Strachan & Holdsworth, 1988). A weak axial planar schistosity is associated with an intersection lineation that plunges sub-parallel to the mullions.

In the gorge section, the Moine rocks strike ESE and are generally inclined steeply to the SSW; cross-bedding indicates that the beds also young in that direction. The rocks show abundant small-scale and meso-scale folding and lie on the northerly long limb of a kilometre-scale, tight, asymmetrical, ESE-plunging antiformal F_2 fold. The opposing SSE-dipping overturned limb is exposed along the River Einig to the west with the major hinge coincident with the confluence of the rivers Oykell and Einig. Note that in the Oykel Bridge area, the Moine rocks are inverted and face downwards and the antiform is informally termed the Einig Syncline (Leslie *et al*, 2010). The mullions are developed on the steep eastern limb of this fold a short distance from the hinge zone, but are absent from the moderately-dipping short limb.

Fig. 10.2
Well-developed mullions plunging parallel to the regional L_2 lineation at Oykell Bridge (Locality 10.1).

The origin and tectonic significance of these mullion structures were controversial for many years. Were the mullions developed parallel to the 'b-axis' of monoclinic symmetry, that is normal to the kinematic transport direction according to the German 'symmetrological' school, or parallel to the tectonic 'a-axis' defined by the regional stretching direction and the Moine Thrust movement direction (Howarth & Leake, 2002)? It is now clear that the ESE-plunging linear fabrics in the Moine rocks define the principal extension direction of the main Caledonian (Scandian?) deformation (e.g. Holdsworth & Grant, 1990 and references therein). The regional D_2 folds and the locally developed mullions were formed within this regional strain/displacement field. Statistical parallelism of the fold axes and the lineation is thought to be due to passive rotation of the axes towards the extension direction during NW-directed overthrusting (e.g. Strachan & Holdsworth, 1988; Holdsworth, 1989a; Alsop & Holdsworth, 2004a & b). The bulk finite D_2 deformation at Oykell Bridge was constrictional, within an overall prolate (cigar-shaped) strain ellipsoid. The precise mechanism by which the mullions developed into discrete but interlocking structures is not, however, fully understood.

Locality 10.2 [NC 3399 0512 to NC 3457 0613]

Glen Oykell (Fig. 10.1). Polyphase deformation and mullion structures within Moine psammites and semi-pelites.

Drive west along the track that leaves the main road just east of Locality 10.1 and runs along the north side of the River Oykell valley. Parking is available for two to three cars at [NC 3407 0520]. Allocate 2 hours for this locality which involves 2-3 km of walking. Walk east to 2A in the river bed beneath the suspension bridge at [NC 3399 0512]. Moderately thickly bedded Morar Group psammites are deformed by close, non-cylindrical D_3 folds with Z-geometry that plunge mainly steeply to the SE, sub-parallel to the regional L_2 lineation. Weakly developed mullion structures are present in some psammite layers. Within semi-pelitic layers it can be seen that a schistosity (S_2?) is folded and crenulated in the hinges of the folds.

Walk north along the eastern bank of the River Oykell, noting shallowly-plunging mullions within psammites at [NC 3416 0573]. Keep following the east bank until the Allt Rugaidh is reached at [NC 3415 0606]. Locality

10.2B comprises a traverse up this stream section. Follow the stream upwards (wellingtons useful, small waterfalls can be bypassed on the banks). At [NC 3241 0605] incipient mullions can be studied in psammite and semi-pelite. In sections normal to the mullions, an early fabric (S_2?), axial planar to cm-scale minor (D_2?) folds is seen, especially along the contacts between psammite and semi-pelite. This early fabric is deformed by a younger set of (D_3?) folds that are clearly associated with the mullions. Traverse further upstream, noting excellent mullions at [NC 3424 0604]. At [NC 3457 0613] siliceous psammite shows excellent cross-bedding indicating that strains are still overall relatively low, despite the presence of variably developed mullions. Study of semi-pelitic layers again provides evidence for polyphase deformation. An early (S_2?) mica fabric is emphasised by concordant quartz segregations; these are tightly folded and the early fabric crenulated and variably transposed. Minor folds of the quartz segregations plunge parallel to the mullions. In contrast to Locality 10.1, the development of mullion structures at these exposures is apparently associated with F_3 folding. Elsewhere within the western Moines and Moine Thrust Zone of Sutherland (see Excursions 11 and 13), D_2 and D_3 folds are regarded as having developed during continuous progressive deformation associated with NW-directed overthrusting. The association of similarly-oriented fold mullions with both sets of folds in the Glen Oykell/Oykell Bridge area is consistent with this interpretation.

Locality 10.3 [NC 5219 1542 to NC 5297 1291]

Airde of Shin (Figs 10.1, 10.3). Infolded Lewisianoid basement within Moine psammites.

Parking is available for two to three cars by the side of the A838 road NW of Lairg at [NC 5231 1600]. Alternatively, parking for a coach and four to five cars is possible at [NC 5281 1543]. Allocate 3-4 hours minimum for this locality which involves 7km of walking, some of it over rough ground. Walk to the gate at [NC 5249 1579]. Go through the gate and head SSW across rough ground to cross another gate at [NC 5232 1551]. Keeping the fence to the left, walk down to the stream at the base of the valley.

Exposures in the stream at 3A [NC 5219 1542] (Fig. 10.3) are of Morar Group micaceous psammites with cm-scale clasts of vein quartz. Possible

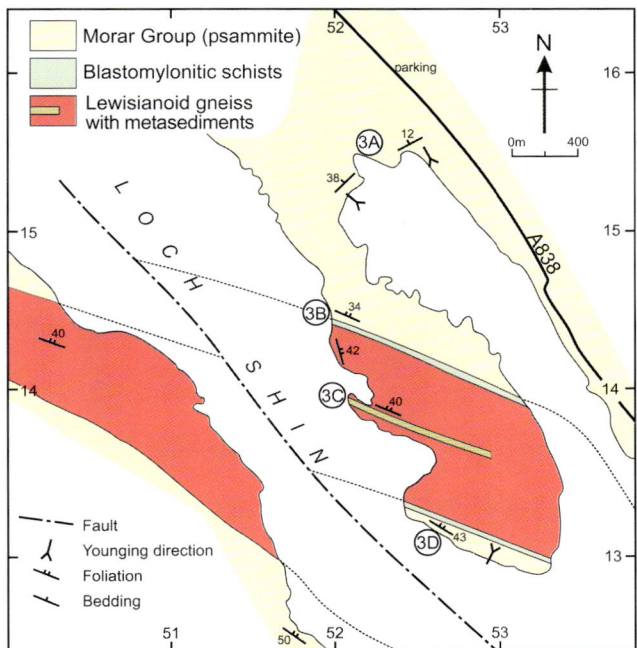

Fig. 10.3 Geological map of Locality 10.3, the Airde of Shin.

cross-bedding indicates that the psammites are right way-up. A strongly developed S_2 schistosity dips north-northeastwards more steeply than bedding. These psammites are interpreted to lie structurally above the Loch Shin basement inlier.

Head southwards to reach the shore of Loch Shin. Cross a low-lying fence and traverse SE along the shore to reach the exposures at 3B [NC 5194 1439] (Fig. 10.3). These are strongly foliated, platy blastomylonitic Moine psammites with numerous quartz veins. Most are attenuated parallel to schistosity and apparently highly deformed, while others are cross-cutting. An intense grain shape fabric is defined by quartz ribbons, quartz-feldspar aggregates and aligned muscovite grains. Cross a small beach to [NC 5196 1433] to view exposures of gently-dipping muscovite schists with abundant lunate quartz segregations and attenuated quartz veins. These are interpreted as blastomylonitic 'tectonic schists' derived from the intense shearing of basement lithologies (Peacock, 1975; Strachan & Holdsworth, 1988). A strong mineral and extension lineation plunges eastwards. The schists contain numerous tight-to-isoclinal D_2 folds that deform a strong S_1 schistosity, so that the dominant fabric is a composite $S_0/S_1/S_2$ foliation. The schists are underlain by banded Lewisianoid horn-

Fig. 10.4 Boudinaged mafic pods within strongly deformed intermediate to felsic gneiss of the Loch Shin basement inlier at Locality 10.3B.

blende gneisses; distinctive dark, hornblende-rich layers range in thickness from a few cm to over a metre. Metre-scale pods of amphibolite and ultrabasic lithologies are common and in places form elongate trains of boudins (Fig. 10.4). The majority of D_2 folds of gneissic banding have Z-geometry and S_2 is anticlockwise of S_0/S_1 (Strachan & Holdsworth, 1988). Traverse eastwards to the limit of exposure at [NC 5209 1409].

Continue eastwards, walking around the small bay and noting the old lime kiln on the hillside to the southeast. The headland at 3C [NC 5208 1388] (Fig. 10.3) exposes layers of white marble, calc-silicate rocks and rusty-brown mica schists that are interpreted as metasedimentary components of the basement inlier. The calc-silicate rocks consist mainly of tremolite, diopside, calcite, quartz and titanite (Read *et al.*, 1926). These metasediments are underlain by hornblende-garnet gneisses that contain concordant, foliated quartz-feldspar pegmatites, varying in thickness from 10-20cm to just over a metre. A strong mineral and extension lineation plunges to the east. Tight-to-isoclinal D_2 folds have S-geometry and S_2 is clockwise of S_0/S_1. The change of vergence of D_2 folds, and the relationship of S_2 to S_0/S_1 across the Loch Shin inlier implies that it occupies the core of a major D_2 fold (Strachan & Holdsworth, 1988). Traverse eastwards across more basement gneisses that extend as far as [NC 5214 3181] and then extensive boulder fields with no exposure.

A traverse through the Moine psammites that structurally underlie the basement inlier commences at 3D [NC 5258 1313] (Fig. 10.3) at the back of the beach where platy, high strain Moine psammites contain numerous deformed quartz veins and are very similar to those immediately above the inlier at 3B. These continue to the SE with numerous surfaces showing high strain, platy blastomylonitic fabrics that wrap elongated and boudinaged veins of granitic pegmatite. Despite the generally high strains, deformed but readable cross-bedding is preserved at three localities [NC 5280 1293; NC 5282 1292; NC 5297 1291] where in all cases the psammites are inverted. The location of the Loch Shin basement inlier in the core of a major (D_2) fold is thus clearly demonstrated on both structural and sedimentological grounds. Exposures at [NC 5309 1285] are notable for the development of layers of pseudo-conglomerate, apparently as a result of the boudinage and extreme flattening of quartz veins.

Return along the coast to [NC 5214 3181] and then head across the hillside, going through the gate at [NC 5217 1489], along the north coast of the Airde of Shin to Locality 10.3A, and back to the vehicles.

Locality 10.4 [NH 6400 8839 to NH 6504 8802]

Creich Peninsula (Figs 10.1, 10.5). Infolded Lewisianoid basement within Morar Group psammites; Sgurr Beag Thrust; Glenfinnan Group gneisses.

This locality requires low tides, and parties are advised to take particular care on the seaweed-covered rocks. Allocate 3-4 hours for this locality which involves *c.*4km of walking. From Bonar Bridge, follow the A949 eastwards parallel to the north shore of the Dornoch Firth. Turn off the A949 at [NH 6433 8910] onto a small track. Parking is available for up to three cars at the first bend in the track at Creich Mains [NH 6424 8879]. If the group is larger than can be transported in three cars, extra vehicles could be parked at the cemetery beside the A949 [NH 6354 8933] and members of the party ferried to Creich Mains. From the vehicles, walk along an overgrown track that leads southwards. After 350m the track branches to the right; follow it down through the woods to the shoreline, emerging by an old ruin at [NH 6418 8813].

Walk NW along the shoreline to the prominent outcrops at 4A [NH 6400 8839] (Fig. 10.5) that comprise blastomylonitic Morar Group psammites. These psammites lie within the wide ductile shear zone associated with the structurally overlying Sgurr Beag Thrust (Strachan & Holdsworth, 1988; Grant & Harris, 2000). Sedimentary structures are absent, presumably as a result of the high strains, although common orange K-feldspars may represent original detrital grains. A pervasive mm-scale schistosity defined by aligned micas and quartz plates dips moderately east-

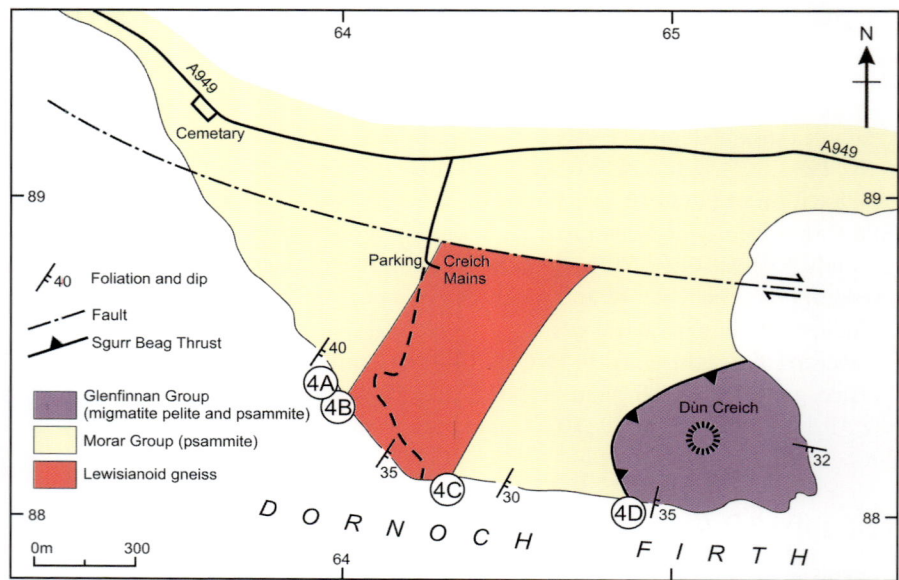

Fig. 10.5 Geological map of Locality 10.4, the Creich Peninsular (from Strachan & Holdsworth, 1988 & Grant and Harris, 2000).

wards and is accompanied by an ESE-plunging mineral and extension lineation. The schistosity is axial planar to mesoscopic tight to isoclinal folds that plunge sub-parallel to the lineation. Numerous cm-scale concordant quartz veins are present and are mostly strongly foliated and/or boudinaged.

Retrace your steps southeastwards to 4B [NH 6402 8835] (Fig. 10.5) where at the back of the beach Morar Group psammites are separated by a thin layer of highly tectonized pelite from banded blastomylonitic Lewisianoid gneisses (Grant & Harris, 2000). These are composed of quartz, feldspar and biotite; the banding is less continuous than that within the Moine rocks and is interpreted as a highly sheared metamorphic segregation fabric. Traverse SE, noting metre-scale sheets of pink, mylonitic pegmatite and/or acid gneiss [NH 6406 8830] and metabasic sheets now mostly composed of retrogressive biotite [NH 6407 8830]. Similar lithologies are exposed east of the ruin, although here there is a higher proportion of variably retrogressed metabasic sheets and pods, some of which preserve hornblende. Strongly foliated hornblende gneisses occur at [NH 6432 8811]. Throughout the basement inlier, the dominant foliation dips moderately eastwards and the associated mineral and extension lineation plunges to the ESE.

At 4C [NH 6437 8812] (Fig. 10.5), the eastern margin of the basement inlier is exposed under the low branches of a large tree. The contact with the highly deformed Morar Group psammites to the east is sharp and concordant. The psammites display essentially the same lithological and structural features as those listed above for Locality 10.4A. The symmetrical disposition of Moine lithologies either side of the basement inlier suggests that it might lie within the core of an isoclinal Caledonian (D_2) fold, and thus has a similar structural setting to the Loch Shin basement inlier. However, this is difficult to demonstrate conclusively due to the lack of minor fold structures and the high tectonic strains that have obliterated sedimentary structures within the Moine rocks. An alternative interpretation is that the lower boundary of the basement sheet is a ductile thrust that was responsible for interleaving the gneisses with the Moine cover (Grant & Harris, 2000).

Traverse eastwards to 4D [NH 6483 8805] (Fig. 10.5) where the Sgurr Beag Thrust is exposed as a sharp, concordant, east-dipping contact between Morar Group psammites and strongly foliated pelitic schists of the Glenfinnan Group. The regional metamorphic contrast across the thrust is not immediately apparent as a result of intense strain and retrogression of the Glenfinnan Group lithologies. A mineral and extension lineation plunges to the ESE, and shear bands indicate a general top-to-the-west sense of overthrusting parallel to the lineation. Continue eastwards, noting garnets as well as occasional lenticles and layers of quartzo-feldspathic material within the pelites that are interpreted as the sheared remnants of an older migmatitic segregation fabric. The pelites pass transitionally eastwards into coarse, striped psammitic and semi-pelitic gneisses with cm-scale quartzo-feldspathic migmatitic layers, well exposed at [NH 6504 8802]. Only a few km to the SW, on Ben Wyvis, migmatization of the Glenfinnan Group has been assigned to a Neoproterozoic prograde event as it pre-dates emplacement of the *c.*730 Ma Carn Gorm pegmatite (Hyslop, 1992). On that basis, migmatization within the Glenfinnan Group rocks of the Creich Peninsula is also tentatively assigned to the Knoydartian event. Return westwards along the shore to the ruin, and then back to the vehicles.

Locality 10.5 [NC 5324 2712]

Vagastie Bridge (Figs 10.1, 10.6). Syn-tectonic granite sheets intruding Moine psammites in the footwall of the Naver Thrust.

Park adjacent to the A836 in the large parking place immediately south of Vagastie Bridge [NC 5324 2712] (Fig. 10.1). There is sufficient space for a coach if necessary. Allocate ½-1 hour for this locality. The Vagastie Bridge granite represents one of the larger members of a series of deformed igneous intrusions that occur in Central Sutherland (Read, 1931), the Vagastie suite (Soper, 1971; Soper & Brown, 1971). The granite is not a single body, but a series of sheets that intrude Morar Group psammites. A detailed description of the petrology and field relationships of these granitic sheets is presented by Holdsworth & Strachan (1988) and only the most important points are covered here.

Walk to the bridge and examine the outcrops of a pink, coarse-grained augen granite by the banks of the stream on the east side of the bridge. A strong tectonic foliation defined by recrystallized feldspar augen dips gently to the ESE and carries a SE/SSE-trending mineral and extension lineation. NW-dipping shear bands are occasionally present and demonstrate a general top-to-the-NW sense of displacement parallel to the lineation. A sample of the augen granite collected a few metres NE of the bridge yielded a U-Pb zircon age of 424 ± 9 Ma, interpreted as dating intrusion of the igneous protolith (Kinny *et al.*, 2003b). It therefore follows that the deformation that affects the granite must have occurred either during or after granite intrusion.

Cross the road and descend to the outcrops in the stream immediately beneath the bridge to examine the relations between the granite and its host Moine psammites. The psammites are deformed by a metre-scale, open-to-close, SW-overturning D_2 fold pair, the upper hinge and short limb of which are cut by the base of the lowest granite sheet (Fig. 10.6). Contacts between the granite and its psammitic host are sharp and well-defined. Three important observations can be made: (1) the granite sheet is not folded; (2) it carries a foliation that is oblique to its margins and sub-parallel to D_2 axial planes and S_2 in the psammites; and (3) the SE/SSE-trending lineation within the granite is sub-parallel to the dominant (L_2) lineation within the psammites. Granite-Moine contacts elsewhere display the same structural relations (e.g. [NC 5329 2726]) (Fig. 10.6).

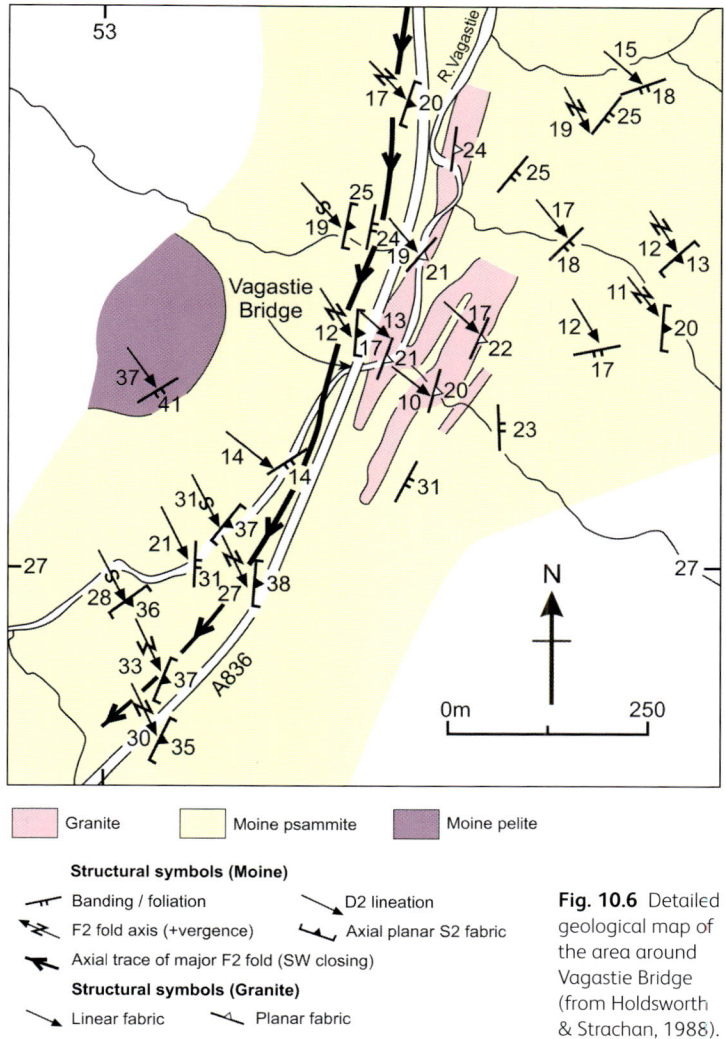

Fig. 10.6 Detailed geological map of the area around Vagastie Bridge (from Holdsworth & Strachan, 1988).

Given that the granite sheets are not folded by D_2, but nonetheless carry S_2 and L_2, the simplest interpretation is that granite intrusion occurred during D_2 (Holdsworth & Strachan, 1988). Intrusion must have post-dated folding, but predated the later stages of fabric development. The U-Pb zircon age obtained for the Vagastie Bridge Granite thus demonstrates that D_2 and correlative structures in this part of Central Sutherland occurred during the Scandian phase of the Caledonian orogeny (Kinny et al. 2003b).

Locality 10.6 [NC 6288 4048 to NC 6537 3925]

Loch Naver (Figs 10.1, 10.7). Interfolded Morar Group and Lewisianoid basement; Naver Thrust; Naver Nappe migmatitic gneisses.

If driving north from Vagastie Bridge, turn right at Altnaharra onto the B873 signposted to Syre and Bettyhill. Drive eastwards for *c.*13km and park by the roadside either side of the bridge at [NC 6405 3867]. There is space for four to five cars; allocate 6-7 hours for this locality, which involves *c.*14km of walking over hillsides and rough ground.

Follow the Allt Gruama Beg NW as far as Loch Morlach. Walk around the north shore of the loch to the small stream that enters the loch at [NC 6305 3925] and head northwards across moorland to Cnoc Liath. Low-lying exposures at 6A [NC 6288 4048] (Fig. 10.7) are massive to coarsely-foliated garnet-pyroxene gneisses of the Naver basement inlier. Although the igneous protoliths of this inlier are undated at the time of writing, a late Archaean age seems likely by comparison with the Borgie inlier just a few kilometres to the north (Friend *et al.*, 2008). The high-grade metamorphic event that resulted in formation of the garnet-pyroxene gneisses has been correlated with the Scourian event in the Caledonian foreland (Moorhouse, 1976), but a much younger age cannot be ruled out (see also Locality 13.7, Excursion 13). Late retrogressive shear zones, defined by narrow bands of hornblende schist, cut these high-grade gneisses. Other exposures within a radius of 100m are of banded hornblende gneisses with concordant amphibolite sheets, more typical of large tracts of the Naver inlier. Gneissic banding dips to the ESE and carries a down-dip mineral lineation commonly defined by aligned hornblende. The hornblende gneisses were probably derived, at least in part, from retrogression of the garnet-pyroxene gneisses. Additional outcrops of relic garnet-pyroxene gneisses are present a little further to the north on a small hillock at [NC 6276 4070].

Traverse NE across numerous other low-lying outcrops of hornblende gneisses. The Morar Group rocks that structurally underlie the Naver basement inlier are exposed at 6B [NC 6356 4197] (Fig. 10.7) on a series of north-facing crags. These lithologies are banded, fine-grained psammites with thin semi-pelite bands. Note the lack of any evidence for high-grade metamorphism and migmatization, in contrast to those to be seen later above the Naver Thrust. The banding is interpreted as tectonically modified bedding; sedimentary structures are absent, but detrital feldspar grains are

South and Central Sutherland

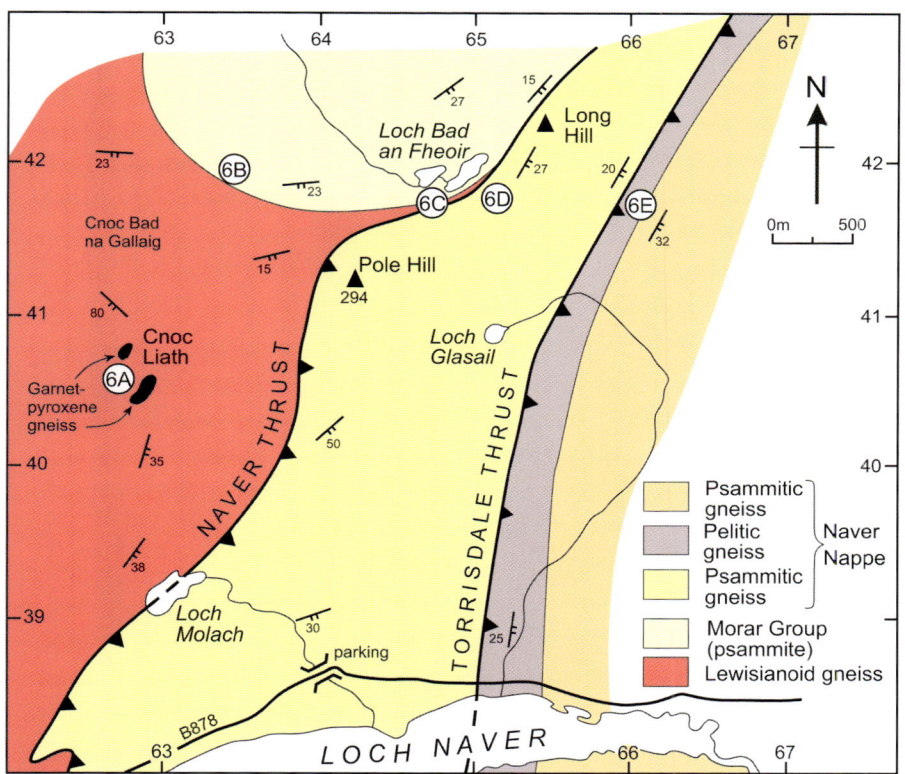

Fig. 10.7 Geological map of Locality 10.6, the Loch Naver area.

common. A mesoscopic, reclined D_2 fold with S-geometry plunges to the SSE, parallel to a mineral and extension lineation defined by aligned micas and quartz-feldspar aggregates. Walk eastwards along strike, noting several metre-scale, foliated and boudinaged sub-concordant granite sheets, possibly members of the same suite as those seen at Vagastie Bridge.

Head eastwards across the moorland to the two lochs labelled 'Loch Bad na Fheoir' on the 1:25,000 scale OS map (Fig. 10.7). On the east side of the smaller of the two lochs at 6C [NC 6481 4191] are exposed very flaggy Morar Group psammites and semi-pelites. The high-strain fabric dips gently to the SE and carries a down-dip mineral and extension lineation that is inferred to lie parallel to the direction of tectonic transport along the overlying D_2 Naver Thrust. Walk around to the SE side of the larger loch to

189

[NC 6495 4184] where similar high-strain psammites are cut by thin, discordant pegmatites. A sharp contact separates these psammites from overlying biotite schists that are interpreted as highly retrogressed basement gneisses near the northern termination of the Naver inlier (Fig. 10.7). A series of low-lying exposures to the south (e.g. [NC 6495 4181]) show rather less-retrogressed, banded hornblende gneisses.

Walk upslope to 6D [NC 6501 4178] (Fig. 10.7), crossing the unexposed Naver Thrust, and traversing into extensive outcrops of banded psammitic, migmatitic gneisses of the Naver Nappe. These are deformed by highly attenuated isoclinal D_2 folds of banding and open, asymmetric D_3 folds, both plunging sub-parallel to the SSE-trending mineral and extension lineation. The gneisses are intruded by numerous concordant and boudinaged granitic pegmatites. Migmatization in Central Sutherland is thought to have occurred during the Ordovician Grampian phase of the Caledonian orogeny (*c.*470-460 Ma; Kinny *et al.*, 1999). Traverse to the top of Cnoc Bad an Fheòir through similar lithologies, and then head east from the summit of the hill for *c.*1 km to the next set of exposures on the west side of an unnamed hill.

A prominent set of outcrops at 6E [NC 6602 4166] (Fig. 10.7) comprise flaggy garnet-biotite schists that overlie the Torrisdale Thrust, a tectonic break within the Naver Nappe (Fig. 10.7). The thrust is not exposed here, but near the summit of Ben Klibreck to the SW is associated with interleaving of basement orthogneisses with Moine metasedimentary gneisses of the Naver Nappe (Strachan & Holdsworth, 1988). The garnet-biotite schists are intruded by a series of pegmatites that vary from early, foliated types to late, discordant and undeformed sheets. Traverse upslope through SE-dipping, banded psammitic and semi-pelitic gneisses with numerous concordant sheets and pods of foliated amphibolite, and abundant pegmatite. A SSE-plunging mineral and extension lineation is present locally. Head southwards from the top of the hill to reach the Allt Dail a'Thuraich.

Follow the stream southwards to 6F [NC 6537 3925] (Fig. 10.7) where coarse migmatitic pelites with abundant concordant pods and layers of coarse pegmatitic material are exposed in the stream bed. Follow the stream southwards to the main road, and then walk westwards for *c.*1 km back to the vehicles.

Excursion 11

The Moine Thrust Belt at Loch Eriboll

Rob Butler

Purpose: Four transects to examine aspects of thrust belt geometry and the evolution of fault rocks.

Aspects covered: Lewisian gneisses, Cambro-Ordovician stratigraphy, imbricate thrusts, duplexes, piggy-back stacking and breaching sequences of thrusts, mylonites and other fault rocks.

Maps: OS: Explorer series (1:25,000) sheets 445, 446 and 447; BGS 1:50,000 Scotland sheet 114W, Loch Eriboll.

Type of terrain: Rocky coastlines, moorland, wet valley bottom, steep grassy slopes, scree, mountain crags and bare rock exposures.

Distance and time: 2 full days: transect 1 (1 day); transect 2 (2 hours); transect 3 (4 hours); transect 4 (2-3 hours). Allow a further 30-45 minutes for the drive from Durness or Tongue.

Short itinerary: Transect 2 for a major thrust, the Arnaboll Thrust and associated deformation features; transect 1 for a fuller range of geology if only one day is available.

Note: all the localities described in this chapter are Sites of Special Scientific Interest and as such are protected. No outcrops should be hammered nor specimens collected, even from float.

The Loch Eriboll area (Fig. 11.1) is hallowed ground not only for Highland geology but also for the discovery of thrust tectonics. It was here in the early 1880s that Charles Lapworth demonstrated that the sequence of rocks was not a simple stratigraphic order, but was repeated by folds and faults (Lapworth, 1883). In accepting Lapworth's geological interpretations, it was for this area that Geikie (1884) first coined the term 'thrust', inspired by the structures on Ben Arnaboll. Lapworth (1885) went on to describe how large shear strains associated with thrusting generate new types of rocks, coining the term 'mylonite' for these 'intensely milled' rocks. This

EXCURSION 11

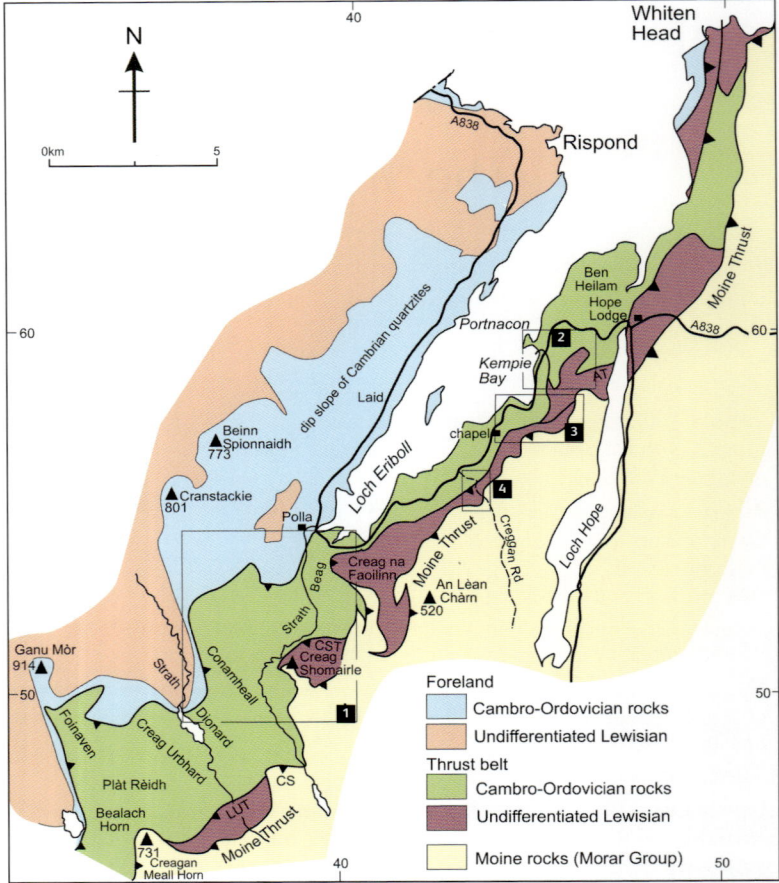

Fig. 11.1 Simplified geological map of the northern part of the Moine Thrust Belt, in the vicinity of Loch Eriboll, Foinaven to Whiten Head.

The sites of the four transects are indicated by boxed areas (1-4).
AT = Arnaboll Thrust;
CST = Creag Shomhairle Thrust.

work not only forms the foundation for much modern work (see Snoke & Tullis, 1998; Law & Johnson, 2010), it also established clearly that the layered structures in metamorphic rocks could be formed by deformation rather than simply mimic original sedimentary bedding. This opened the way for better interpretations of the geological structure of the interiors of mountain belts around the world. Lapworth's key sites are described here together with spectacular sections across these units at the Moine Thrust. The region is also important internationally for understanding the structure and geometric evolution of thrust belts. Early attempts to model the formation of thrusts using analogue materials were inspired by the spectacular imbricate structures on Foinaven (Cadell, 1888; see Butler, 2004a). A century later the same outcrops provided the type example for duplex structures (Boyer & Elliott, 1982). The whole area is protected as Sites of Special Scientific Interest, recently described by Butler (2009).

The overall aim of the combined transects described here is to introduce the key structural elements of thrust belts at a variety of scales, from the processes that form fault rocks that permit tens of kilometres of displacement on relatively narrow zones of movement, to the ways in which many faults can interact to create wonderfully complex stacks and piles of thrust sheets. Depending on the structural level within the thrust belt, the structures formed vary from simple localised fault zones to the broader zones of mylonites that characterise the higher structural levels of the thrust belt. The thickness of thrust sheets and slices varies considerably also, from a few centimetres where individual beds within the Pipe Rock are imbricated to several hundred metres of gneisses where Lewisian basement has been displaced. Indeed the Moine Thrust Belt is a classic place to study basement involvement in thrust belts (see Butler *et al.*, 2006). Remarkably for such an intensively studied and relatively well-exposed thrust belt, there are still fundamental disagreements between researchers on such matters as the correlation and kinematic significance of certain thrusts and the protoliths of various mylonitic lithologies. The main issues are outlined at the appropriate points in these transects.

Transect 1: South Eriboll

The mountainous country south of Loch Eriboll (Fig. 11.2) contains some of the finest examples of thrust geometry in the British Isles. Although the area can be visited for a few hours, by hiking up the ridge of Conamheall (*c.*480m OD, [NC 363 514]) for panoramic views, especially of Foinaven, a full day is required to get an appropriate appreciation of the geology. The excursion described here visits the key structural domains of the Moine Thrust Belt, from the lowest-level imbricates made by slices of Cambrian quartzites, to major refolded duplex systems, thrust sheets of Lewisian basement and culminates in the Moine Thrust with its associated mylonites. Apart from examining the large-scale structural geometry of thrust systems, there are also sites that provide excellent outcrops of key types of fault rock.

The excursion involves a full day out in terrain that includes rugged

EXCURSION 11

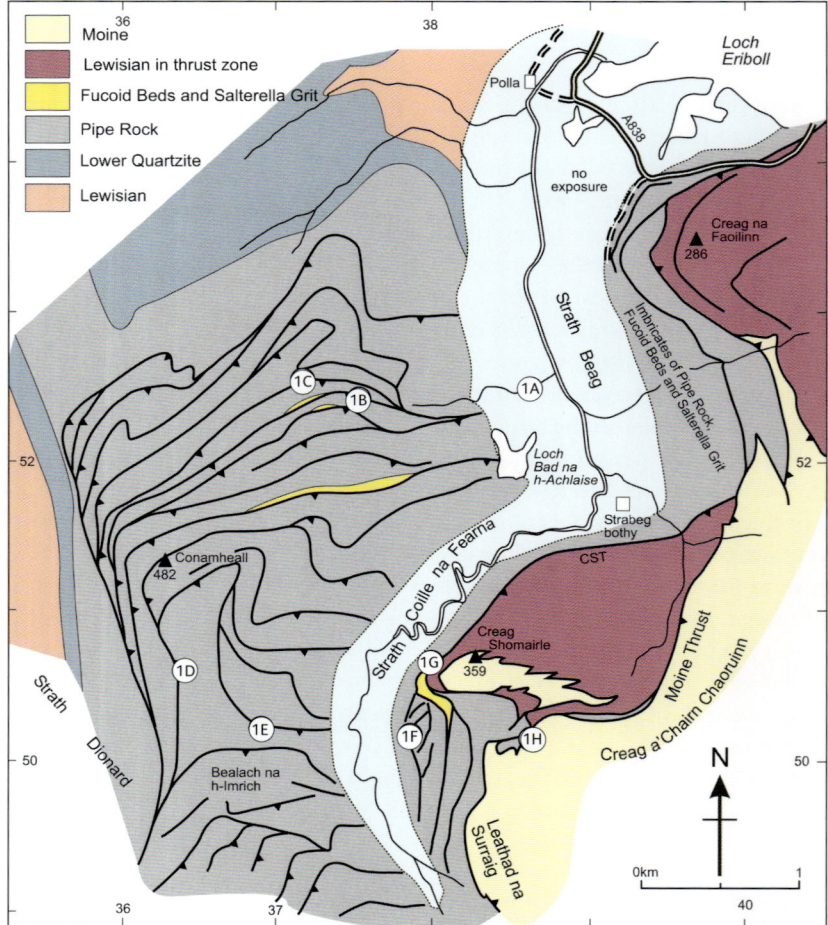

Fig. 11.2 Simplified map of the south Eriboll district, showing the positions of sites visited on transect 1. Located as 1 on Fig. 11-1. CST = Creag Shomhairle Thrust.

upland bounder fields, boggy valleys and steep hillsides. As described, the route involves fording the Strath Beag river which should not be attempted in times of flood. The area is particularly prone to harsh weather and river levels can rise quickly. Regrettably, and in common with many parts of the Scottish Highlands, once-open countryside in upper Strath Beag is being fenced to restrict deer movement. These high obstacles also restrict people – please take care if you need to cross any fences.

Park at Polla [NC 3865 5458], taking care not to obstruct access to the farm and associated activities. Further parking is available adjacent to the

Moine Thrust Belt at Loch Eriboll

A838 [NC 3908 5535]. Regardless of the choice of parking place, pause to look up the valley. The view is described in a counter-clockwise direction, starting on the western side of the valley. Here lies the gently-inclined slope of Cambrian quartzites dipping down from the summit of Cranstackie (800m OD, [NC 3505 5560]). These rocks constitute part of the foreland but, in tracing their continuity leftwards along the ridge leading to Conamheall (482m OD), the bed structure appears disrupted. This transition into more complex structure is across the most western thrusts in this part of the thrust belt, with the Sole Thrust (structurally lowest thrust) lying on the dip slope. With diligence a number of imbricate thrust faults and related folds can be identified from this distance. The imbricate slices incorporate the Pipe Rock, repeating this unit many times. Butler & Coward (1984) estimate that this stack of imbricates originally represented a distance of over 50km, now telescoped into about 6km on the Conamheall ridge. The first part of this excursion will examine some of these structures and their most dramatic expression on Creag Urbhard, Foinaven.

On the east side of Strath Beag the geology is rather different to that on the west. The large cliff seen due south of Polla is part of Creag Shomhairle. The cliffs are in Lewisian gneiss, part of a thrust sheet emplaced onto the now-imbricated Cambrian quartzites described above. The base of the Creag Shomhairle Thrust sheet lies in the lowest part of the crag, but is better seen further up the valley to be visited during the transect. Up behind Creag Shomhairle is a corrie overlooked by the dark cliffs of Creag a' Charn Chaoruinn. These are mylonites, although whether they are derived chiefly from Moine metasediments (Butler, 1982) or Lewisian gneisses (British Geological Survey, 2002; Holdsworth *et al.*, 2006) has been debated. They lie above the structure interpreted by the present author as the Moine Thrust that runs through the obscured ground behind Creag Shomhairle. Continuing the panorama to the left, the cliff of Creag na Faoilinn lies to the SE of Polla. The structure of this hill is complex but in many ways comparable to that on Shomhairle. The northern part of the main cliff is made from Lewisian gneisses thrust onto Cambrian sediments that form the grassy lower slopes. The southern part of the Faoilinn cliff is made of more Cambrian strata, including Pipe Rock quartzites, imbricated many times beneath the Lewisian sheet. In summary, the Strath Beag transect provides a complete cross-section through the Moine Thrust Belt and contains examples of all of its main structural elements.

From Polla it is possible simply to head up onto the foreland quartzites

EXCURSION 11

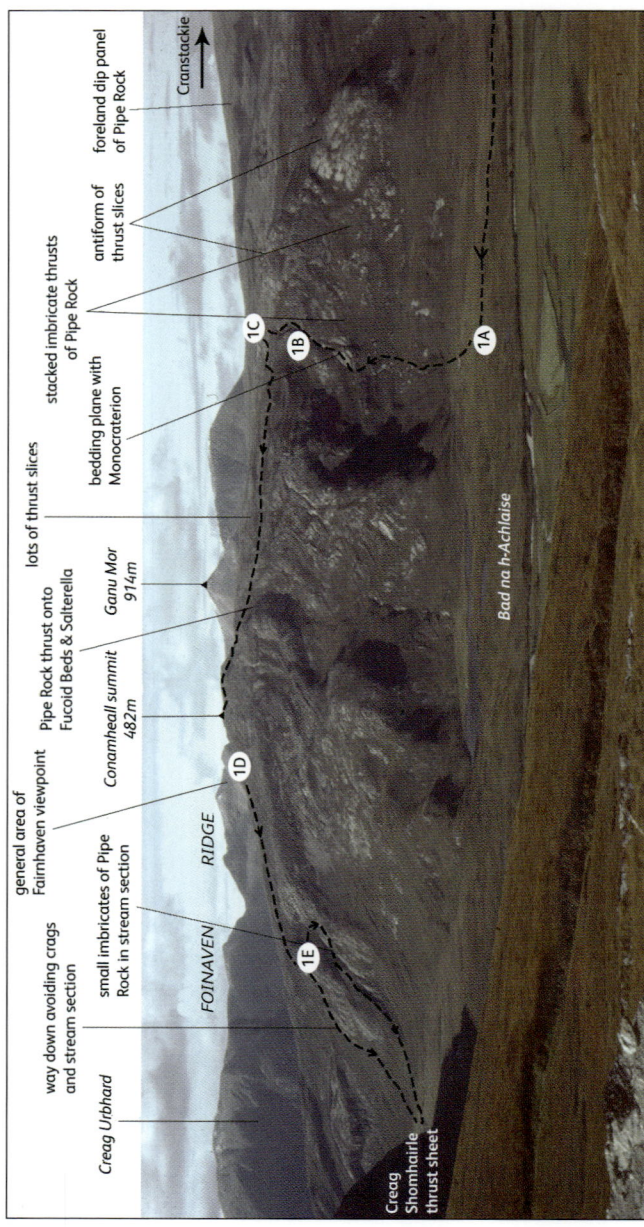

Fig. 11.3 Annotated photograph view SW from slopes of An Lean Charn (e.g. [NC 406 526]) onto Conamheall – showing some of the outcrop locations from transect 1. (Note that this viewpoint is not part of the excursion, but can be visited readily by hiking up from Strath Beag bothy or from the southern end of the Creag na Faoilinn crags. For photographs the best lighting is generally in the morning).

and then walk out the Conamheall ridge. However, it can be difficult to find some of the key outcrops by this route. Consequently the excursion as described follows the valley for about 2 km to where a small stream (Allt a' Chuilinn, [NC 3865 5250]; Locality 11.1A) runs down from the quartzites. The hillside to the west (Fig. 11.3) contains numerous imbricate slices that incorporate at most about 15 m of Pipe Rock. These thrust slices bunch up to form a prominent antiformal structure seen towards the WNW. The stream section gives access to one of these thrust slices with superb bedding-plane exposures of *Monocraterion* trace fossils (trumpet pipes; e.g. at [NC 3765 5240]). The northern branch of the stream leads steeply up onto the plateau. For much of its length the stream contains excellent exposures of cataclasites, formed by strongly fractured quartzites. These fault rocks can be either pasty cream or bluish in colour (colloquially termed 'bruised'). These textures show evidence for multiple fracturing and grain suturing processes with both brittle and ductile processes recorded in thin section. Through these sections imbricate thrusts are intraformational, carrying Pipe Rock onto Pipe Rock. However, at higher levels the thrusts climb up-section into the Fucoid Beds and Salterella Grit that tend to form gullies in the hillside (e.g. Locality 11.1B; [NC 3730 5245]). The Fucoid Beds are particularly easy to identify as they generate yellow screes with lush vegetation. Yet there are more thrusts present in this area than those that involve Fucoid Beds. About 200 m to the north of the gullies underlain by Fucoid Beds lies the antiformal stack of thrust slices exclusively in Pipe Rock. A short diversion to examine some of these structures is rewarding, with exceptional exposures of small thrust-related folds and bedding planes containing weakly flattened *Skolithos* and *Monocraterion* burrows (Locality 11.1C; [NC 3730 5257]). The strains are however barely detectable in *Skolithos*, but *Monocraterion* are generally visibly distorted presumably reflecting a greater propensity for the host gritty beds to deform. However, in profile these burrows are generally perpendicular to bedding. In general it may be deduced that thrust stacking has been accomplished with barely any internal distortion of the thrust slices.

So far the examination of thrust structures on Conamheall has focussed on relatively small scale examples. However, the entire ridge is made of thrust repetitions. This can be appreciated by gaining a view of Foinaven's Creag Urbhard which provides an equivalent, but historically more important, profile through the thrust belt. Walk SE onto the summit area of Conamheall, then move a few hundred metres further south to obtain

EXCURSION 11

Fig. 11.4 Annotated photograph (a) and cross-section (b) of the Fcinaven-Meall Horn ridge, illustrating the imbricate structure of Cambrian quartzites. (after Butler, 2004b.)

LQ = Lower Quartzites; PR = Pipe Rock; MT = Moine Thrust

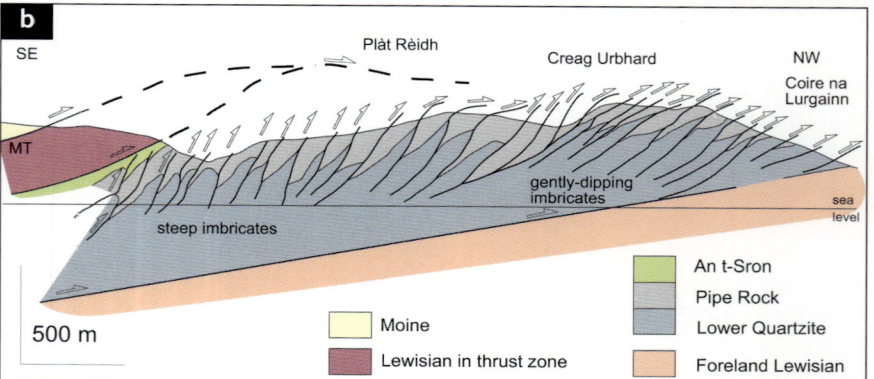

uninterrupted views (Fig. 11.4) into Strath Dionard and across onto Foinaven (Locality 11.1D, e.g. [NC 3612 5105]). Large numbers of imbricate slices can be identified in Creag Urbhard section, repeating the Cambrian quartzites (see Butler, 2004b, figure 6). The crags are aligned such that the view is nearly parallel to the inferred direction of thrust sheet emplacement. It was this view that inspired Cadell (1888) to model thrust tectonics (see Butler, 2004a; and Boyer & Elliott, 1982) to formalise the duplex model of imbricate thrusting. In apparent contrast to Conamheall, the Foinaven section involves repetition of the lower part of the quartzites (note that all BGS maps since those of the 1880s have wrongly shown Creag Urbhard to be made exclusively of Pipe Rock, although Cadell's field slips indicate both quartzite units) suggesting that the Sole Thrust cuts laterally down-section to the south to incorporate more stratigraphy in the imbricate slices.

Strath Dionard ends in a headwall that exposes the thrust sheets that cap the imbricated quartzites of Foinaven (Fig. 11.4). The upper part of the headwall forms Meall Horn (777m OD) with its north facing crags. These are mylonites with the Moine Thrust lying below. Beneath this thrust, and separating it from imbricate quartzites seen at the head of Strath Dionard,

Moine Thrust Belt at Loch Eriboll

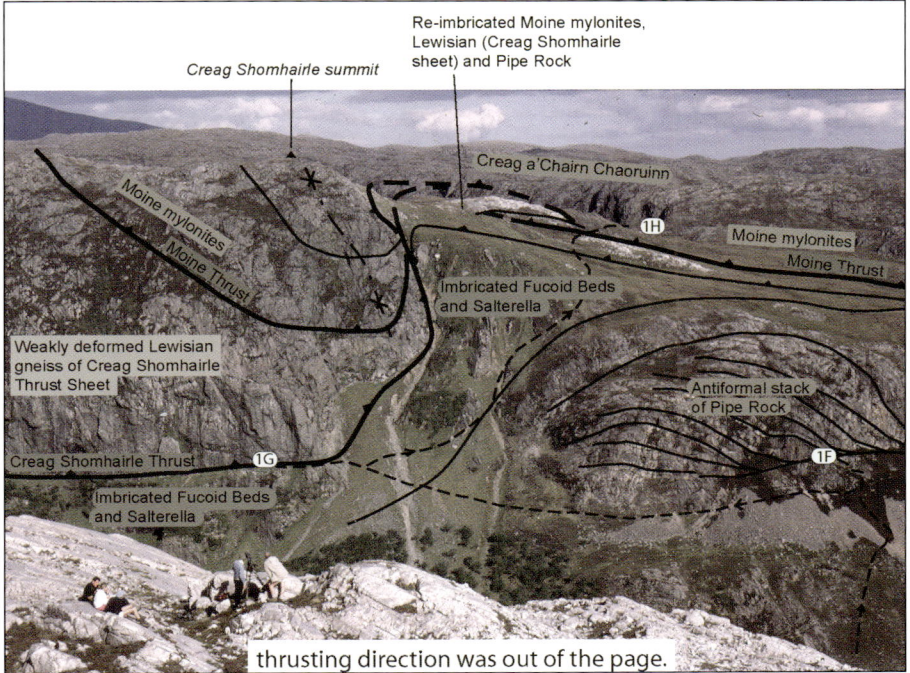

Fig. 11.5 Annotated photograph of Creag Shomhairle (from Location 11.1E; after Butler, 2004b) showing the culminations in imbricated Cambrian strata that fold higher thrust sheets (Creag Shomhairle and Moine).

is a slice of Lewisian gneisses equivalent to but not continuous with those of Creag Shomhairle and Creag na Faolinn in Strath Beag. From the viewpoint over Dionard return eastwards across the shattered quartzites of Conamheall and descend carefully towards Strath Beag. A small bluff overlooking the valley provides a useful landmark (Locality 11.IE, [NC 3718 5095]). This vantage point gives spectacular views onto the west face of Creag Shomhairle (Fig. 11.5) and into upper Strath Beag. Repetitions of Pipe Rock, similar to those seen on Foinaven, form craggy ridges running from the ridge (Beallach a' Chonnaidh) down leftwards (NE) into Strath Beag. Further to the SE lies Creag Staonsaid (454m OD), whose crags are made of mylonites associated with the Moine Thrust. These are broadly continuous with the mylonites to Meall Horn and Creag a' Chairn Chaoruinn, permitting the observer to trace the Moine Thrust through the landscape.

The main purpose of the viewpoint is to consider the structure on Creag

Fig. 11.6 Photograph of the antiformal stack duplex at Creag Shomhairle (after Butler, 1987).

Shomhairle (Fig. 11.5). This view looks directly back down the inferred direction of thrust transport and therefore gives an ideal perspective on how thrust sheet geometry may vary perpendicular to displacement. The lowermost slopes of Creag Shomhairle contain Pipe Rock, the upper parts of the imbricate structures that are found on Conamheall. These imbricates form a small culmination, about 100m high, on the southern end of the cliffs. The internal structures of thrust slices are very well-exposed within the culmination (Fig. 11.6; see Butler, 1987 for discussion). Separating the culmination of Pipe Rock from the main cliffs of Creag Shomhairle is a complex gullied area that is partly wooded. This ground is made of steeply dipping and thrust-repeated slices of Fucoid Beds and Salterella Grit. These imbricates are folded over the Pipe Rock culmination and beneath the main cliff of Creag Shomhairle return to a subhorizontal attitude. The main cliffs are largely composed of Lewisian gneisses of the Creag Shomhairle Thrust sheet (Fig. 11.5). The basal thrust of this sheet is also folded by the culmination of Pipe Rock so that, as with the Fucoid Beds and Salterella Grit imbricates, it is sub-vertical in the gully area. However, the summit area of Creag Shomhairle is made of mylonites that occupy the core of a synform. The interpretation favoured here (Butler, 1982, 2004b) is that these mylonites are derived from the shearing chiefly of Moine psammites. Thus the Moine Thrust marks the lower boundary of these mylonites with the less sheared

Lewisian of the Creag Shomhairle Thrust sheet. However, a Lewisian parentage has been suggested (British Geological Survey 2002; Holdsworth *et al.*, 2006) which sites the Moine Thrust at structurally higher levels. Regardless of this controversy, the summit mylonites on Creag Shomhairle are carried by a major thrust. In detail the structure is still more complex (Butler, 1982), with strongly mylonitised quartzites, presumably derived from the Cambrian, lying between the Moine and Lewisian units. All three units are interleaved by small thrusts that are themselves folded around the synform.

The field relationships on Creag Shomhairle were important for demonstrating the sequence of thrust sheet emplacement in NW Scotland (Butler, 1982, but see Butler, 2004b). As higher thrust sheets are either re-imbricated or folded over underlying ones, it implies that these were emplaced first, with underlying ones developed sequentially in turn. This is classically termed piggy-back thrusting and is now widely assumed to be the normal way in which thrust belts develop. The outcrops on Creag Shomhairle and their access are discussed shortly. However, the descent into Strath Beag provides opportunities for observing some thrusts in the Pipe Rock.

The stream that flows steeply into Strath Beag on the north side of the bluff of Locality 11.1E provides arguably the finest exposures of small-scale thrusts and their related fault rocks in the region. However, great care is needed to access these, especially in the normal, damp conditions. If in doubt the best way down hereabouts is to remain on the south side of the stream (see Fig. 11.3). However, the watercourse follows a thrust surface that can only be appreciated from its north side. The bedding plane of the stream bed is also a thrust plane and is decorated by photogenic examples of fault breccia and other cataclasites. The southern retaining wall of the stream contains imbricated Pipe Rock on the bed-scale. These include slices approximately 2 m thick together with a 'microduplex' [NC 3628 5098] that incorporates about 10 cm of stratigraphy. More description is unnecessary and the reader is left to discover these informative outcrops themselves.

Continue down into the boggy valley bottom, choosing a fording point for the river that is appropriate to the conditions. In the event of high water levels, discretion is strongly advised. The remainder of the excursion should be abandoned and road regained by a 5 km walk back along the river's west bank. However, if possible, the next goals are outcrops of fault rocks on the western slopes of Creag Shomhairle that were seen from Locality 11.1E. The first target lies within the culmination of Pipe Rock at the southern end of the cliffs. Recently erected deer fencing has hindered access here. Scramble up the heathery and scree-covered slopes to the foot of the Pipe Rock cliffs at Locality 11.1F ([NC 3785 5024]; see Fig. 11.5). The floor thrust to this culmination is well-exposed with excellent sections through the cataclasites that decorate it. In many places these fault rocks are foliated. In the cliffs above there are numerous examples of both footwall and hanging-wall ramps of the culmination's constituent imbricate thrusts.

To observe outcrops of the Creag Shomhairle Thrust it is necessary to access a narrow ledge beneath the main cliff to the north (Locality 11.1G; [NC 3800 5059]; see Fig. 11.5). With care it is possible to follow animal tracks, contouring around beneath the Pipe Rock culmination, traversing below outcrops of Fucoid Beds and Salterella Grit, and crossing some unstable scree shoots. It may be preferable to return to the valley bottom and re-ascend to the ledge, thus avoiding the traverse. However, while exposed, the ledge provides comfortable access to the Creag Shomhairle Thrust in its sub-horizontal attitude, overlying strongly sheared Fucoid Beds. The Lewisian immediately above the thrust plane is strongly retrogressed with dark, chloritic seams and epidotic veins. Cataclasitic seams are also evident. However, penetrative forms of this thrust-related damage are restricted to within only one or two metres of the thrust plane. Otherwise the gneisses are largely unaffected by thrusting, apart from having experienced a presumably substantial displacement. More intrepid visitors may choose to access the steeply inclined parts of the Creag Shomhairle Thrust that are found in the steep gully ([NC 3804 5048]; see Fig. 11.5) to the south of the ledge. High in this gully the thrust zone includes a slice of strongly deformed quartzites that increases dramatically in size to the south, away from Creag Shomhairle. However, this is very steep, potentially hazardous ground and access is not encouraged. The lower parts of this approach is up scree shoots largely formed of blocks of the Moine (*sensu* Butler, 1982) mylonites from high on the cliff. Many samples contain thin seams of iron oxides that cross-cut the mylonitic foliation. Veining and fracturing presumably happened when these mylonites, after their emplacement, were folded into the major synform seen on Creag Shomhairle (Fig. 11.5).

The upper structural levels of the Moine Thrust Belt in Strath Beag may be accessed in the ground SSE of Creag Shomhairle, reached via a steep path above the culmination of Pipe Rock (Fig. 11.5). The grassy slopes at the top of the gully lead eastwards up towards Creag a'Chairn Chaoruinn. As outcrop improves, this route crosses low crags of variably sheared Cambrian quartzites and mylonites. Good examples are found on the hillsides overlooking the col between Shomhairle and Chairn Chaoruinn (Locality 11.1H; [NC 3860 5022]). These relationships are interpreted as representing imbrication of the rocks on either side of the original Moine Thrust (Butler, 1982). If time permits, it is instructive to trace out some of these slices. They are not laterally persistent. Further north the slices incorporate tracts of sheared Lewisian, at least some of which can be traced into the Creag Shomhairle sheet. Although in detail these geological relations are very complex, they show how major thrusts (in this case the Moine) can be strongly disrupted during later parts of their history by other thrusts. Consequently, in many places it is very difficult to trace out continuous thrust surfaces. A feature of the outcrops at Locality 11.1H is the intense, polyphase ductile deformation. In many cases the strongly mylonitic foliation, found in all units, is crenulated. While the mylonitic fabrics presumably reflect penetrative shearing associated with the emplacement of the Moine Thrust sheet, the crenulations have been related to a combination of the re-imbrication and the refolding by underlying culminations (Butler, 1982).

The Strath Beag transect has taken the visitor from the Sole Thrust to the Moine Thrust with increasingly ductile deformation and structural complexity. This trend is taken to show that the structurally over-lying thrusts (e.g. the Moine and its re-imbrications) formed at relatively deep crustal levels ($c.$ 10-15 km) at warm temperatures ($c.$ 300-350°C). As thrusting continued, the tectonic overburden presumably eroded so that later structurally underlying thrusts formed at shallower crustal levels, at progressively lower temperatures, faster strain rates and correspondingly more 'brittle' conditions. Although this is a common motif in thrust belts, these variations are not uniform and thick thrust sheets can be incorporated with very little internal distortion. The Creag Shomhairle sheet is one such example. If time and enthusiasm permit, it is informative to walk over the summit of Creag Shomhairle ([NC 3814 5052]; excellent views onto Conamheall) and down onto the rocky slopes to the NE. Abundant clean outcrops of Lewisian gneiss are to be found, preserving structures and metamorphic states otherwise encountered in the foreland. The easiest way back to the road is to walk down into the corrie between Shomhairle and Chairn Chaoruinn, crossing deer fences and descending to Strathbeg bothy. A path then continues down the valley, passing beneath Creag na Faoilinn and reaching the A838 [NC 3940 5391]. This can then be followed back to Polla.

Transect 2: Arnaboll

Along with Glencoul in northern Assynt, Ben Arnaboll is one of the most important geological sites in NW Scotland. It was here that Lapworth (1883) first demonstrated that the rock sequence was repeated by what were to become known as thrusts (Callaway did the same at Glencoul). Exposures of the Arnaboll Thrust in its type area, placing Lewisian gneisses onto Cambrian Pipe Rock, are justly famous and have inspired generations of both student and professional geologists for well over a century. Indeed Geikie (1884) first coined the term 'thrust' to describe these relationships. They are also Lapworth's (1885) type-locality for mylonites. The site was important in the 1980s for determinations of thrust sequences and for deducing the geometric evolution of rather complex thrust structures, especially by Coward (1980; 1984, 1988). It has continued to influence discussions of how basement comes to be incorporated into thrust belts (e.g. Rathbone *et al.*, 1983; Ramsay, 1997; Butler *et al.*, 2006). These inspirational outcrops form the cornerstone of this transect. The adjacent geology (Fig. 11.7), chiefly concerned with imbricated Cambrian strata, is described to provide context.

The Arnaboll outcrops can readily be accessed over a period of a couple of hours, although diversions over Heilam can occupy a full day. In contrast with the southern end of the Loch, the weather in northern Eriboll is positively arid. However, the terrain is still rough and includes steep hillsides and cliffs. As ever, these outcrops should be treated with great respect and should not be hammered.

Park in the large lay-by on the A838 [NC 4520 5992], or on a portion of the old road opposite. Before starting the transect it is worth examining the view across Loch Eriboll from the parking area. The hillside on the western shore of the loch displays prominent slabs of Cambrian quartzite that incline down to the sea. This is a dip-slope, inclined at 12° which, although tilted, is essentially undeformed. These distant rocks form part of the foreland. The dip-slope can be traced southwards to the head of the loch, running down from the summits of Beinn Spionnaidh (773m) and Cranstackie (801m). The Cranstackie dip slope runs along the ridge towards the hill of Conamheall (482m). At the saddle between the hills the simple dip of bedding in the quartzites becomes disrupted. The rocks here have been imbricated, stacked up on thrusts. The lower edge of the simple dip slope is the approximate position of the Sole Thrust – the outer edge of

Moine Thrust Belt at Loch Eriboll

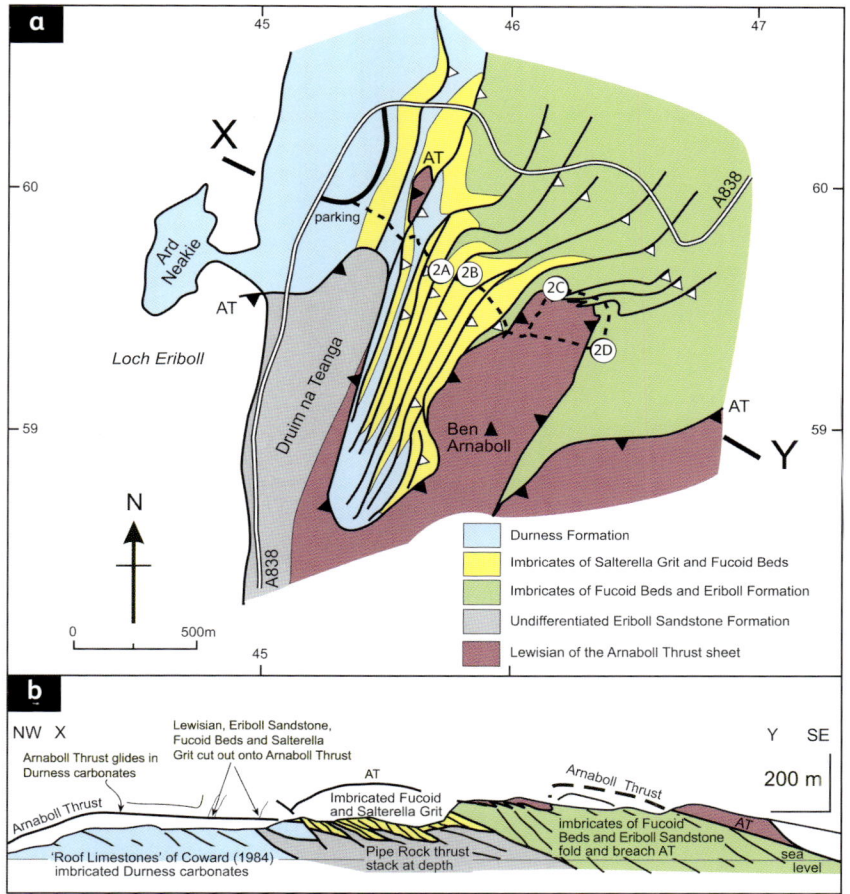

Fig. 11.7 Simplified geological map (a) and cross-section (b) of the Arnaboll hill area. Modified after Coward, 1984 (but see recent interpretation by Wibberley & Butler, 2010, of the structural relations between Lewisian and adjacent quartzites). AT = Arnaboll Thrust.

the Moine Thrust Belt. The rest of this area is described in transect 1 of the Eriboll excursion.

Although the chief objective of this transect is the Arnaboll Thrust, the approach route crosses an important tract of imbricated Cambrian strata (Coward, 1984; Fig. 11.7). These are described on the approach. From the parking area take the old road for a few metres, then head SE across boggy ground to a grassy gully [NC 4568 5971] that leads up a short escarpment.

205

The southern (right) side of the gully includes a cliff of Fucoid Beds. These are thrust onto Durness carbonates seen to the north of the gully. At the top of the gully the terrain opens up, revealing low outcrops and some rather boggy ground. Ahead, to the southeast, the upper crags of Ben Arnaboll are visible. Between the gully and these crags lie imbricated Cambrian strata, chiefly Fucoid Beds and Salterella Grit together with a few metres of the uppermost Pipe Rock. The imbricates have an across-strike width of several hundred metres. Exploring geometry of these structures can be instructive. Directly ahead of the gully, across some marshy ground, lies a small outcrop of Pipe Rock, about 1m high, sandwiched between Fucoid Beds (Locality 11.2A, [NC 4574 5969]). The Pipe Rock slice can be traced to the north, on the western flank of a knoll. The thickness of the slice gradually increases to about 8m, implying that the imbricate thrust that carries it changes its stratigraphic position. Indeed all of the imbricate thrusts in this area display the same trend, carrying more Pipe Rock in the north than in the south.

After this diversion, scramble to the top of the knoll (Locality 11.2B, [NC 4581 5971]) for the view onto the western cliffs of Ben Arnaboll. This view is generally down the inferred direction of thrust transport so the variations in the structure seen in the cliff are lateral. The highest crags are Lewisian gneiss, with their characteristic massive appearance, that form the hanging-wall to the Arnaboll Thrust. They lie on a near-continuous cliff of bedded, cream-coloured rocks – the Pipe Rock. The bedding is not continuous, but terminates to form features that appear to be sedimentary channels. Closer inspection reveals that these are thrust structures, with the bedding terminations representing lateral ramps. The Arnaboll Thrust is clearly visible at the base of the Lewisian gneisses. It is folded by the underlying imbricates, a feature best appreciated at the small cave [NC 4596 5938]. In the foreground, forming the gentle ground beneath the cliffs, are imbricates of Fucoid Beds and Pipe Rock, readily identified from the characteristic vegetation and outcrop of these units.

The excursion now visits the exposures of the Arnaboll Thrust. From the knoll, head down the heathery slope to the southeast and then ascend the steep slope between the crags to the Lewisian of the Arnaboll Thrust sheet. This ascent is aided by an indistinct track, but beware slippery rocks amongst the heather. A brief diversion can be taken to the north to find exposures of the Arnaboll Thrust plane ([NC 4617 5951]; see White 1998). However, great care should be taken here on this steep ground. Otherwise continue onto the plateau and walk northeast along the top of the cliffs.

Moine Thrust Belt at Loch Eriboll

The outcrops hereabouts are Lewisian gneisses that show little sign of Caledonian deformation. Structures, including the gneissic banding, developed under and preserve amphibolite facies metamorphic assemblages. They are cross-cut by, presumably, late Laxfordian (Palaeoproterozoic) granitic pegmatites. At the northern end of the plateau [NC 4611 5951] there is a spectacular viewpoint north up the coast towards Whitten Head and, in the immediate ground, onto Ben Heilam (Fig. 11.1). The general 'grain' of the geology, picking out the trend of the imbricate thrusts of Pipe Rock and Fucoid Beds discussed above, can be traced across Heilam. To the NE lies typical Moine outcrop.

From the plateau drop carefully down to the west for a few metres. Here are the outcrops of the Arnaboll Thrust (Locality 11.2C, [NC 4615 5958]; Fig. 11.8). This internationally-important site should be treated with the utmost respect. According to Teall (see White, 1998), the site was important for Lapworth's (1885) first descriptions of mylonites. Lewisian gneisses rest tectonically upon a footwall formed of Pipe Rock. Bedding is clearly visible within these quartzites, parallel to the thrust contact. *Monocraterion* burrows show the Pipe Rock to be the correct way up, demonstrating that the presence of Lewisian gneisses above requires the contact to be tectonic, rather than, say, an upside-down unconformity. Despite the recognition of sedimentary structures, the Pipe Rock has been sheared with the prominent Skolithos burrows inclined with respect to bedding. Given the assumption that these burrows formed perpendicular to bedding, they constitute excellent markers by which the deformation can be quantified (see Fischer & Coward, 1982). This deformation penetrates for at least 2m below the thrust. The sense of deflection clearly indicates a top-to-the-WNW shearing which presumably reflects the tectonic transport direction of the Arnaboll Thrust sheet.

The Arnaboll Thrust plane itself is a discrete, knife-sharp surface, although it is locally warped by minor thrust structures within the Pipe Rock below. Its hanging-wall is marked by a narrow zone of dark, chlorite-rich mylonite about 50cm wide. More than about 3m above the thrust plane the Lewisian shows few effects of Caledonian deformation (as seen on the plateau). So the thrust zone shows a dramatic gradient, not only in deformation but also in the intensity of the greenschist overprint (Wibberley, 2005). The deformation can be tracked using the pegmatites embedded within the gneisses that are progressively more streaked out and deflected towards the thrust.

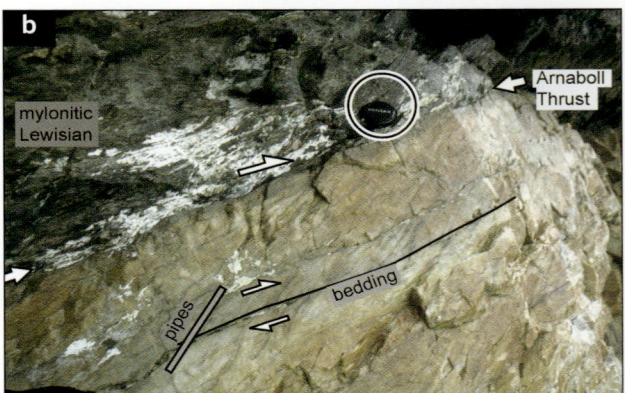

Fig. 11.8 The classic location (2C) of the Arnaboll Thrust and of Lapworth's (1883) mylonites:

(a) general aspect of the thrust, carrying Lewisian gneisses onto Pipe Rock; hammer (circled) for scale

(b) detail of the Arnaboll Thrust, showing deflected Skolithos (pipes) that indicate the sense of shear (top to WNW); scale from lens cap (circled)

If time and the inclination permit, it is instructive to walk out the Arnaboll Thrust across the slopes to the SE of Locality 11.2C as far as Locality 11.2D [NC 4626 5933]. The thrust is cut repeatedly by thrusts that climb out of the underlying Pipe Rock and up into the overlying Lewisian (Fig. 11.7b). This type of geometry is described as 'breaching' (Butler, 1987) and gives a clear indication of the relative sequence of thrust development: the Arnaboll Thrust sheet was emplaced onto the Pipe Rock before the imbricates within it developed. One of the clearest examples of this geometry is to be found in the small natural amphitheatre [NC 4622 5953]. Further along the outcrop trace, the Arnaboll Thrust can be inferred to have been folded (Fig. 7b) so that it maps out as a vertical contact. At the end of the short walk (Locality 11.2D) outcrops of the Pipe Rock, from the footwall to the Arnabol Thrust, can be found with similarly sub-vertical dips. Careful searching reveals Monocraterion burrows that, as at Locality 11.2C, young towards the Lewisian rocks of the Arnaboll Thrust sheet.

Transect 3: Kempie

Parking is available in the large lay-by on the A838 [NC 4441 5800] where there is room for about six cars. The transect running up the hillside from Kempie follows in the footsteps of Lapworth (1883). It covers the transition from folded Cambrian quartzites and their Lewisian basement through a dramatic deformation gradient that increases upwards into mylonites associated with the Moine Thrust. The lower parts of the transect are concerned with less deformed units and include excellent introductory outcrops for parts of the Cambrian stratigraphy. Indeed the coastal outcrops include the type section for the An t-Sron Formation units. This transect lies exclusively within the hanging-wall to the Arnaboll Thrust sheet. Therefore it provides a continuation from transect 2 into progressively higher structural levels within the Moine Thrust Belt. The whole section contains a number of informative outcrops that can be used to build up a structural interpretation from first principles. Consequently the area has been much used for student training exercises. It has been described most recently by Butler *et al.* (2006). This transect provides an introduction to some of the critical elements and is illustrated by a sketch geological map (Fig. 11.9) and cross-section (Fig. 11.10). It is described so that visitors can appreciate the geological interpretations based on individual outcrops that can build up into a coherent structural model through this part of the thrust belt. As described, this transect takes about 4 hours, involving ~2 km of walking and ~200 m of ascent with some steep terrain.

Start on the shore below the parking area, on the headland of An t-Sron (Fig. 11.9). Walk down to the headland to reach clean outcrops of Cambrian Pipe Rock (Locality 11.3A, [NC 4431 5817]). These quartzites lie in the core of a large anticline. Moving west along the coast (passable at all but the highest part of the tide) provides an excellent introduction to the stratigraphy of Cambrian strata. The stratigraphic top of the Pipe Rock, passing

EXCURSION 11

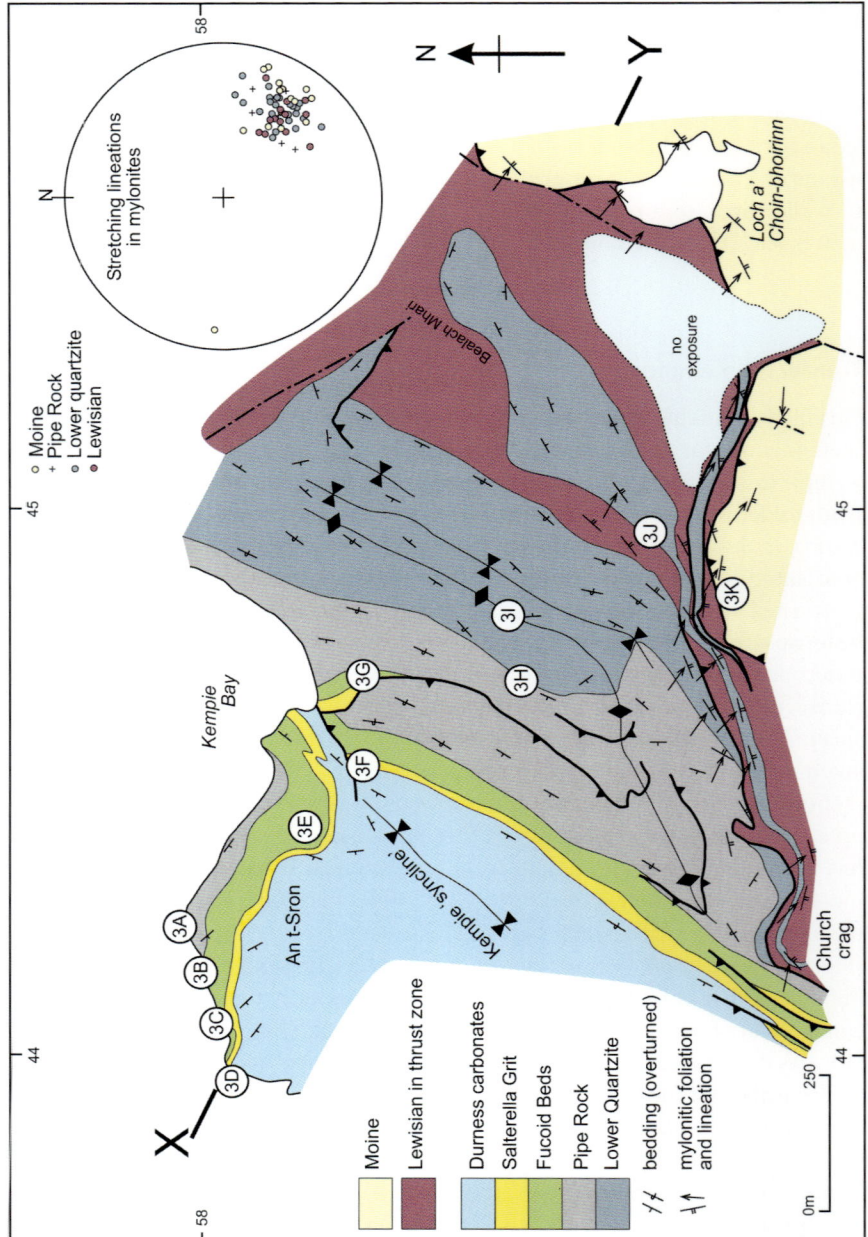

Fig. 11.9 Geological map of the Kempie area (modified after Butler *et al.*, 2006). X-Y is the section line of Fig. 11.10.

Moine Thrust Belt at Loch Eriboll

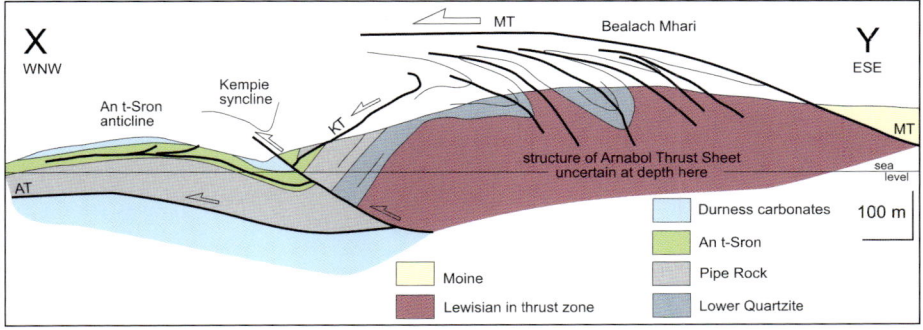

Fig. 11.10 Cross-section through the Kempie area (modified after Butler et al., 2006). KT = Kempie Thrust; AT = Arnaboll Thrust; MT = Moine Thrust.

abruptly up into the Fucoid Beds, is found at the base of the low sea-cliff (Locality 11.3B, [NC 4417 5814]). In detail, this boundary is deformed. Bedding planes show evidence for slip and some of the finer-grained layers in the Fucoid Beds are cleaved. Within the Eriboll area, the base of the Fucoid Beds can be inferred to have acted as a detachment horizon from which imbricate thrusts have splayed to repeat Fucoid Beds and the overlying Salterella Grit. Here on the An t-Sron shore line there are no such imbricates, but examples are encountered on the Arnaboll transect described below. There are, however, minor thrust structures, chiefly directed towards the east ('back-thrusts') that cut up from the Pipe Rock into the Fucoid Beds (see Coward, 1988).

Continuing along the wave-cut platform (*beware slippery rocks*) is a journey up stratigraphic section. There are excellent outcrops of the sedimentary structures within the Fucoid Beds that are commonly difficult to study in the sparse outcrops of this formation inland. Towards the top of the Fucoid Beds at Locality 11.3C [NC 4405 5809], there are some spectacular yet enigmatic folds and thrusts with trends that are strongly oblique to the regional orientations.

The Salterella Grit Member on An t-Sron has a significantly lower thickness compared to other parts of the NW Highlands (<5m as opposed to the usual 8-10m). These clean quartzites are well exposed in the low sea cliff section and capped by the Ghrudaidh Formation – the gritty dolostones that form the basal unit of the Durness Group carbonates. Bedding within the dolostones is markedly disturbed – well exposed examples lying on the low rocky promontory, Locality 11.3D [NC 4393 5807]. These structures have been variously interpreted as formed by extensional faults or by backthrusts. The latter of these is preferred here, as there are examples of such structures along the foreshore section and many other contractional structures can be proven within the dolostones in the ground above the shore. The development of thrust structures within the dolostones of the Ghrudaidh Formation, but not in the Salterella Grit, requires the presence of a detachment horizon near their geological boundary. Evidence for this in outcrop include cleavage development (foreland-vergent) and local slip surfaces.

From the shore, return to the roadside of the A838. A useful exercise is to trace the geological boundary between the dolostones and quartzites of the Salterella Grit. This is picked out by vegetation and confirmed by scattered outcrops. The final climb to the road (at a crash barrier) breaks out opposite a small spring and water trough that lie between cuttings with dolostones (west) and Salterella Grit (east: [NC 4437 5799]). For those wishing to find more evidence for deformation within the dolostones, the road cuttings to the west along the A838 provide sections through complex thrust stacks. However, unless experienced with carbonates, it may be difficult to distinguish between depositional bedding surfaces and tectonic surfaces in these sections.

The remaining objectives for this transect lie to the east and up to the Moine Thrust. This route is described in detail so as to identify the necessary evidence for the structural relationships depicted on Fig. 11.10. The road section opposite the parking lay-by (Locality 11.3E, [NC 4439 5798]) is made of Fucoid Beds that are variably thickened up by thrusts. These structures lie in the crest of the An t-Sron Anticline, the western limb of which has formed the focus of attention so far. The excursion now moves into its eastern limb. Regrettably this is rather poorly exposed at the level of the road, but can be proven down on the coast (below the cottages of Kempie). The first outcrop to the east along the road, next to the small stream next to the woods (Locality 11.3F, [NC 4453 5787]) are quartzites partly covered in moss that are interpreted to be Salterella Grit. Bedding is sub-vertical. The adjacent road outcrops to the east are Fucoid Beds [NC 4455 5787]. Consequently the interpretation here is that the Salterella Grit and Fucoid Beds young to the west and therefore lie on the eastern limb of a synform whose axis runs through the dead-ground between these outcrops are the lay-by. This structure is the Kempie Bay Syncline. Cleavage within the Fucoid Beds in the road section here verges east, consistent with the syncline interpretation.

Moine Thrust Belt at Loch Eriboll

It is worth working carefully eastwards along the low road cutting below the trees, watching carefully for traffic. For the most part the section consists of steeply-dipping Fucoid Beds with inclined, east-vergent cleavage. After 40m [NC 4460 5786] there is a screen of vertically-bedded quartzites with prominent Skolithos burrows, deduced to be the Pipe Rock that stratigraphically underlies the Fucoid Beds. The burrows are sheared with a sense consistent with the vergence of the cleavage in the Fucoid Beds. Similar vergence can be determined for cleavage in the shaley units in the Pipe Rock here. For enthusiasts it is possible to trace the boundary between Pipe Rock and Fucoid Beds, and the Fucoid-Salterella boundary encountered to the west up the wooded hillside to the south of the road.

Given the stratigraphic thickness (>150m) of the Cambrian quartzites, the expectation moving east along the road from the previous outcrops at the stratigraphic top of the Pipe Rock should be to remain in quartzites. However, within a few metres the road section returns to Fucoid Beds. This unit can be found in a small damp gully (Locality 11.1G, [NC 4470 5784]). Just to the east, at the sharp bend in the A838 [NC 4472 5785], is a return to Pipe Rock. As with the Pipe Rock found to the west on the road, the *Skolithos* burrows here are also sheared with an eastward vergence. Consequently the panel of Fucoid Beds caught between the Pipe Rock outcrops is interpreted as having been incorporated by thrusting rather than in a syncline hinge (Fig. 11.10). This is called the Kempie Thrust. It is inferred to lie on the western side of the Fucoid Beds at Locality 11.1G, while their eastern side is interpreted as the original stratigraphic contact at the top of the Pipe Rock, now tilted to be vertical. Note that the inferred Kempie Thrust has also been tilted to vertical, an important corollary that will be explored at the end of this excursion. Thrusting within the Cambrian stratigraphy deduced from consideration of these road sections means that the Kempie Syncline is difficult to prove as there are no simple geological boundaries that can be traced around this fold. Further structural com-plexity can be investigated on the rocky shore at Kempie [NC 4464 5795].

The plan now is to follow the structure eastwards, working back across strike. In doing this, the route is predicted to run down stratigraphic section, through steeply-dipping Cambrian quartzites on the steep, eastern limb of the Kempie Bay Syncline. The thin slice of Fucoid Beds described above (Locality 11.3G) cannot be followed far away from the road and consequently will not add to the structural complexity! To access this ground, return back along the road a few metres to the most western slice of Pipe Rock and follow an indistinct path up into the trees. A number of different routes can be taken up the steep heathery hillside, but the target is the narrow, north-south-trending hanging valley that ends at an elevation of about 130m. It is worth checking the outcrops of quartzites, confirming the general orientation of bedding (subvertical) and the presence of *Skolithos* burrows (confirming the status of these rocks as being Pipe Rock). The burrows display elliptical bedding sections, with the long axes sub-parallel to strike, implying significant bed-parallel shortening strains. With care, upon entering the hanging valley, it is possible to pick the stratigraphic boundary with the underlying Lower Quartzite. The key distinction lies in identifying mm-scale depositional lamination and therefore the absence of the intense burrowing that characterises the Pipe Rock. Good examples can be found at Locality 11.3H [NC 4478 5771]. It is worth studying these rocks in some detail. Cross-bedding can be identified here and used to determine the westward younging of these strata. Individual sedimentary grains can be readily identified either with the naked eye or through a hand-lens. They are undeformed and show a good granular texture. It might be instructive to collect a small sample of these quartzites from surface detritus for comparison with equivalent rocks further up the section.

Lapworth (1883) interpreted the structure further up the hanging valley in terms of fold structures, the western of which are the An t-Sron Anticline and Kempie Bay Syncline described above. The next fold hinge is found in the cliffs on the east side of the hanging-valley at Locality 11.3I (best seen from [NC 4472 5757]). This anticline shares its steep western limb with the Kempie syncline. Quartzites of its eastern limb dip gently eastwards, indicating that the fold axial surface is inclined to the east (Fig. 11.10). The interlimb angle (the measure of fold tightness) is about 100 degrees.

Continue up the hanging-valley consulting outcrops, especially on its northern side. Towards the top of the valley (e.g. [NC 4478 5743]) bedding in the Lower Quartzite dips moderately towards the WNW, with younging determined by cross-lamination towards the WNW. Consequently a synclinal fold axis may be inferred to have been crossed (Fig. 11.10; it can be mapped through the adjacent ground, see Fig. 11.9). Good clean outcrops of quartzites are to be found at Locality 11.3J [NC 4488 5743]. Here the beds dip at about 50° to the ESE but young westwards, indicating that they are upside-down. Careful inspection reveals that the sedimentary grains are flattened, creating a weak to locally intense protomylonite deformation fabric that is approximately axial planar with respect to the main folds that were encountered on the traverse. A few metres to the east are out-crops of pegmatite-rich Lewisian basement (e.g. [NC 4492 5741]) which also show a weak schistosity defined by chlorite and epidote that is sub-parallel to the foliation in the quartzites. It may be deduced that the contact between the Lewisian and the quartzites is an overturned unconformity. A slight diversion along the plateau reveals this contact in outcrop [NC 4501 5756].

Returning to Locality 11.3J, the next objective is to work carefully up section, best achieved walking south. The small knoll (219m OD) about 200m SSW of Locality 11.3J forms a useful landmark, with outcrops lying along a small escarpment facing the plateau area (Locality 11.3K; [NC 4487 5726]). The upper part of the escarpment consists of distinctive folded mylonites, and again the interpretation (Barber & Soper, 1973; Butler *et al.*, 2006) put forward here is that these were chiefly derived from Moine psammites, although others favour a foreland Lewisian protolith (British Geological Survey 2002; Holdsworth *et al.*, 2006). They contain a strong linear fabric defined by elongate quartz aggregates that plunges towards the ESE. The base of these mylonites is considered by Butler *et al.* (2006) to be the Moine Thrust. Below lie more mylonites of distinct compositions, arranged in bands of about a few metres thickness. One type of mylonitic layer is highly quartzitic. Others are essentially chloritic phylonites with thin feldspathic seams. Where evident, these units also show stretching lineations that plunge ESE.

The derivation of the mylonites beneath the Moine Thrust can be established by briefly tracing out a deformation gradient. Return to Locality 11.3J. The plan is to walk out these quartzites and the neighbouring Lewisian for a few hundred metres to the WSW, along the strike (Fig. 11.9). As seen previously, at Locality 11.3J, the quartzites retain visible bedding but also display moderate protomylonitic deformation fabrics. Further WSW the deformation increases (e.g. [NC 4470 5731]) to become fully mylonitic with the same ESE-plunging stretching lineation as seen at Locality 11.3K. Thus these mylonites are products of progressive deformation that, at lower strain states, involves folding of the Cambrian quartzites and their Lewisian basement (Fig. 11.10). Elsewhere (e.g. [NC 4417 5713]) the shearing focuses onto a discrete thrust that carries mylonites derived from the Cambrian quartzites and their Lewisian basement onto more outlying parts of the fold belt crossed on this transect. Further description of these forms of structural relationship is reserved for the next transect. Return to vehicles by carefully descending the slopes to the A838.

Many interpretations of structural evolution in crustal-scale shear zones, such as the Moine Thrust Belt, assume that there is a simple progression from ductile deformation, manifest by mylonites development into brittle deformation and cataclasis as rocks become progressively exhumed. However, the transition from ductile to brittle deformation need not be controlled simply by depth (or temperature), but also by strain rate. The structural evolution on this transect illustrates this complexity. The main folds (An t-Sron Anticline, Kempie Bay Syncline and un-named folds seen higher on the transect) face WNW. Their axial surfaces become increasingly inclined up-section as the deformation state increases, culminating in mylonite formation directly beneath the Moine Thrust. Therefore these folds formed before or during the latest ductile movements on the Moine Thrust (Butler *et al.*, 2006). Yet the folds deform earlier thrust structures such as those that are now found facing downwards on the steep eastern limb of the Kempie Bay Syncline. These earlier structures represent periods when deformation was strongly localized. It is not clear what the timing of the folding is relative to slip on the Arnaboll Thrust, although it is plausible that this structure is folded by the Kempie Bay Syncline (e.g. Coward, 1984). However, there were periods in the structural evolution of this part of the thrust belt when ductile shear was partitioned strongly onto the Moine Thrust, then distributed across the fold belt, then onto the Moine Thrust again. Thus the deformation has alternated between ductile and brittle styles.

Transect 4 – Creagan Road
(contributed by Rob Strachan, Bob Holdsworth and Ian Alsop)

This is a short traverse to examine the internal tectonostratigraphy of the mylonite belt (Fig. 11.11) that occupies a high structural level in the Moine Thrust Belt (Barber & Soper, 1973; Soper & Wilkinson, 1975; Evans & White, 1984; Law *et al.*, 1986; Holdsworth *et al.*, 2006). The description here makes an interesting counterpoint to that for the Kempie transect in that the naming of thrusts and correlative approaches are different. This debate has continued since the work of Peach *et al.* (1907; see Barber & Soper, 1973). The traverse involves 2-3 hours of moderate walking on tracks and hillsides.

Park (with permission) adjacent to the entrance to Eriboll Estate at [NC

4323 5630], taking care not to block any gateways or entrances. There is sufficient space for a coach or four to five cars. Take the track (the 'Creagan Road') that leads southwards from the telephone box through a gate. Pause further on at a second gate to look northeastwards to the crags in the trees of steeply dipping Durness Limestone on the overturned limb of the Kempie Bay Syncline. After going through the gate, follow the fence southwards and then walk across the hillside to the low-lying crags that form Locality 11.4A [NC 4301 5577]. Here is exposed the tectonic lower boundary of the mylonite belt which is termed the Lochan Riabhach Thrust by Holdsworth *et al.* (2006). They propose that it is a late, out-of-sequence brittle structure that everywhere underlies the mylonite belt and is entirely distinct from the Moine Thrust that is exposed at a higher structural level. It is a sharp, gently-dipping fault that emplaces intensely deformed quartzo-feldspathic mylonites that are here interpreted to be of Lewisian origin onto largely undeformed and probably inverted Cambrian Salterella Grit. Small pips of carbonate lie along the thrust plane and are interpreted as detached slices of Durness Limestone. The thrust cuts obliquely across the inverted limb of the Kempie Bay Syncline, to rest discordantly on Durness Limestone in the stream section a few hundred metres to the southwest. Walk from here northeastwards, cutting uphill to the first hairpin bend in the Creagan Road. Continue along the track, passing outcrops of Cambrian Pipe Rock, the unexposed Lochan Riabach Thrust and, above that, further outcrops of Lewisian-derived mylonite between the second and third hairpin bends. Pass through a gate and continue along the track.

Locality 11.4B [NC 4356 5534] is by the track, where 'Oystershell Rock' mylonites are well exposed. These are platy, white mica-chlorite phyllonites with numerous lunate quartz segregations – the superficial resemblance of the latter to fossil shells gave rise to the informal term that has continued to be used for this lithology (Peach *et al.*, 1907; Soper & Wilkinson, 1975; Holdsworth *et al.*, 2001a, 2006). The protolith for the Oystershell Rock has long been considered as Lewisian (e.g. Barber & Soper, 1973; but see also Soper & Wilkinson, 1975). Holdsworth *et al.* (2001a) have confirmed a metamorphic protolith and a Lewisian one seems most likely. The phyllonites carry well-developed shear band fabrics (McClay & Coward, 1981) that indicate a top-to-the-west sense of displacement parallel to a locally developed mineral and extension lineation. Continue uphill, passing further outcrops of the Oystershell Rock, some containing early syn-mylonitization isoclinal folds ('F_2') that are refolded by asym-

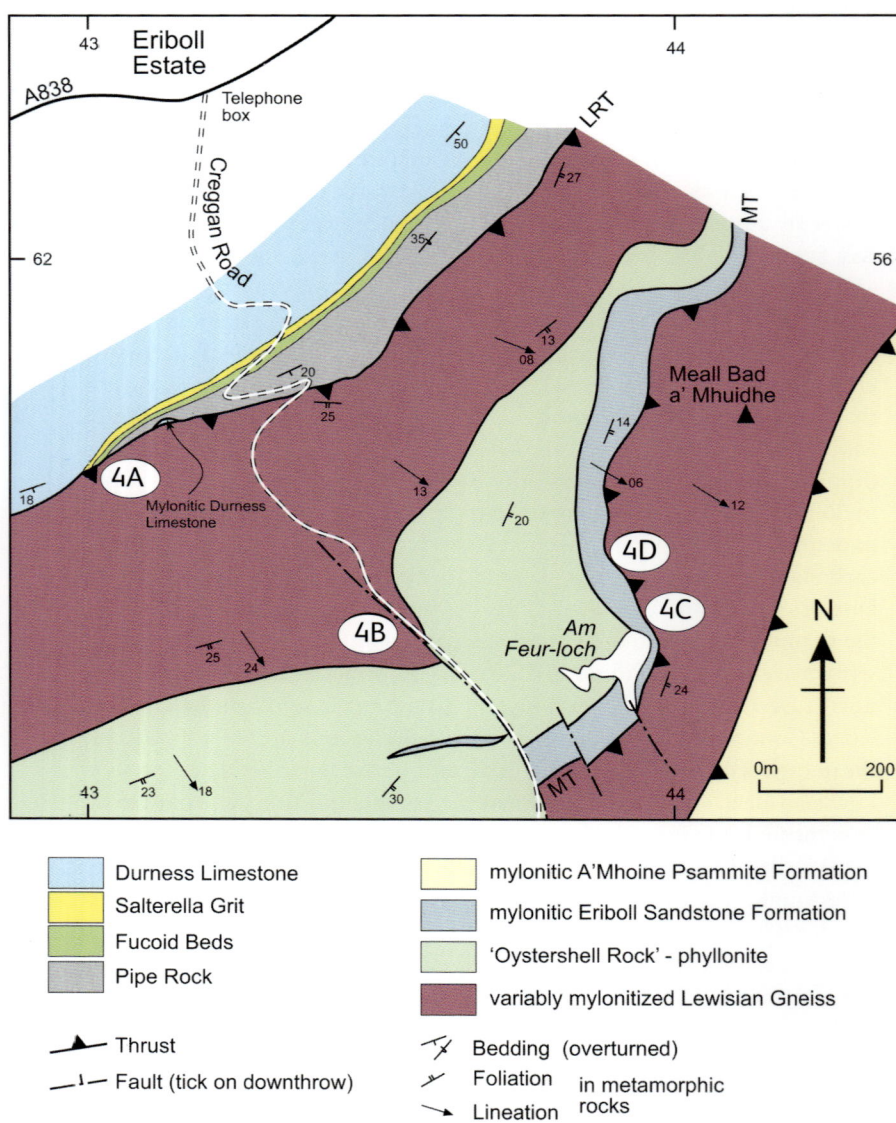

Fig. 11.11 Geology of transect 4 through the mylonite belt along the Creagan Road.

metric folds ('F_3'). Just beyond the telegraph poles, the track starts to flatten by low outcrops of Oystershell Rock [NC 4367 5525]. At this point, head across the hillside due east towards Am Feur Loch, walking over more low outcrops of Oystershell Rock. Pause at [NC 4388 5530] to view the loch and surrounding outcrops. The steep crags on the east side of the loch comprise in their lower part quartz mylonites derived most probably from deformation of the Eriboll Sandstone Formation. These are separated by what Holdsworth *et al.* (2006) and others before have interpreted to be the Moine Thrust from overlying mylonites derived from Lewisianoid basement. Exposures of the latter have a characteristic blotchy pale colouration due to greater amounts of lichen cover compared to the more homogeneous grey quartz mylonites.

Whilst there is agreement on the protolith of these quartz mylonites, the problem of their structural setting is central to current debate on the structural evolution of this part of the Moine Thrust Belt. According to Holdsworth *et al.* (2006), the quartz mylonites overlie an original (albeit highly tectonized) unconformity with the Lewisian protoliths of the Oystershell Rock, both of which are assigned to the Caledonian foreland (British Geological Survey 2002; Holdsworth *et al.*, 2006). An alternative view, consistent with Butler's (1982) interpretation on Creag Shomhairle (see transect 1), is that the quartzites have been imbricated into the Oystershell Rock. This could mean that the Moine Thrust and the Lochan Riabhach Thrust described here are essentially the same structure repeated by displacement on a breach thrust. If this is the case, the Oystershell Rock need not be derived from the foreland, but could form part of the far-travelled Moine Thrust sheet.

Alternatively some or all of these contacts could be minor structures associated with distributed shearing, as found at the top of the Kempie area (transect 3).

Walk to Locality 11.4C [NC 4395 5525] at the low crags to the north of the loch. The lowest exposures are spectacular quartz mylonites; an intense mylonite fabric dips gently to the ESE and carries a mineral and extension lineation that plunges down-dip. Asymmetrically sheared quartz veins indicate a top-to-the-west sense of shear parallel to the lineation. West-verging, open F_3 folds deform the mylonite fabric and associated lineation. Quartz veins vary from intensely mylonitic to apparently undeformed. Some discrete thrusts are associated with late folds. Higher parts of the crags expose the Moine Thrust as a sharp contact between the white quartz

Fig. 11.12
Quartz mylonites and associated quartz veins deformed by F_3 folds at Locality 11.4D.

mylonites and grey to cream coloured Lewisianoid-derived mylonites. The latter contain quartz, feldspar and mica, and lack the continuous banding that is a prominent feature of the quartz mylonites. The contrast between the evidently ductile nature of the Moine Thrust here, located within a broad belt of mylonites, and the brittle Lochan Riabhach Thrust, is the rationale for Holdsworth *et al.* (2006) regarding the structures as entirely separate and not the same thrust repeated by breaching.

Head northwestwards upslope to small outcrops 50m away of quartz mylonite at Locality 11.4D [NC 4388 5545]. These expose excellent examples of F_3, tight-to-open, asymmetric S-folds of the mylonite fabric and associated lineation (Fig. 11.12). Fold hinges are variably oriented: some are almost parallel to the lineation, others are normal to the lineation (Evans & White, 1984). Also visible are isolated F_2 isoclinal folds that formed during mylonitization; note that the mylonite fabric is appreciably more intense on the fold limbs than in the hinges. Numerous quartz veins are present, some elongated parallel to lineation. The F_2 and F_3 folds are interpreted as resulting from continuous, progressive deformation within the evolving mylonite belt (Holdsworth *et al.*, 2006).

Splendid views may be had of Ben Hope and Ben Loyal to the east, and Cranstackie and the dip-slope of the Cambrian quartzites to the west. Return to the lochside and walk back down the Creagan Road to the vehicles, leaving all gates as you find them.

Excursion 12
Durness and Faraid Head

Bob Holdsworth and Rob Strachan

Purpose: To examine down-faulted segments of the Moine Thrust sheet and Moine Thrust Zone in the Caledonian foreland.

Aspects covered: Moine and basement-derived mylonites; Cambrian quartzites and limestones, Caledonian thrusts and other low-angle faults; late to post-Caledonian faults.

Maps: OS: 1:25,000 sheet 446 Durness and Cape Wrath; BGS: 1:50,000 sheet 114W Loch Eriboll.

Useful information: Accommodation may be obtained in Durness, Rhiconich and Tongue.

Type of terrain: Almost entirely coastal exposures.

Distance and time: If staying in Durness, driving is minimal; the excursion involves c.6-7km walking and could easily be completed in one day. Low tide is advantageous for the Durness part of the excursion. Durness is c.30 minutes drive from Rhiconich and c.1 hour from Tongue.

Short itinerary: The Durness part of the excursion is very accessible, could be covered in 2-3 hours and demonstrates the most important aspects of the geology.

This is a short excursion designed to examine unique occurrences on the Caledonian foreland of segments of the Moine Thrust sheet and Moine Thrust Zone that were down-faulted during the late Palaeozoic and/or the Mesozoic (Fig. 12.1; Peach *et al.*, 1907; Hippler & Knipe, 1990; Holdsworth *et al.*, 2006, 2007). These outcrops are of historical importance because Peach *et al.* (1907) were able to deduce from them a minimum displacement of *c.*15km along the Moine Thrust – one of the first times that this approach had been used to constrain large-scale horizontal movements in an orogenic belt.

If driving to Durness from Tongue, the route passes up the west side of Loch Eriboll, along the strike of the easterly-dipping Cambrian quartzites

EXCURSION 12

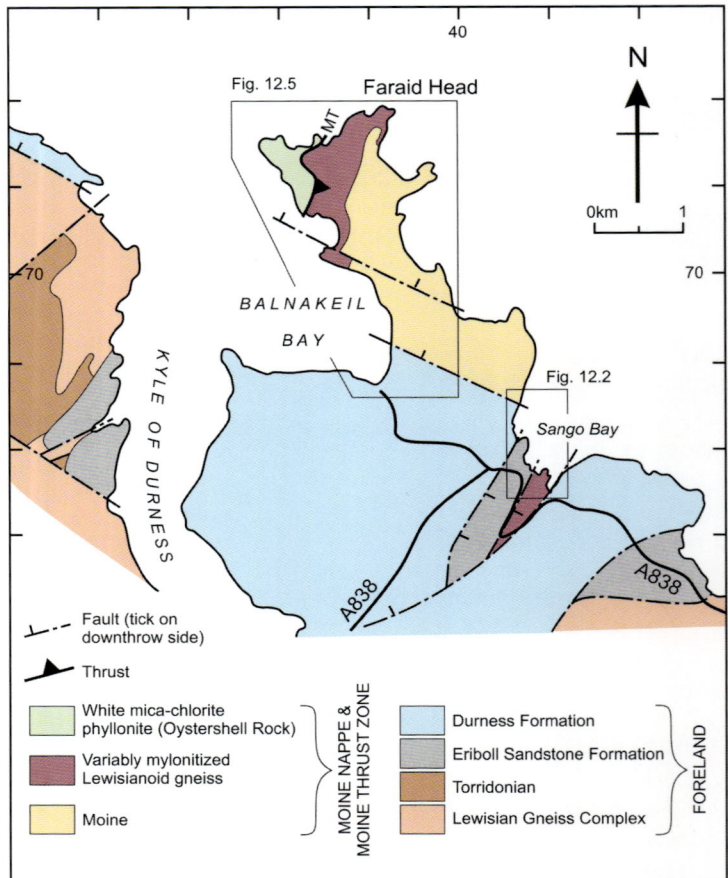

Fig. 12.1 Simplified geological map of the Durness-Faraid head area (modified from British Geological Survey, 2002). MT = Moine Thrust.

of the Caledonian foreland. There are numerous opportunities along this road section for a short stop to examine these rocks. Groups may also wish to stop *c.* 4 km ESE of Durness at Traigh Allt Chailgeag (parking at [NC 4435 6537]) to view steeply-dipping Lewisian basement gneisses that were strongly reworked in the Laxfordian event. Kinny & Friend (1997) reported a U-Pb zircon age of *c.* 2680 Ma for the igneous protolith of a dioritic gneiss from this locality. The road climbs slightly to the west, and it is worth pausing briefly at one of the various passing spaces in order to look north-eastwards across Loch Eriboll to the spectacular cliffs of Whiten Head that expose the northernmost part of the Moine Thrust Zone. The white cliffs comprise imbricated Cambrian Pipe Rock, and the darker rocks at a structurally higher level are Lewisian gneisses of the Arnabol Nappe (Holdsworth *et al.*, 2007; see Excursion 11).

Locality 12.1 [NC 4100 6740 to NC 404 685]

Sango Sands, Durness (Fig. 12.2). Down-faulted basement and Moine Thrust sheet and Moine Thrust Zone mylonites overlying the Cambrian Durness Limestone.

Parking is available for coaches and cars at the Tourist Information Centre in Durness [NC 4070 6775] overlooking Sango Sands. From the car park, view the steep wall of Durness Limestone *c.* 400m to the SE; this lies along the Sangobeg Fault, one of the main bounding normal faults of the Durness outlier (Fig. 12.2). Walk *c.* 300m east along the road to roadside exposures in the Durness Limestone at Locality 12.1A [NC 4100 6740] that lie in the immediate footwall of this fault. Here, a series of carbonate-cemented red sandstone-breccia infills and carbonate veins are preserved in sub-vertical fractures trending NNE-SSW, approximately parallel to the trend of the adjacent normal fault. The sedimentary material is thought to have infilled tectonically active open fractures in the limestone that formed synchronously with normal faulting activity. This suggests that this phase of extension was associated with sedimentation, although most of the basin infills have subsequently been eroded. Similar red-bed infills are common in the region between Durness and Cape Wrath (e.g. see Locality 12.1F for further details). Clast types are mainly Durness Limestone, but isolated examples of mylonite and Cambrian quartzite are also preserved. Climb the grassy slope above the exposure, look northwestwards across Sango Bay and compare the annotated view presented in Fig. 12.3 with the geological map (Fig. 12.2).

Return to the car park and walk down onto the beach via the wooden steps. East of the base of the steps is a prominent headland (1B) that comprises outcrops of banded quartzo-feldspathic and amphibolitic Lewisianoid basement gneisses (Fig. 12.2). These are thought to lie close to the base of the Moine Thrust sheet; the underlying Moine Thrust is not exposed in the bay, but is seen on Faraid Head (see Locality 12.2). Creamy-pink acidic

EXCURSION 12

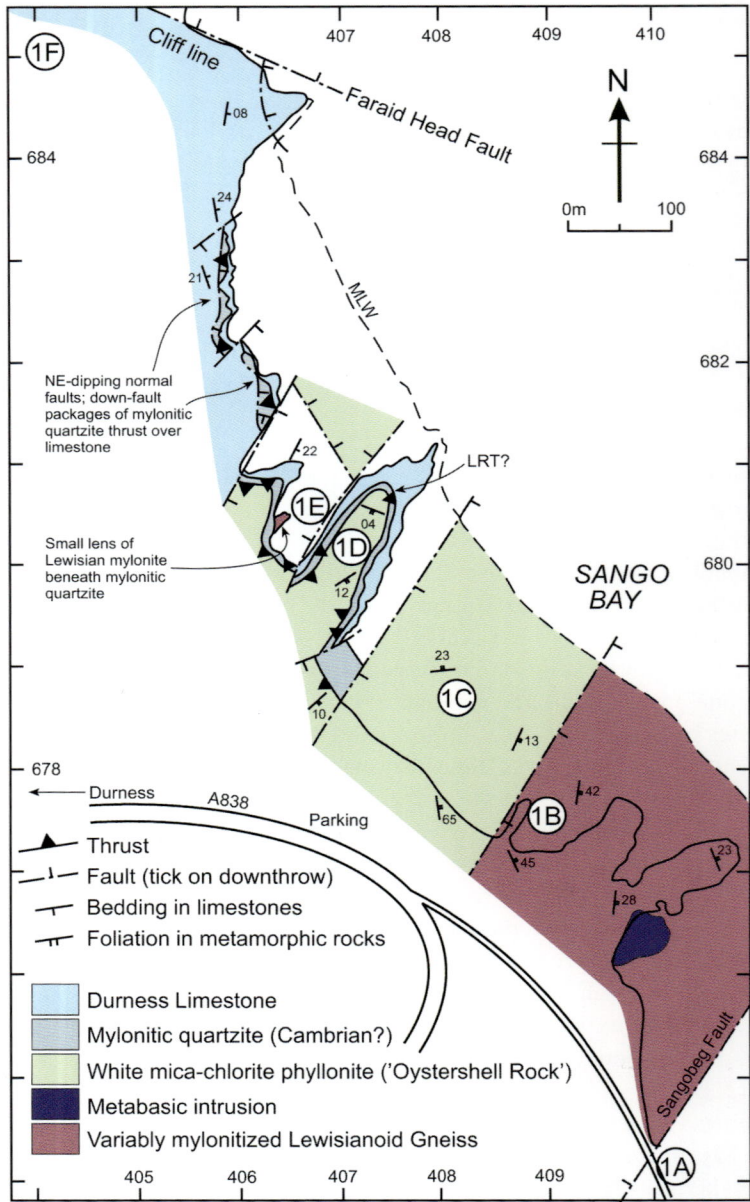

Fig. 12.2 Geological map of the coastal area around Durness showing Locality 12.1. LRT = Lochan Riabhach Thrust.

Durness and Faraid Head

Fig. 12.3 View from above Locality 12.1A, looking NW across the Durness outlier and towards the Faraid Head peninsula in the distance. Locations of Lewisianoid rocks of the Moine Nappe, Oystershell Rock (OSR), Lochan Riabhach Thrust (LRT) and the Faraid Head Fault (FHF) are also shown.

gneisses and dark green metabasic sheets are cut by pegmatitic and quartz veins. The gneisses contain greenschist-facies mineral assemblages (chlorite, actinolite, epidote) indicative of retrogression; the dominant banding dips east and carries an ESE-plunging mineral and extension lineation.

The gneisses are probably bounded to the west by a normal fault that separates them from upstanding outcrops (1C) in the central part of the bay of a green chlorite phyllonite (Fig. 12.2). These phyllonites are identical to the 'Oystershell Rock' identified within the mylonite belt of the Moine Thrust Zone at Loch Eriboll (see transect 4, Excursion 11). They contain numerous lunate quartz segregations and pervasive shear bands that indicate a top-to-the-west sense of displacement parallel to an E-W-trending lineation that is particularly well developed in the quartz-bearing layers.

Walk towards 1D, the rocky headland at the NW limit of the beach (Fig. 12.2). Look up towards the cliffs to the left to see further outcrops of the Oystershell Rock. These contain thin mylonitized pegmatitic veinlets; a set of intrafolial isoclinal folds can be identified as well as later folds that deform the mylonite fabric. The structurally lowest rock unit on the headland is rather fractured, pink-purple weathering Durness Limestone. This is separated by a gently-inclined detachment fault from a 2-3m-thick slice of quartz mylonite, that is itself overlain by another low-angle fault, above which is the Oystershell Rock. Walk up onto the headland to examine these detachments in detail, and then follow them around to the west into the next bay. Note that the low-angle faulted contacts are very much disrupted by the effects of later carbonate veining, located mainly in the footwall of the lowermost detachment, and also due to offsets along numerous, steeply-dipping normal faults (Hippler & Knipe, 1990). Care should be taken on this path if conditions are wet. An alternative route into this bay is to retrace the route back to and up the wooden steps, and walk westwards along the

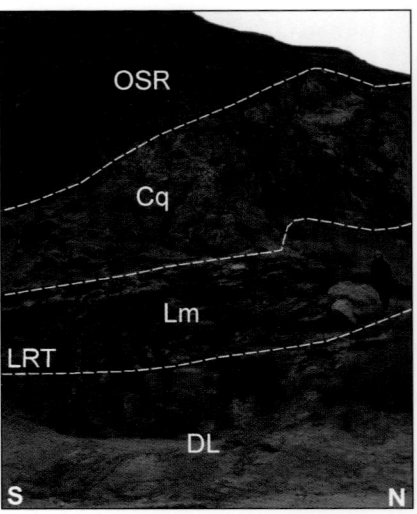

Fig. 12.4 View of cliff at Locality 12.1E at NW end of Sango Bay, showing Moine Thrust Zone rocks Oystershell Rock (OSR), Cambrian quartzite (Cq), Lewisian (Lm) and Lochan Riabhach Thrust (LRT) directly overlying totally unmylonitized, autochthonous Durness Limestone (DL) in the uppermost part of the Caledonian foreland sequence.

main cliff-top, parallel to the boundary fence of the campsite. The interpretation favoured here is that the lowermost detachment corresponds to the Lochan Riabhach Thrust that is thought by Holdsworth *et al.* (2006) to underlie the mylonite belt at Loch Eriboll (see transect 4, Excursion 11) and interpreted by them as a separate structure from the Moine Thrust exposed at a higher structural level at Faraid Head (see below). The alternative view is that the lowermost thrust does in fact correspond to the Moine Thrust (R. W. H. Butler, *pers. comm.*).

Locality 12.1E (Fig. 12.2) is bounded to the SE by a steep face of Durness Limestone that is thought to lie very close to another NNE-trending normal fault. Slickenlines on the fault pitch steeply. Carbonate fault breccias are extensively developed in places and there is abundant evidence for carbonate veining and cementation. Low tides expose further outcrops of Oystershell Rock (Fig. 12.2). A small headland at the back of the beach exposes brecciated and fractured Durness Limestone that is overlain by the Lochan Riabhach Thrust (Fig. 12.4). However, at this locality the rocks that immediately overlie the thrust appear to be a slice of Lewisian-derived mylonite. These are in turn overlain by quartz mylonites and Oystershell Rock. Walk further west to see a steep face of Durness Limestone cut by NE-dipping brittle faults. White-weathering quartz mylonites (presumably underlain by the Lochan Riabhach Thrust) are visible in the upper parts of the cliff, but are inaccessible.

Return to the top of the cliff and, following the fence, walk approximately 500m NW along the cliffs of Durness Limestone towards Locality 12.1F which lies south of Geodha Brat (Fig. 12.2; [NC 404 685]). The steep cliffs here define the trend of the WNW-ENE-trending Faraid Head Fault that down-faults the main Faraid Head outlier of Moine rocks and the Moine Thrust Zone mylonites that outcrop to the north. Most of the cliff comprises variably brecciated Durness Limestone, but at its western end where the upper parts of the cliff can be accessed from the sand dunes, carbonate-cemented red sandstone-breccia infills are preserved in a series of sub-vertical fractures trending parallel to the main fault. These are virtually identical to those exposed at Locality 12.1A, and are also thought to represent sedimentary material that has infilled tectonically open fractures in the limestone formed synchronously with normal faulting activity. The dominant clast types are limestone, but clasts of Moine psammite, mylonitized Lewisian gneiss and quartzite mylonite are also present. At least two units of infill are recognized based on differences in grain-size and sorting.

The ages of the sedimentary infills at Localities 12.1A and 12.1F – and hence the age of extension – are unknown, but a Devonian (Old Red Sandstone) or Permo-Triassic age seems likely, given the timing of sedimentary basin formation in the West Orkney Basin that lies immediately offshore and to the north (Coward & Enfield, 1987; Enfield & Coward, 1987; Stoker *et al.*, 1993). Detailed studies of the normal faulting along the north coast in the Durness-Cape Wrath area (Beacom, 1999; Wilson *et al.*, 2010) suggests that the NNE- and WNW-trending normal faults are likely to be contemporaneous, forming a complex transfer zone that defines the southern margin of the West Orkney Basin. Return to the vehicles, retracing your steps along the cliff-top.

Locality 12.2 [NC 3925 6965 to NC 3785 7135]

Faraid Head (Fig. 12.5). Down-faulted Moine, Lewisianoid basement and 'Oystershell Rock'.

Drive a few hundred metres west and at the sharp bend in Durness village take the minor road signposted to Balnakeil. Parking is available for coaches and cars on the south side of Balnakeil Bay at [NC 3915 6870]. Walk northwards across the beach to exposures of Moine psammites at 2A [NC 3925 6965]. These carry a strong mylonitic fabric that dips shallowly

EXCURSION 12

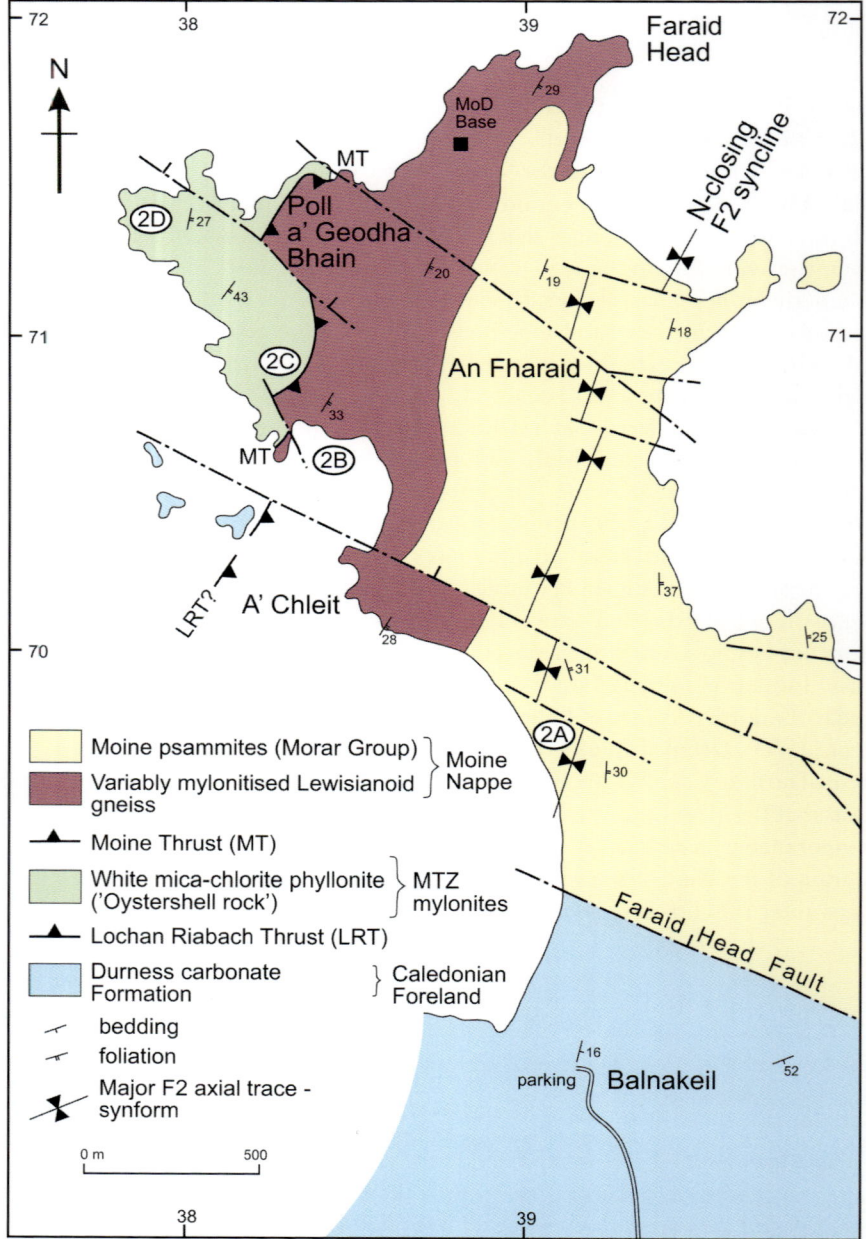

Fig. 12.5 Geological map of Faraid Head showing Locality 12.2 (modified from Holdsworth et al., 2007). LRT = Lochan Riabhach Thrust.

to the ESE and is associated with a down-dip mineral and extension lineation. The mylonitic fabric is axial-planar to rare F_2 isoclinal folds that plunge ESE sub-parallel to the mineral lineation. Thin layers of garnetiferous semi-pelite contain shear bands that indicate a top-to-the-WNW sense of displacement parallel to the lineation. Rare lenticular bands of garnetiferous pelite up to 10cm thick carry a quartz segregation fabric that is folded around the hinges of F_2 minor folds; garnets up to 7-8mm in diameter are strongly wrapped by the S_2 fabric that is axial planar to these folds.

Walk northwards to beach outcrops at 2B [NC 3855 7070]. If tide is high, good outcrops are also present above the high water mark. The spectacular exposures here are of basement-derived mylonites that exhibit a wide variety of features typical of mid-crustal shear zones. The mylonites vary from creamy-pink types derived from acid gneiss to strips and pods of chlorite-actinolite schist that may represent boudinaged and highly retrogressed amphibolites. Relict gneissic layering is represented by colour banding in the acid types. The mylonite fabric dips to the ESE and a strong lineation plunges down-dip. Classic examples of shear criteria such as shear bands, asymmetrically-wrapped porphyroclasts and boudins all indicate a top-to-the-WNW sense of displacement parallel to the lineation. Locally, cm-scale, close to tight minor folds are present plunging at low to moderate angles to the mineral lineation. Numerous quartz-chlorite-epidote veins are preserved; early types are concordant and mylonitic, later types cross-cutting, often in boudin necks, and little deformed. In more feldspathic units, mylonite is associated with pale yellow-green, epidote-rich cataclasite seams, many of which are concordant with the foliation. These are examples of semi-brittle behaviour typical of greenschist facies fault rocks in which feldspar-rich layers deform in a brittle fashion whilst adjacent quartz and phyllosilicate-rich layers undergo dynamic recrystallization. In places just above the high water mark to the west, strain is less intense and the mylonites resemble more closely the Lewisianoid rocks at the same structural level at Sango Sands (Locality 12.1B).

Traverse inland to 2C [NC 3825 7084], a series of crags composed of the Oystershell Rock just below the unexposed trace of the Moine Thrust. Gently-dipping Oystershell Rock includes numerous deformed quartz veins; a strong lineation plunges to the ESE. Pervasive shear bands and 10cm-scale shear zones again indicate a top-to-the-WNW sense of displacement. The Oystershell Rock is notable here for the presence of 10-15cm-thick bands of brown-weathering marble that are continuous for several metres in some cases. F_3 S-folds of the mylonite fabric verge northwest to north-hinges are markedly curvilinear, plunging between the northeast and east. A 10cm-thick E-W-trending, steeply-dipping basic dyke of possible Permo-Carboniferous age cross-cuts the upper part of these crags at [NC 3828 7093].

Walk 700m NW across the raised beach in which isolated crags of Oystershell Rock are exposed until the far end of the headland is reached at Locality 12.2D [NC 3785 7135]. Look ENE towards the Ministry of Defence buildings and the steep cliffs on the northern coast of Faraid Head to view the ductile Moine Thrust. This is the flat-lying boundary exposed in the western cliffs of Poll a Geodha Bhain that separates stripy, multi-coloured Lewisianoid mylonites in the hanging-wall from more uniform dark-green-grey Oystershell Rock in the footwall. If time permits, trace this boundary south to the coast at [NC 3725 7080] where this boundary is exposed. Note, however, that it is difficult to distinguish precisely the location of this contact as it separates mylonites that are both derived from Lewisianoid protoliths.

Return to the vehicles, noting the excellent views to the west of Cape Wrath, and to the south of the east-dipping Durness Limestone succession on the south side of Balnakeil Bay.

Excursion 13
North Sutherland

Rob Strachan, Bob Holdsworth,
Clark Friend, Ian Burns and Ian Alsop

Purpose: A general traverse across the Caledonian thrust nappes that outcrop between the Moine Thrust and the sedimentary cover of the Devonian Orcadian Basin.

Aspects covered: Various metasedimentary lithologies, metamorphic minerals and migmatites, Lewisianoid basement gneisses, Precambrian and Lower Palaeozoic (meta)igneous intrusions, Caledonian ductile structures, late to post-Caledonian brittle faults, Devonian and probable Permo-Triassic sedimentary rocks.

Useful information: Hotel and B&B accommodation and camping are available at Talmine, Tongue and Bettyhill.

Maps: OS: 1:25,000 sheets 446 Durness & Cape Wrath, 447 Ben Hope, Loch Loyal & Kyle of Tongue, 448 Strath Naver & Loch Loyal, 449 Strath Halladale & Strathy Point; BGS: 1:50,000 sheets 114W Loch Eriboll, 114E Tongue and 115W Strathy Point.

Type of terrain: Rocky coastline, moorland, quarry and roadside exposures.

Distance and time: The excursion is best followed from a base in either Tongue or Bettyhill, taking 4 days. See each locality for suggested times.

Short itinerary: Localities 13.1, 13.4 and 13.8 could be accomplished in one day.

The traverse along the well-exposed north coast of Sutherland (Fig. 13.1) provides the best opportunity to examine the complex regional structure of the Moine and Naver nappes, as well as a number of the Archaean basement inliers that crop out in the area. Particular points of interest include the nature of the relationships of these inliers with the Moine, and the evidence for their pervasive reworking during the Caledonian orogeny. These basement inliers have often been correlated with the Lewisian gneisses of

EXCURSION 13

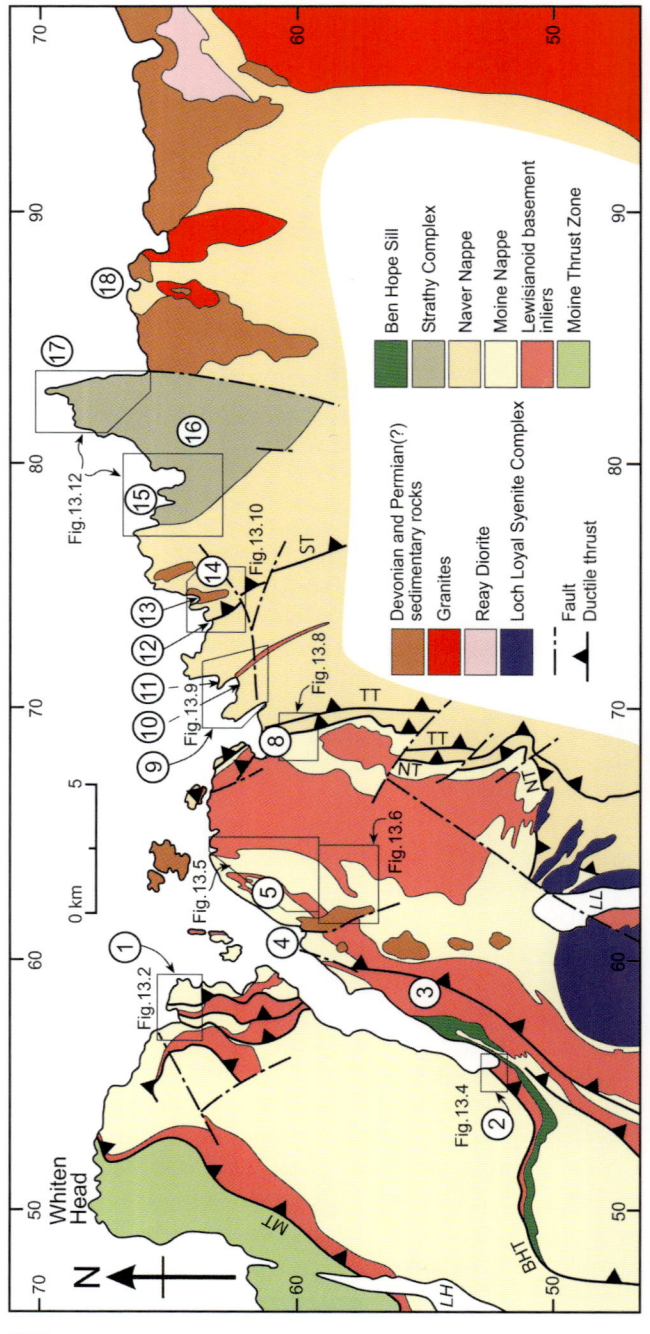

Fig. 13.1 Simplified geological map of north Sutherland together with the localities for the excursion.

LL = Loch Loyal; LH = Loch Hope;
MT = Moine Thrust; NT = Naver Thrust;
ST = Swordly Thrust; TT = Torrisdale Thrust.

232

the foreland, although unambiguous correlation remains to be demonstrated. Accordingly, use of the less specific term 'Lewisianoid' is employed here to denote basement that is of similar age and lithology to that of the foreland. Both the basement inliers and the Moine have undergone a complex polyphase deformation history. Complications in attempting to erect a consistent chronology across the traverse arise from the recognition that in the Moine Nappe the earliest structures and metamorphic fabrics recognizable in the field are Neoproterozoic in age, whereas in the Naver Nappe they are Ordovician (Grampian). It is assumed that Neoproterozoic fabrics were originally present within the Naver Nappe, but were extensively reworked during Ordovician high-grade metamorphism and migmatization (Table 1). The Grampian event has so far not been recognized within the Moine Nappe. In the deformation chronology presented in Table 1, we therefore recognize D_1 (Moine Nappe) and D_{1N} (Naver Nappe) events that are of different ages. Both thrust nappes record a similar polyphase Silurian (Scandian) history: sets of structures assigned to D_2, D_3 and D_4 episodes are thought to have resulted from a continuous, progressive deformation.

Locality 13.1   [NC 580 643 to NC 5850 6501]

The Melness area (Fig. 13.2). A comparison of low and high strain Moine metasedimentary rocks of the Moine Nappe; Lewisianoid basement rocks; amphibolite of the Ben Hope Sill Suite; the Strathan basal conglomerate.

From Tongue, turn north off the A838 on the west side of the Kyle of Tongue, continue through Talmine for 1.3km and park where the road forks [NC 580 643] just south of Loch Vasgo. Parking is available for a coach or four to five cars. Allow 4 hours for this locality.

Examine the low cliffs (1A, Fig. 13.2) to the east of the fork. Here, relatively low-strain, cross-bedded gritty psammites are folded into a mesoscopic recumbent, north-facing D_2 fold pair that plunges sub-parallel to the ESE-trending L_2 lineation. The foresets dip north, sedimentary transport is from the south, and the lower limb is right-way up. Variations in the state of strain can be estimated by measuring the angles of cross-beds around the fold. The apparent strain increases dramatically on sections parallel to L_2. These psammites are interpreted to rest with modified unconformity upon the basement rocks of the Achinahaugh inlier that are

Table 1: Table of structural events

	MOINE NAPPE	NAVER NAPPE
SCANDIAN (c.435-425 Ma)	Emplacement of the **Loch Loyal Syenite Complex**. D_4 formation of the transpressive Torrisdale Steep Belt, as well as the major synform-antiform pair within the Kirtomy gneisses and the Strathy Complex. D_3 tight-to-open, asymmetric recumbent folding. D_2 NW-directed ductile thrusting (e.g. Swordly, Naver and Ben Hope thrusts) accompanied by widespread tight to isoclinal interfolding of Moine and Lewisianoid basement; recumbent folds commonly curvilinear about NNW- to NW-trending mineral and extension lineation.	
GRAMPIAN (c.470-460 Ma)	Not recorded.	D_{1N} upper amphibolite facies metamorphism and migmatization; minor tight development of S_{1N} foliation and N-S-trending L_{1N} mineral and extension lineation.
LATE NEOPROTEROZOIC	Intrusion of alkaline **Loch a' Mhoid Metagabbro Suite**.	Not recorded.
KNOYDARTIAN? c.820-800 Ma	D_{1M} garnet-staurolite grade metamorphism, S_{1M} foliation, minor isoclinal folding.	Amphibolite facies metamorphism and fabric development inferred but not observed due to intensity of Grampian overprint.
EARLY NEOPROTEROZOIC (c.1000-870 Ma)	Deposition of Moine sediments in an extensional basin within the Rodinia supercontinent, accompanied by emplacement of mafic igneous intrusions such as the **Ben Hope Sill**.	

NB: (1) It is envisaged that D_2-D_4 structures formed essentially as a continuum, accompanied by amphibolite facies metamorphism – garnet and hornblende stable; (2) D_2 and D_3 structures are generally only found in structurally lower parts of the Naver Nappe; (3) deformation was accompanied by intrusion of a range of granite and pegmatite sheets, particularly within the Naver Nappe and in the vicinity of the Naver Thrust.

North Sutherland

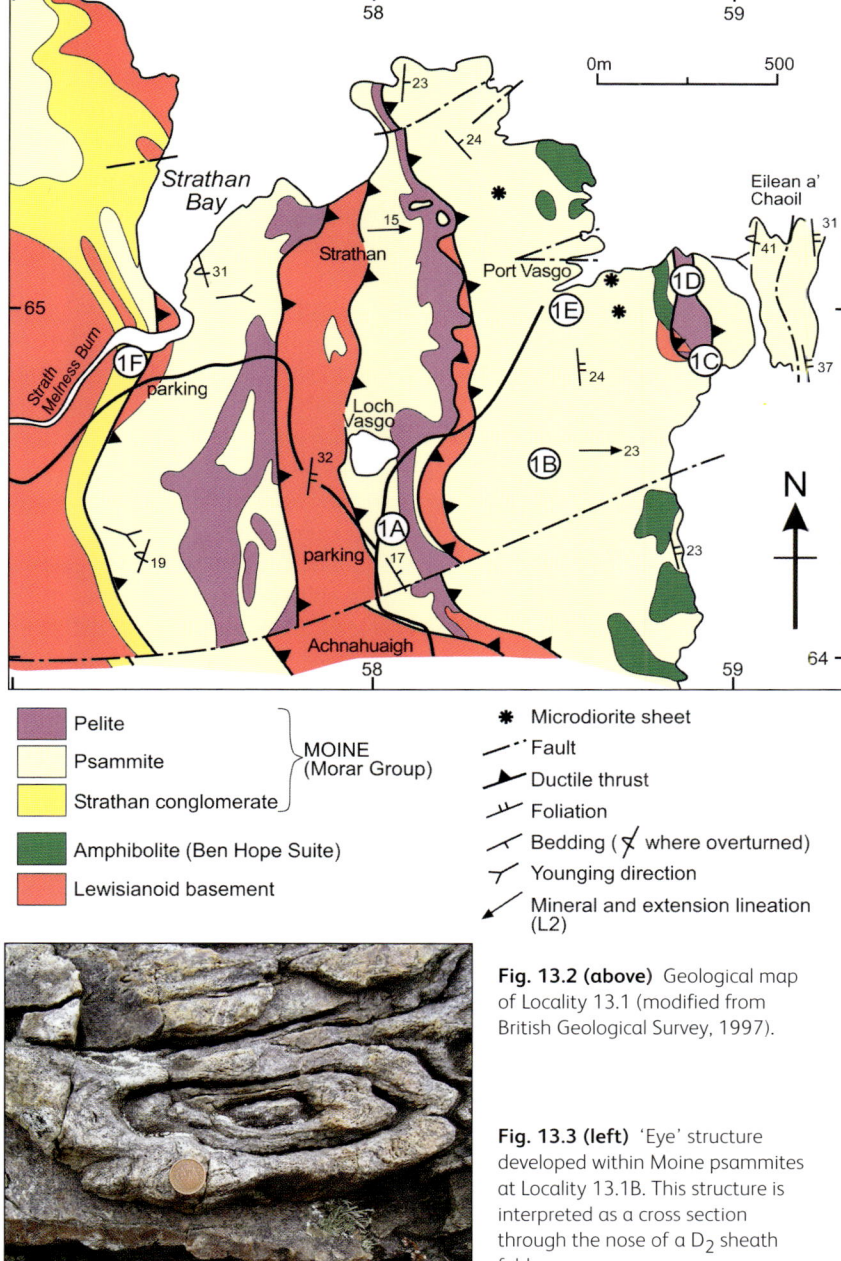

Fig. 13.2 (above) Geological map of Locality 13.1 (modified from British Geological Survey, 1997).

Fig. 13.3 (left) 'Eye' structure developed within Moine psammites at Locality 13.1B. This structure is interpreted as a cross section through the nose of a D_2 sheath fold.

235

exposed on low-lying outcrops immediately west of Loch Vasgo (Alsop & Holdsworth, 2004a, figure 6). On top of the cliffs, strain and white mica content increase towards a minor ductile thrust that cross-cuts the fold pair. A prominent pelite horizon above the thrust contains garnets up to 5mm in diameter. Small inclusion trails within these porphyroblasts define S_{1M}; the garnets are wrapped by the dominant S_2 mica fabric and are associated with pressure shadows that are elongate parallel to L_2. The evidence thus suggests that garnet growth occurred post-D_{1M} and pre-D_2. Formation of the S_{1M} fabric and garnet growth are tentatively assigned to the Knoydartian orogenic event. Quartz veins are lineated and elongate parallel to L_2. Shear bands within the pelites indicate a top-to-the-west sense of displacement parallel to the lineation. S_2 and the quartz veins are folded by cm-scale, open to close, asymmetric D_3 folds (e.g. [NC 5811 6431]).

From here, head northeastwards across the intermittently well exposed, undulating ground. Spectacular rodded quartz veins are present at [NC 5820 6440] on the west side of a large hummocky outcrop. Traverse across psammites and occasional pelite bands, displaying good examples of D_2 and D_3 folds. 1B (Fig. 13.2, [NC 5847 6455], 25m on a bearing 195° from the main summit cairn) is marked by a small cairn above a west-facing, 3m high surface displaying D_2 and D_3 folds, including numerous spectacular 'eye' structures (Fig. 13.3) that represent cross-sections through the noses of D_2 sheath folds (i.e. not interference patterns). *Please do not hammer these localities*. Detailed analyses of both D_2 and D_3 folds in this area are presented by Alsop & Holdsworth (1999, 2002, 2004a & b). A little further to the east (25m on a bearing of 140° from the main summit cairn) another west-facing surface marked by a small cairn [NC 5851 6454] displays various D_2 eye structures that are refolded by tight, asymmetric D_3 folds. Then walk northeastwards downslope towards the coast.

1C [NC 5892 6487] is by the beach on the west side of the gully (Fig. 13.2), where pelites contain prominent 1-2cm diameter garnets. These show obvious zonation with orange cores (some with aligned inclusion trails) and darker rims wrapped by S_2. Prominent shear bands indicate a top-to-the-west sense of displacement parallel to the locally developed L_2 lineation. To the east of the gully, above a thin garnet amphibolite of the Ben Hope Sill Suite, east-dipping surfaces of highly deformed psammites display intersections of S_2 with S_0/S_{1M} curving about the E-W-trending L_2 mineral and extension lineation. This reflects the progressive rotation of fold hinges during west-directed D_2 thrusting. Quartzofeldspathic clasts within pebbly horizons show a marked elongation parallel to the intersection lineation. Still further to the east, the L_2 mineral lineation disappears, leaving a uniformly N-S-trending stretching direction thought to represent a relict L_{1M} lineation (Holdsworth, 1989a).

Head northwestwards, over inclined surfaces of garnet pelite, with isoclinally folded and rodded quartz veins elongate parallel to L_2. Further excellent examples of garnets wrapped by S_2, showing cores and rims and inclusion trails, are present at [NC 5885 6507]. At 1D [NC 5884 6504], at the south end of the gully (Fig. 13.2), an outcrop-scale brittle detachment fault cuts across upright, brittle kink folds of fabric in its footwall. On the west side of the gully are steeply-inclined slabs of garnet amphibolite of the Ben Hope Sill Suite. Garnets are wrapped by the S_2 fabric and some have pressure shadows that are elongate E-W parallel to L_2. Walk westwards across undulating ground towards Port Vasgo. At [NC 5867 6499], pause to look ENE to view steep rock slabs that show mesoscopic tight to isoclinal D_2 folds.

Follow the path down into Port Vasgo (Fig. 13.2) to a stony beach at 1E [NC 5850 6501], where NE-dipping, mylonitic psammites define a zone of high D_2 strain. If tides are low, cross the rocks on the east side of the bay to access a pale-weathering lenticular pod of psammite associated with a microdiorite sheet exposed on a SW-dipping cliff at [NC 5862 6508; see Holdsworth *et al.* 2001a, plate 8). The pale eye-shaped area of psammite is a low-strain augen preserving cross-laminations folded by D_2 folds that appear to be cross-cut by a little-deformed, fine-grained sheet of microdiorite, with well-defined chilled margins. On careful inspection, however, it is apparent that the sheet pinches out and passes laterally into bedding-parallel shears that locally preserve irregular isolated pods of microdiorite, often in equivalent small-scale low-strain augen. Other sheets of microdiorite can be traced laterally into the high-strain, platy psammites where they become schistose and carry fabrics parallel to S_2 and L_2 in the adjacent psammites. Although clearly still discordant in both anticlockwise and clockwise senses, they appear to represent intrusions that were emplaced as curviplanar units that have subsequently been deformed during D_2. Similar microdiorite intrusions are common around the Kyle of Tongue and have been grouped as the Port Vasgo Microdiorite Suite by Holdsworth *et al.* (2001a) who interpreted them as being syn-tectonic with respect to the D_2 deformation. It is equally possible, however, that the intrusions pre-date D_2, with some or all of the folds occurring close to the discordant sheets representing flanking folds (Passchier, 2001). If the intrusions are pre-D_2, they may be the same age as the Loch a' Mhoid Metagabbro Suite (see Locality 13.6).

Retrace your steps back to the beach, and follow the road up out of Port Vasgo back to the vehicles. Take the left fork and drive to Strathan Bay (Fig. 13.2), parking at the east end of the crash barrier at [NC 5743 6480]. Walk down the slope to the shore on the south side of the bay to 1F [NC 5734 6489] to view exposures of the Strathan Conglomerate (Mendum, 1976), interpreted as a basal Moine conglomerate overlying Lewisianoid basement (Holdsworth, 1987, 1989a; Holdsworth et al., 2001a). The conglomerate lies within a high strain zone and contains numerous highly flattened 'clasts' that are up to 30cm in length and elongate parallel with L_2. 'Clast' types are, in order of decreasing frequency and degree of strain: pale grey quartzite, fine to medium-grained quartzo-feldspathic gneiss, white quartz, dark grey quartz-magnetite rock, and granite/pegmatite. It is doubtful, however, that all these types were genuinely incorporated as clasts within the Moine sediments prior to D_{1M}. In low strain zones, the grey 'quartzite' and white quartz 'clasts' appear to represent tectonically disrupted, isoclinally folded metamorphic segregations (compare with the segregated pelites and possible conglomerate layers seen at Kinloch Broch, Locality 13.2 below). According to Holdsworth (1987), the only lithologies that can be identified confidently as pebbles in areas of low strain are those of quartz-magnetite rock and pink granite/pegmatite which typically display lower amounts of strain. At low tide and low sand levels, highly strained hornblendic material within the conglomerate could represent early infolds of the Lewisianoid basement. Upstream, the conglomerate is underlain by flaggy, pink acidic basement gneisses that are interleaved with mafic sheets at [NC 5725 6483].

Locality 13.2 [NC 5500 5282 to NC 553 531]

Kinloch Broch (Fig. 13.4). The D_2 Ben Hope Thrust, highly strained Lewisianoid basement and Moine psammite, complex D_2 and D_3 folds.

Two to three hours should be allocated to this locality, at the south end of the Kyle of Tongue. Parking is available for a coach or several minibuses in the small gravel pit [NC 5516 5337], 150 m east of the Allt Ach' an t-Strathain.

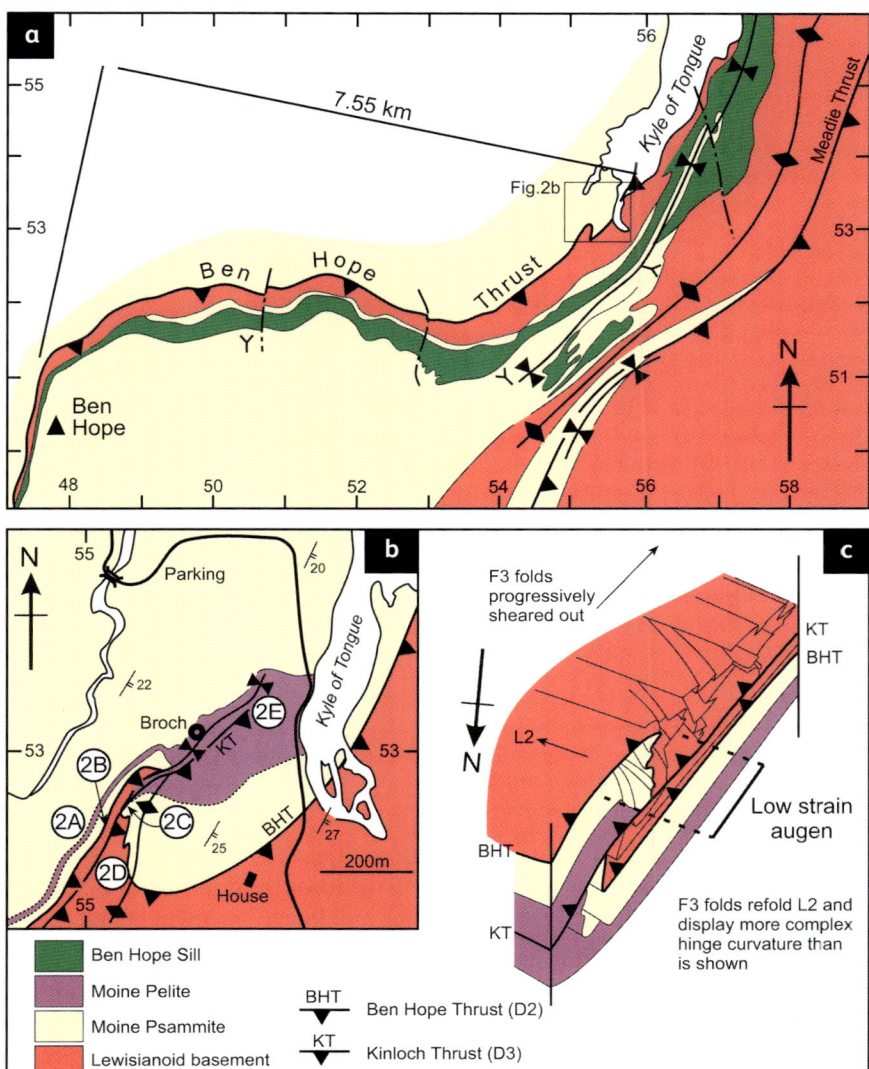

Fig. 13.4 Locality 13.2.
(a) the Ben Hope Thrust and basement-cover relations in the Ben Hope to Kinloch area of the Moine Nappe;
(b) detailed map of the Kinloch area;
(c) interpretative sketch of D_3 structures in the Kinloch area (from Moorhouse et al., 1988).

Follow the burn for 550m SSW, until it turns sharply west [NC 549 629]. Traverse ESE uphill for 100m, noting the flaggy to platy, highly strained, sparsely gritty psammites. At 2A (Fig 13.4b, [NC 5500 5282]), thin (1m) units of what has been interpreted as an intensely deformed, intraformational conglomerate are interbanded with psammites a few metres below a laterally extensive pelite (cf. stop 1A). The pebble 'clasts' are mainly quartzose, resembling closely those in the Strathan Conglomerate (stop 1F), and hence similar questions arise as to their true origin. Psammite adjacent to the overlying pelite displays a marked increase in white mica content. Traverse ESE for 30m over extremely platy psammite, containing the ESE-trending L_2 mineral lineation and isolated small, strongly flattened feldspar clasts, elongated parallel to this fabric. At the top of the slope at 2B [NC 5503 5282], these psammites are overlain by hornblende-biotite schists of the Kinloch basement sheet; this boundary is the D_2 Ben Hope Thrust (Figs 13.4a, 13.4b). This thrust, and the overlying basement sheet, has been traced for at least 25km (Fig. 13.1) from south of Ben Hope to Strathan Bay where it is associated with the zone of intense D_2 strain affecting the Strathan Conglomerate (Locality 13.1F). The marked asymmetry across this D_2 thrust contact is demonstrated by the main (pre-D_{1M}) Ben Hope Sill intrusion occurring above the accompanying basement body, but never below the lower contact with the platy psammites (Fig. 13.3a). The outcrop pattern of the Ben Hope Thrust suggests a minimum WNW displacement of 7.55 km (Fig. 13.4a).

Walk NNE for some 150m towards the summit of the hill to 2C. Here a number of open to isoclinal, south- to ESE-plunging folds form a 'Z' geometry pair, distorting the lower Moine-basement contact (i.e. the Ben Hope Thrust) and clearly refolding the D_2 platy fabric and associated lineation. Thus these folds belong to the local third phase, D_3. Note the apparent repetition of the upper pelite and lower basement contacts (Fig. 13.4b) by a thrust which must lie within the pelite to the north (Figs 13.4b, 13.4c) as the lower boundary of the pelite is unaffected by thrusting or the D_3 folding. The common limb of the fold pair is a zone of relatively low strain where D_{1M}-D_2 refolded folds occur in basement rocks [NC 5507 5288] along with migmatitic fabrics and ultramafics pods; 50m further south at 2D [NC 5507 5282], D_3 folds display up to 100° of curvature of their hinges and clearly deform the L_2 lineation.

Traverse SSW for 200m within the basement gneisses. The D_3 folds tighten, becoming progressively smeared out into the foliation and indistinguishable from D_2 structures. The exposed D_3 fold pair appears to form the southern part of a large fold pair of the same age that initially overturned WNW before being modified into highly curved sheath fold geometry by continued shearing (Holdsworth, 1990). The northern part of this structure has been largely removed by erosion, but a remnant of an intensely curved synformal infold of psammite within pelite is still preserved to the northeast at 2E (Figs 13.4, 13.4c) [NC 553 531]. Within the Moine Nappe, many D_3 folds are similarly associated with zones of high D_2 strain and so may be genetically related to ductile thrusting processes (Holdsworth, 1990; Alsop & Holdsworth, 2007). This implies that a direct correlation of all D_3 structures may not be valid, with the structures forming at various times during a protracted ductile displacement event.

Locality 13.3 [NC 5618 5258 to NC 5860 5455]

Ribigill (Fig. 13.1). Lewisianoid basement gneisses of the Ribigill inlier, the metabasic Ben Hope Sill.

This locality comprises three short stops (each 15-30 minutes), all close to the road between Kinloch Broch and Tongue. Driving from south to north, first of all stop at 3A [NC 5618 5258] where parking is available for three to four cars at intervals along the road. Examine the crag exposures immediately east of the road. The lowest outcrops are of banded Moine psammite with locally well-preserved, inverted cross-bedding. The beds are deformed by mesoscopic, tight to isoclinal D_2 folds with a strong axial-planar S_2 fabric, as well as asymmetric, open D_3 folds. Both sets of folds plunge broadly parallel to L_2. The crags upslope expose planar banded gritty psammites; ductile strain is apparently higher as no sedimentary structures are preserved. Their contact with overlying platy, banded hornblendic gneisses of the Ribigill East basement inlier is concordant and interpreted as a tectonically modified unconformity on the inverted limb of a major D_2 fold (Holdsworth, 1989a). A $^{40}Ar/^{39}Ar$ age of $c.416$ Ma obtained from muscovite within the psammites is interpreted to date cooling through a closure temperature of $c.350°C$ either during or after D_2 (Dallmeyer *et al.*, 2001).

Return to the vehicles and drive northwards to [NC 5667 5334] where roadside parking is available for a minibus or four to five cars. Head westwards over the hillside for 150m or so to find west-facing slabs of foliated garnet amphibolite of the Ben Hope Sill at 3B [NC 5660 5338]. The amphibolites carry an intense S_2/L_2 fabric; thin, concordant quartzo-feldspathic layers and quartz veins are often boudinaged. The garnets are wrapped by the dominant schistosity and show well-developed pressure shadows that are elongate parallel to the lineation, indicating that they formed pre-D_2. Inclusion trails of S_{1M} that are highly oblique to S_2 are present occasionally. The evidence therefore indicates that garnet growth occurred post-D_{1M} and pre-D_2.

Return to the vehicles and drive northwards to park in the entrance to Ribigill Quarry [NC 5860 5455]. This quarry is in intermediate, banded hornblende-biotite gneisses of the Ribigill East basement inlier. At the back of the quarry at 3C, a unit of homogeneous, finely-banded gneiss is exposed. U-Pb SHRIMP dating of zircons from this unit sampled in the main face of the quarry indicates a late Archaean age of $c.2850$ Ma for the igneous protolith (Friend $et\ al.$, 2008). Mm-cm scale concordant granitic veins may have formed during migmatization of the basement prior to intense reworking during the Knoydartian and Caledonian events. The dominant planar and linear fabrics within the gneisses are essentially parallel to the S_2/L_2 fabric in nearby Moine rocks. A $^{40}Ar/^{39}Ar$ age of $c.433$ Ma obtained from hornblende within the gneisses is interpreted to date cooling through a closure temperature of $c.500°C$ either after or during D_2 (Dallmeyer $et\ al.$, 2001). High in the quarry face, the gneisses are cut discordantly by pink granitic to syenitic veins that may have been intruded at the same time as the late-Caledonian Loch Loyal Syenite Complex some 5km to the south (Holdsworth $et\ al.$, 1999).

Locality 13.4 [NC 6124 6053]

Coldbackie Bay (Fig. 13.5). D_2 deformation in the Moine Nappe; brittle deformation, faulting, possible New Red Sandstone (Permian) conglomerates and unconformity.

Extensive roadside parking space is available by the A836 at Coldbackie [NC 612 601]. Allow 1 hour for this locality. In the roadside cutting on the south side of the road, 4A (Fig. 13.5), are well-developed mullion structures that are parallel to the hinges of prominent 'Z' geometry D_2 folds. These

North Sutherland

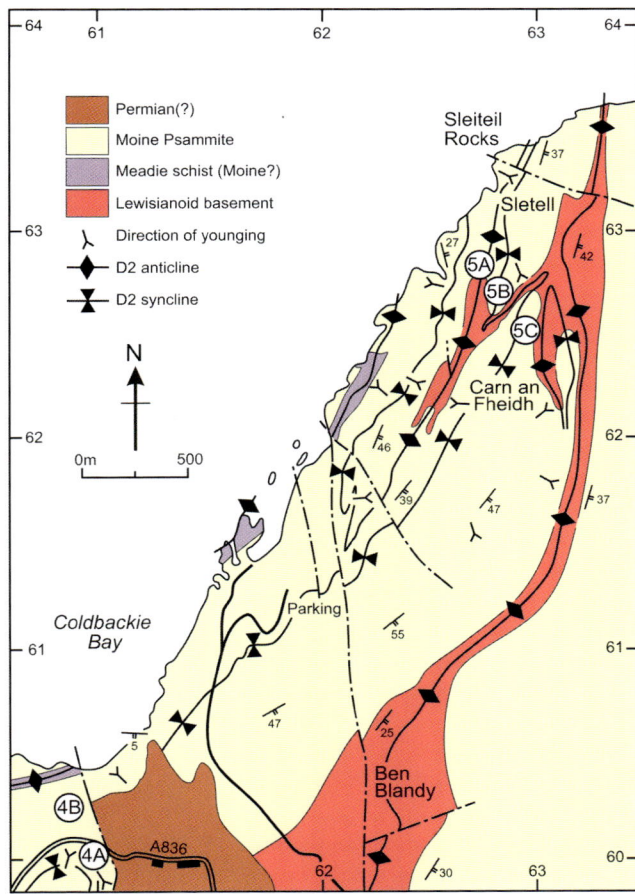

Fig. 13.5 Localities 13.4 and 13.5: basement-cover relationships in the Coldbackie Bay and Sletell area of the Moine Nappe (from Moorhouse et al., 1988).

deform a well-defined, bedding-parallel D_{1M} planar fabric. Cross-bedding is well preserved in a fold hinge close to the road surface [NC 6102 6003] and indicates that these folds face southwest and lie in the normal limb of a major D_2 synform lying just to the southeast. Some 30m to the south and east of these exposures, psammites with gritty bands lie in the inverted limb of this fold as shown by cross-bedding [NC 6105 5997] and 'S' geometry minor D_2 folds.

Descend through the gate to the beach, 4B, to examine Moine psammites with tight-to-isoclinal D_2 folds and a strongly developed S_2/L_2 fabric. On the east side of the beach, numerous folded, lineated and boudinaged pegmatites and quartz veins were clearly deformed during D_2 (e.g. [NC 6124 6053]). A $^{40}Ar/^{39}Ar$ age of $c.419$ Ma obtained from muscovite within

243

the psammites is interpreted to date cooling through a closure temperature of $c.350°C$ either during or after D_2 (Dallmeyer et al., 2001). Undeformed red conglomerates and occasional sandstone layers unconformably overlie the Moine rocks on the beach. The sediments are thought to have been deposited in a series of alluvial fans and braided channels. Rounded to subrounded clasts within the conglomerates are mostly of Moine psammites, basement gneisses and late Caledonian syenite plutons and are therefore essentially locally derived (Blackbourn, 1981). The age of the conglomerates and sandstones is uncertain, with both Devonian (Old Red Sandstone) and Permian ages proposed (see Holdsworth et al., 2001a and references therein). However, recent studies suggest that strata of both ages may be present, with the NNW-SSE bounding fault, lower conglomerate and sandstone units (only exposed inland) being of Devonian age, and the upper, syenite-bearing conglomerates (seen at this locality) being of Permian age (Wilson et al., 2010).

Locality 13.5   [NC 6281 6283 to NC 6180 6150]

Sleteil & Skullomie Harbour (Fig. 13.5). D_2 infolds of Lewisianoid basement gneisses in unconformable contact with Moine psammites preserving sedimentary structures; late folds and associated brittle detachments.

At Strathtongue on the A836, turn north on the minor road to Skullomie; parking is available for three to four cars in the lay-by at the end of the road [NC 6191 6132]. Allow 2-3 hours for this locality. Walk north through a gate, pass a house on the right and over a footbridge. Follow the footpath through a gate, turn sharp right parallel to a fence and head NE past a ruined croft and up a steep hillside to Carn an Fheidh (Fig. 13.5) [NC 6275 6285] where excellent examples of D_2 folds and sedimentary structures occur in the psammites. A further 700m north at 5A [NC 6281 6283] is a well exposed boundary with locally gritty, feldspathic Moine psammites underlying sheared hornblende, biotite and migmatitic feldspathic basement gneisses. This locality is 250m NE of a prominent coastal inlet and north of three large syenite erratics.

This boundary can be followed around the hinge of an east-plunging D_2 fold; less than 100m to the north [NC 6282 6291], Moine psammites contain poorly preserved cross-bedding younging away from hornblendic basement schists, thus the D_2 fold faces north. Follow the upper boundary of the basement towards 5B [NC 6282 6268] where the Moine psammites young away from the underlying basement schists. In this area, the psammites contain soft-sediment structures and pebbly horizons; crude fining-upwards cycles and cross-bedding indicate the direction of younging and enables the disposition and facing of the D_2 folds to be determined even though up to 180 degrees of fold hinge curvature is developed about the ESE-plunging L_2 mineral lineation. Traverse some 200m SE to 5C that lies just within the inverted limb of a syncline (Fig. 13.5). Around here, examples of D_2 'eye' structures and along-strike changes in fold plunge, sense of vergence and facing are well displayed, with hinge curvature occurring about a weakly developed, ESE-plunging L_2 mineral lineation. This inverted fold limb underlies basement schists exposed further to the east, with the fold closing to the north and facing SW (Fig. 13.5).

The Sleteil basement body lies in the core of complex, en-echelon, anticlinal D_2 folds with curved hinges that appear to close and face both to the north and to the south, and to thus form 'tongue-shaped' sheath structures that originally faced and closed upwards to the WNW; D_2 folds with sheath-like geometry are present on all scales within this area (Fig. 13.5) (Holdsworth, 1988; Alsop & Holdsworth, 1999). The boundary of the Sleteil inlier is believed to be a slightly modified Moine-basement unconformity (Holdsworth, 1989a).

Return to the vehicles, drive south for 500m and take a sharp turn right at [NC 6155 6086] to follow the track down to Skullomie Harbour. 5D [NC 6180 6150] is located along the low cliffs east of the harbour and best approached at low tide. Here the Moine psammites are deformed by a series of minor brittle-ductile folds that are related kinematically to a series of detachment faults that have both top-to-the-SSE and ESE senses of displacement. These structures are developed preferentially in belts of pre-existing D_2 strain in which the foliation is markedly flaggy with few folds developed. The generally eastward-dipping detachment faults lie either parallel to the foliation ('flats') or cross-cut at angles of up to 40° ('ramps'). Examples of ramp-flat geometries are well exposed in the cliffs behind the harbour. These structures are thought to have formed during extensional collapse of the thickened nappe pile at a late stage in the Caledonian orogeny (Holdsworth, 1989b; Holdsworth *et al.*, 1999, 2007).

Locality 13.6 [NC 6234 5858 to NC 6291 5766]

Loch Cormaic (Fig. 13.6). Complex D_2 and D_3 folds, Loch Cormaic Metagabbro.

At Strathtongue on the A836, take the minor road to Dalcharn and park at the end of the track by a gate [NC 6222 5864]. There is space for two minibuses or four to five cars. Allocate 2-3 hours for this locality. Go through the gate and ascend the small knoll on the hillside to the SE. At 6A [NC 6234 5858], banded psammites with concordant quartzo-feldspathic segregations (the onset of melting?) are deformed by numerous complex D_2 folds. These display highly curvilinear axes, resulting in closed 'eye' structures and opposing vergence patterns. Most of the fold hinges plunge to the SE, parallel to a strong mineral and extension lineation (L_2). Coaxial, asymmetric D_3 folds (Fig. 13.7) locally refold D_2 structures.

Descend and follow the track southeastwards to Loch Cormaic. Note the presence by the track just to the north of the loch [NC 6259 5841] of a tiny outlier of red conglomerate of uncertain age. If water levels allow, continue on the track on the east shore of the loch. At 6B [NC 6283 5803] by the lochside, banded psammites with some semi-pelite bands are deformed by a mesoscopic D_2 isoclinal fold pair with 'Z' geometry and a penetrative, axial-planar S_2 schistosity. Although a strong pre-D_2 fabric is commonly present, in places the strain is lower to reveal probable cross-bedding. Hillside exposures above the SE end of the loch (e.g. [NC 6288 5771]) contain examples of cross-bedding within psammites, here deformed by SE-plunging D_3 folds.

Craggy exposures to the southeast are of the Loch Cormaic Metagabbro (a member of the Loch a' Mhoid Metagabbro Suite of Moorhouse & Moorhouse, 1979). It is worthwhile first of all examining the coarse, little-deformed metagabbro that is characteristic of the internal part of the body. At 6C [NC 6291 5766] coarse, randomly oriented hornblende and plagioclase grains define a relict ophitic texture. The boundaries between plagioclase and hornblende clusters are mantled by metamorphic garnet. This may indicate that the mafic domains, now amphibole aggregates, once represented igneous clinopyroxene. Relict layering on the dcm scale is indicated by the alternation of mafic-rich and mafic-poor bands. Near the eastern margin of the body [NC 6292 5758] the metagabbro is noticeably leucocratic and garnetiferous. The igneous and meta-igneous fabrics are

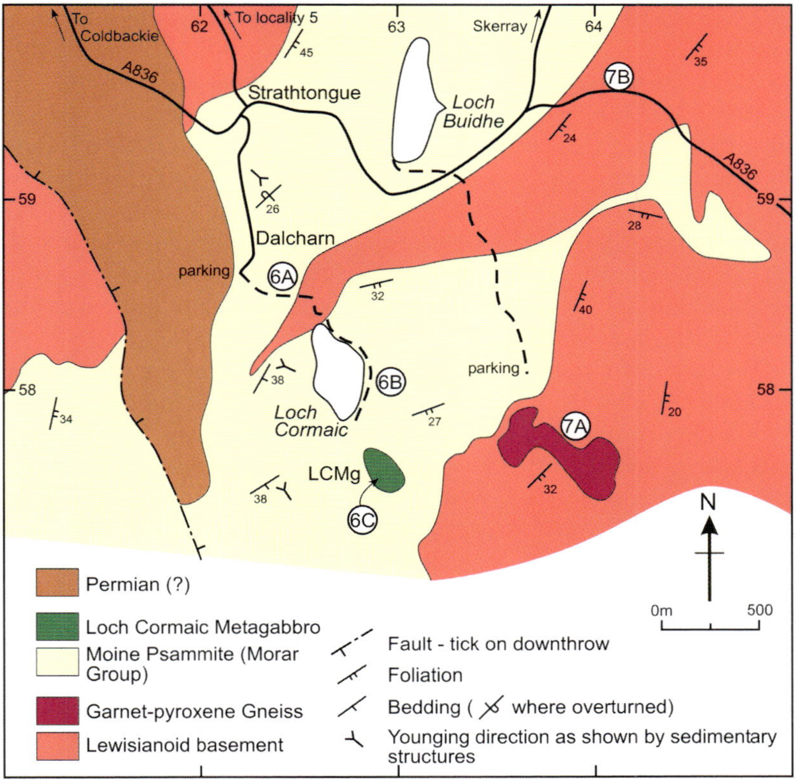

North Sutherland

Fig. 13.6 (above)
Geological map of Localities 13.6 and 13.7 (modified from British Geological Survey, 1997).

Fig. 13.7 (left)
D_3 folds deforming inter-banded psammite and garnetiferous semi-pelite at Locality 13.6A.

commonly reworked within shear zones defined by anastomosing, often curviplanar bands of hornblende and/or actinolite schist. These shear zones typically carry a strong mineral and extension lineation that plunges to the SE, parallel to L_2 in the host Moine rocks. A top-to-the-northwest sense of shear parallel to this lineation can be deduced from the sense of fabric curvature on the margins of some of the southeasterly-dipping shear zones. The margins of the metagabbro body are invariably highly strained, and the intrusion seems likely to have acted in a highly competent manner with respect to its metasedimentary host during deformation. The contact with the adjacent Moine psammites is nowhere exposed, but its location can be narrowed down to within 5-6m.

The field relations are consistent with a pre-D_2 age of intrusion for the metagabbro and it therefore forms an important marker in the regional deformation chronology. U-Pb SHRIMP dating of zircons from a leucocratic part of the Loch Cormaic metagabbro has yielded a late Neoproterozoic age that is interpreted to date intrusion and crystallization of the igneous protolith (Strachan & Kinny, unpublished data). In a regional context, the Loch Cormaic body, and by implication other members of the Loch a' Mhoid Suite, were probably intruded during the period of late Neoproterozoic rifting that resulted in the break-up of Rodinia and formation of the Iapetus Ocean (Kinny *et al.*, 2003a).

Locality 13.7 [NC 6380 5770]

Borgie peat cuts (Fig. 13.6). Relict garnet-pyroxene gneisses within the Borgie basement inlier.

One kilometre east of Strathtongue on the A836, turn south onto an untarred track at [NC 6320 5912]. There is parking for three to four cars at the end of the track. Allocate 1½ hours for this locality. Walk southeastwards across the moor to reach the low-lying outcrops around 7A [NC 6380 5770]. These expose gneisses of the Borgie basement inlier. Lenses of mafic gneiss, up to several hundred metres long, contain largely unretrogressed metamorphic assemblages dominated by garnet and clinopyroxene, indicative of at least upper amphibolite facies conditions. These are cut by narrow shear zones defined by hornblende schist, and at their margins pass imperceptibly into host hornblende gneisses typical of large tracts of the

Borgie inlier. These carry an east- to SE-dipping foliation and a SE-plunging mineral and extension lineation that is parallel to, and correlated with, L_2 in the Moine. Similar relict garnet-pyroxene mineral assemblages are preserved within the Naver basement inlier in Central Sutherland (Locality 10.6, Excursion 10), and they have been compared with the Scourian granulites of the Caledonian foreland (Moorhouse, 1976). However, in the absence of isotopic data, at the time of writing nothing precludes a significantly younger age.

Return to the cars and pay a brief visit (10-15 minutes) to the small quarry adjacent to the A836 at 7B [NC 6413 5958]. This exposes typical banded hornblendic gneisses of the Borgie inlier, intruded by a mafic sheet, now converted to hornblende schist. This has been correlated with the Scourie Dyke Suite of the Caledonian foreland (Moorhouse *et al.*, 1988), but given its state of deformation it could equally be a member of the Ben Hope amphibolite suite or a highly sheared member of the Loch a' Mhoid Metagabbro Suite. U-Pb SHRIMP dating of zircons from a gneiss sampled in the southeast corner of the quarry indicates a late Archaean age of *c*. 2850 Ma for the igneous protolith (Friend *et al.*, 2008). The strong fabric within the orthogneisses is at least partly Caledonian in age, as it carries a mineral and extension lineation that is parallel to the regional L_2 lineation within the Moine rocks. A $^{40}Ar/^{39}Ar$ age of *c*. 421 Ma obtained from hornblende within the gneisses here is interpreted to date cooling through a closure temperature of *c*. 500°C either during or following D_2 (Dallmeyer *et al.*, 2001).

Locality 13.8 [NC 6875 6089 to NC 6896 6169]

Torrisdale Bay (Fig. 13.8). The ductile Naver Thrust separating Moine Nappe psammites from high-grade meta-basic rocks and migmatitic gneisses of the Naver Nappe with Lewisianoid basement gneisses along the thrust; Caledonian granitic bodies emplaced during thrusting; development of the Torrisdale Steep Belt.

Turn off the A836 just west of Borgie Bridge [NC 6675 5872], drive northwards for nearly 3km and park by the roadside at [NC 6807 6111]. There is sufficient space for a coach; allow 3 hours for this locality. Walk down to the river, across the bridge, follow the path towards the raised beach over a

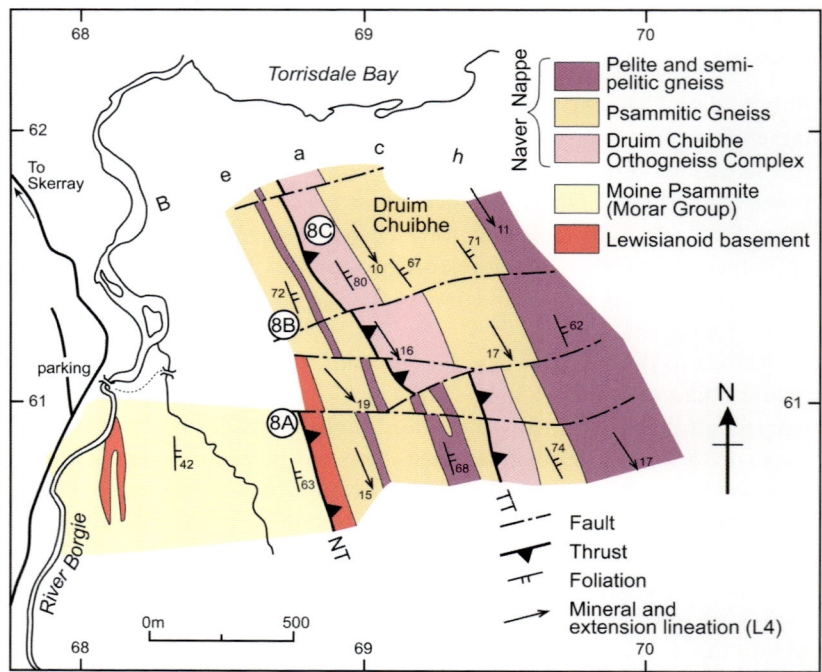

Fig. 13.8 Geological map of Locality 13.8 (from Burns, 1994). Many granite sheets have been omitted for clarity. NT = Naver Thrust; TT = Torrisdale Thrust.

second bridge, and then climb up the gorse-covered bluff in a gully with a small stream to emerge on a large flat grassy area by a wall. Head ESE to the lowest outcrops on the west-facing slope.

At 8A [NC 6875 6089] platy, high strain psammites of the Moine Nappe contain concordant, thin quartz veins and are deformed by asymmetric, minor D_2 isoclinal folds. These psammites lie in the immediate footwall to the Naver Thrust that occurs in unexposed ground along the line of the next gully uphill. The foliation dips rather more steeply eastwards than is the case at most of the localities thus far, and it carries a mineral and extension lineation that plunges gently to moderately to the SSE. The trend of the lineation does not correspond to the direction of tectonic transport during ductile thrusting. Instead, the steepening, composite fabric and associated low-angle lineation are associated with development of the Torrisdale Steep Belt (D_4) that reworks all pre-existing structures, including major ductile thrusts (Burns, 1994; Holdsworth et al., 2001a). Cross the gully containing the Naver Thrust and traverse southeastwards uphill into the next gully and then southeastwards across the next ridge to [NC 6884 6081] to view west-facing exposures of hornblendic basement gneisses that are interpreted to rest as an allochthonous sheet on the Naver Thrust. These gneisses display well-developed tight-to-isoclinal folds and 'eye' structures of probable D_2 age that are interpreted as cross-sections through the noses of sheath folds. The crags above these expose high strain, banded psammitic gneisses of the Naver Nappe, containing migmatitic layering and numerous deformed granitic and quartz veins. Gneissic layering is deformed by isoclinal D_2 and asymmetric D_3 folds. Walk downslope northwards, along the base of the slope, to the first outcrops by the sand dunes.

At 8B [NC 6874 6128] are banded, migmatitic psammitic gneisses of the Naver Nappe. Two generations of granitic rocks are apparent. A series of early, concordant, pink migmatitic leucosomes are attenuated and flattened parallel to the composite gneissic banding; the grey melanosomes that fringe melt layers are dominated by residual quartz. The migmatitic layering is cut by discordant pegmatitic veins of the Torrisdale Vein Complex (Holdsworth et al., 2001a). U-Pb SHRIMP dating of zircons from the early migmatitic phase sampled here yielded an age of 467 ± 10 Ma (Kinny et al., 1999). This is thought to date to mid-Ordovician (Grampian) high-grade metamorphism and migmatization of the Naver Nappe. An L_4 mineral and extension lineation plunges gently to moderately to the SSE.

Traverse northeastwards upslope across psammitic gneisses, including a prominent garnetiferous semi-pelite, across the (unexposed) Torrisdale Thrust and into an interbanded assemblage of strongly foliated amphibolites, hornblendic gneisses (locally with garnet) and augen granites – the Druim Chuibhe Orthogneiss Complex (Burns, 1994; Fig. 13.8). Geochemical data indicates that these lithologies are not part of the pre-Moine basement, and they are thus viewed as most likely constituting a series of possibly contemporaneous intrusions that were emplaced into the Moine rocks of the Naver Nappe. At 8C [NC 6886 6159], the amphibolites contain brown-weathering ovoid cores, 1-2m across, that correspond to a relict, anhydrous garnet-pyroxene metamorphic assemblage from which a P-T estimate of 650-700°C and 11-12kb has been estimated (Friend et al., 2000). These cores are cut by dark, hornblendic rehydration veinlets, and the margins of the cores pass imperceptibly into the host foliated amphibolites. A few similar smaller cores that are more highly retrogressed can be found along strike to the south. These high-pressure granulite-facies rocks were formed on a very different P-T trajectory to that associated with thrust-related (Scandian) deformation and metamorphism. These unusual assemblages are interpreted to be a relic of burial during early crustal thickening, perhaps during the Grampian orogenic event (Friend *et al.*, 2000). The composite gneissic fabrics are deformed by at least two sets of folds: an early set of attenuated isoclines, probably of D_2 age and later asymmetric D_3 structures. Granitic pegmatite sheets of the Torrisdale Vein Complex, ranging in thickness from a few cm to 3-4m, cross-cut the ductile planar fabrics; these sheets are commonly either folded or boudinaged during dextral shear associated with the development of the SSE-plunging L_4 lineation. Those that were apparently intruded clockwise of banding were folded, whilst those that were intruded anticlockwise of banding were boudinaged. A dextral sense of shear parallel to L_4 is apparent from asymmetrically sheared porphyroclasts and granite veins.

Traverse further upslope across extensive exposures that display a variety of intrusive relationships between the granites and host orthogneisses. The percentage of mafic material within the orthogneisses decreases; a common lithology present is a grey, rather homogenous gneiss with feldspar porphyroclasts, and interpreted as a strongly mylonitized granitoid. Refolding of D_2 isoclines and 'eye' structures by asymmetric D_3 folds is apparent at [NC 6896 6169]. The eastern boundary of the orthogneisses with interbanded Moine semi-pelitic and psammitic gneisses is diffuse and apparently intersheeted. Once the presence of Moine rocks is established, it is worthwhile pausing to view the prominent granite sheets of the Torrisdale Vein Complex seen in the cliffs on the northwest and east sides of Torrisdale Bay. These granites, as well as those encountered at outcrop at this locality, are all thought to correlate broadly with the Strath Vagastie Granite (see Locality 10.5, Excursion 10) and similarly to have been emplaced during Silurian (Scandian) displacement along the Naver Thrust. Hornblende and muscovite $^{40}Ar/^{39}Ar$ ages obtained from basement and Moine lithologies across the Torrisdale Bay section described here all fall within the range $c.420$-415 Ma, corresponding to the time of cooling through closure temperatures, most probably shortly after the formation of the Torrisdale Steep Belt and intrusion of the Torrisdale Vein Complex.

Opposite:
Fig. 13.9 Geological map of Localities 13.9, 13.10 and 13.11 (from Burns, 1994).

North Sutherland

Locality 13.9 [NC 6970 6316]

Creag Ruadh (Fig. 13.9). Strongly deformed Moine gneisses and amphibolites of the Naver Nappe within the Torrisdale Steep Belt.

If driving into Bettyhill from the west, turn left at the Post Office, and take the next turn left at a T-junction. At [NC 7025 6213] turn right and drive to the end of the track. Parking for a minibus or three to four cars is available at [NC 7005 6231]. Allow 1-1½ hours for this locality. Walk through a gate, follow the path and then walk northwestwards along the western side of the headland to reach the outcrops at [NC 6970 6316]. These are of psammitic gneisses with concordant bands of amphibolite and augen granite, intruded by numerous sheets and pods of pegmatitic granite. The composite foliation is steep and associated with a prominent L_4 lineation that plunges gently to the SSE – structures typical of the Torrisdale Steep Belt. The dextral kinematic indicators that are associated with L_4 elsewhere (e.g. Localities 13.8, 13.12) are not evident here, and the linear fabric is instead defined by rodding and mullion structures. On a steep, west-facing

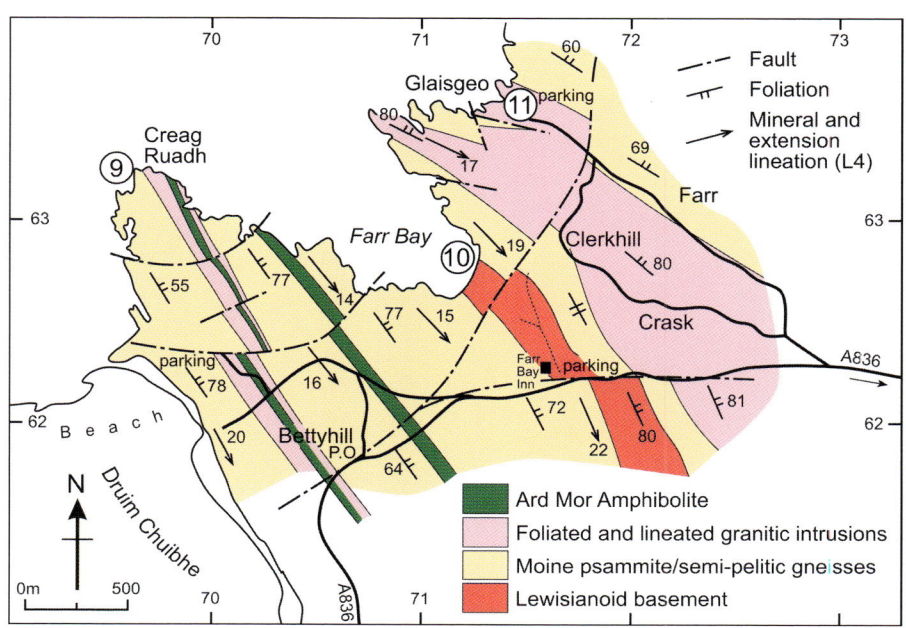

253

outcrop [NC 6971 6314], the gently curvilinear nose of a relict D_3 isocline folds a NW-trending mineral and extension lineation, probably of L_2 age. *Please do not hammer these outcrops.* Cross a small gully to the northwest to see other another example of a NW-trending lineation (L_2?) folded by a tight, D_3 fold that is itself wrapped by L_4 mullions [NC 6970 6317]. Walk out to the end of the headland and look back to the cliffs to the southeast to see prominent pink granite sheets, some folded, intruding complexly deformed gneisses. Similarly, the views to the northwest to Aird Torrisdale show large-scale intersheeting of Moine lithologies by steeply-dipping granite sheets.

Locality 13.10 [NC 7147 6265]

Farr Bay (Fig. 13.9). Farr basement inlier; Moine gneisses and intrusive granites.

Abundant parking space is available on the grassy area adjacent to the Farr Bay Inn at Bettyhill [NC 7163 6223]. Allow ½-1 hour for this locality. Walk up the narrow track alongside the parking area, go through the first gate on the right and head through the sand dunes to Farr Bay. In the east corner of the beach [NC 7147 6265], close to the sand dunes, are banded hornblendic mafic gneisses of the Farr basement inlier that is thought to occupy the core of an early isoclinal fold within the Naver Nappe (Moorhouse, 1979; Moorhouse *et al.*, 1988). U-Pb SHRIMP dating of zircons from the Farr inlier has yielded an Archaean age of *c.*2900 Ma that is interpreted to date crystallization of the igneous protolith (Friend *et al.*, 2008). Gneissic layering is deformed by early 'eye' structures (D_2?) and later asymmetric folds (D_3?); the hinges of both sets of folds are broadly parallel to a gently-plunging L_4 lineation. Low-lying (and somewhat sand-dependent) outcrops a few metres to the NNW show hornblendic gneisses and slightly discordant intrusive amphibolites tightly interfolded by asymmetric folds (D_3?). Outcrops on the east side of the bay are of steeply-dipping, banded psammitic and semi-pelitic gneisses with numerous granite sheets in varying states of deformation from concordant and mylonitic to cross-cutting and essentially undeformed. At least two sets of folds can be identified: early isoclinal folds (D_2?) and later asymmetric (D_3?) types.

Locality 13.11 [NC 7146 6360]

Glaisgeo (Fig. 13.9). Contemporaneous mafic and acid magmas deformed and metamorphosed within the Torrisdale Steep Belt.

Turn off the A836 onto a minor road 2.5km east of Bettyhill; after 250m take the left-hand fork and follow the road for 1.5km. Parking for three to four cars is available at the end of the road by the houses at [NC 7147 6265] (please ask permission). Allocate 1 hour for this locality which is best viewed at low tide. Descend the steep, grassy slope and when on the beach walk east over the first rocky ridge to access a small stony beach. Aim for the notch in a second ridge of rock to the east; scramble over this notch into another stony beach. On the east side of this beach [NC 7146 6360] steeply-dipping psammitic gneisses are interbanded with slightly discordant sheets (10-30cm wide) of a foliated horn-blende-bearing granitoid characterized by distinctive K-feldspar augen. This zone represents the eastern margin of a complex intrusive suite. The augen are interpreted as relict igneous megacrysts that have been modified by metamorphic recrystallization and associated deformation. Numerous boulders on the beach provide excellent examples of this intrusive facies. At the back of the beach are hornblendic mafic gneisses, and flat surfaces on the west side of the beach expose mafic and ultramafic streaks and schlieren within felsic hornblende gneiss. These outcrops are interpreted as deformed mixed and/or mingled magmas. Traverse westwards towards the western margin of the intrusive complex, noting the large-scale alternation of augen granite and mafic sheets, some containing screens of metasediment. Late, discordant pink granite veins and sheets intrude the meta-igneous rocks throughout the section. The steep fabric within these meta-igneous rocks is parallel to the composite fabric within the D_4 Torrisdale Steep Belt, although the associated lineation is absent, perhaps as a result of the partitioning of strain into the host metasedimentary rocks (Burns, 1994). The field relations indicate that the igneous protoliths were intruded after regional migmatization of the host Moine gneisses but prior to formation of the Torrisdale Steep Belt.

These meta-igneous rocks were first described by Cheng (1942, 1943) who thought that they belonged to the pre-Moine basement. U-Pb SHRIMP dating of zircons from a sample of augen granite from this locality has yielded a Silurian crystallization age (Strachan & Kinny, unpublished data). The intrusive suite is thus probably a pre- to syn-tectonic member of the

Caledonian 'Newer Granites'. Muscovite sampled from within Moine psammitic gneiss at Glaisgeo has yielded a ^{40}Ar/^{39}Ar age of $c.$ 419 Ma (Dallmeyer et al., 2001) and again most probably dates cooling after formation of the Torrisdale Steep Belt.

Also worthy of note at these outcrops are the NE-SW trending normal faults and calcite-filled fractures cutting these rocks. Minor structures associated with these faults are consistent with their development as a result of NW-SE extension during the Permian (Wilson et al., 2010).

Locality 13.12   [NC 7354 6355]

Swordly Bay (Fig. 13.10). Swordly Lewisianoid basement inlier; Swordly Thrust and overlying migmatitic pelites reworked within the Torrisdale Steep Belt.

Turn off the A836 onto a minor road $c.$ 5km east of Bettyhill; after 300m take the left-hand fork and follow the road for 1.25km. Park with permission by the farm buildings at [NC 7356 6307] where there is sufficient space for a minibus or three to four cars. Allocate 1 hour for this locality. Walk northwards and turn to the right of a ruined building, passing through a gap in the stone wall and then head across the grass towards Swordly Bay. Cross the fence at a small stone stile and walk down to the beach [NC 7354 6355]. Exposures on the west side are of banded, mafic hornblende gneisses of the Swordly basement inlier that is thought to occupy the core of an early isoclinal fold within the Naver Nappe (Moorhouse, 1979; Moorhouse et al., 1988). It has not at the time of writing been dated isotopically, but its basement affinities have been confirmed by detailed chemical studies (Burns, 1994). The gneissic fabric is deformed by upright, open (D_4?) folds. If the tide and sand levels are low, the eastern contact of the basement with psammitic gneisses is exposed in the centre of the bay. Walk to the east side of the bay, across the unexposed Swordly Thrust, to see the Swordly Pelite, comprising northeast-dipping, pelitic gneisses and schists with numerous sheets and pods of granitic material. This lithology is interpreted as a strongly deformed and mylonitized migmatitic gneiss; the granitic rocks are thought to represent melt layers, consistent with the local presence of garnet and sillimanite, and the abundant muscovite is most likely of retrogressive origin. A strong L_4 lineation plunges gently to the SSE. Pervasive shear

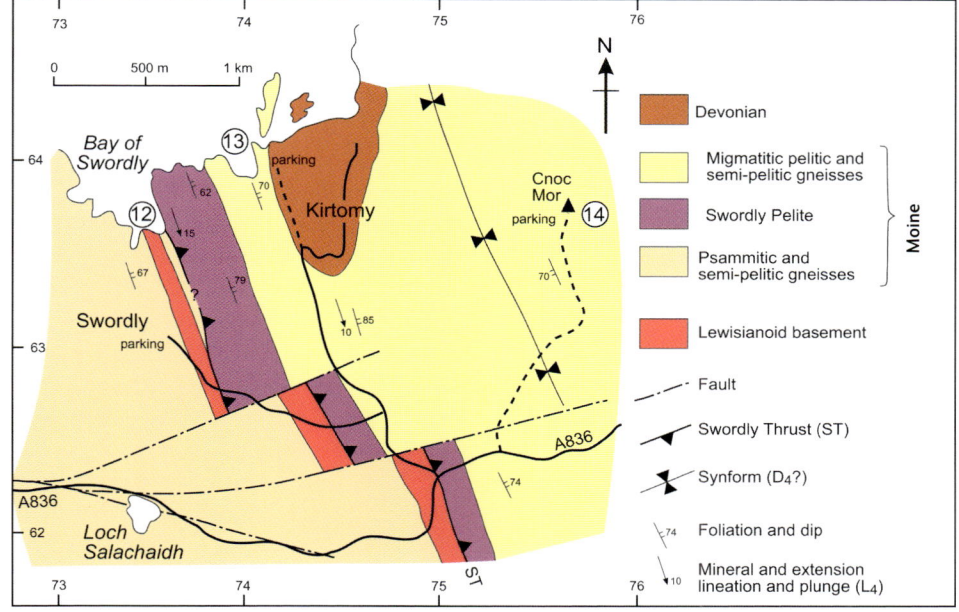

Fig. 13.10 Geological map of Localities 13.12, 13.13 and 13.14 (from Moorhouse et al., 1988 and Burns, 1994).

bands and asymmetrically sheared boudins of granitic material indicate a dextral sense of shear parallel to the lineation. On the west side of a rocky knoll at the top of the beach, a low strain zone preserves a NW-trending mineral lineation (L_2?) that is variably sheared into parallelism with L_4. This early lineation is interpreted to define the direction of tectonic transport along the D_2 Swordly Thrust prior to steepening within the Torrisdale Steep Belt. A $^{40}Ar/^{39}Ar$ muscovite age of $c.$ 423 Ma has been obtained from these mylonitic pelitic gneisses (Dallmeyer et al., 2001).

Locality 13.13 [NC 7413 6408]

Kirtomy Bay (Fig. 13.10). Moine gneisses within the Torrisdale Steep Belt and unconformably overlying Old Red Sandstone (Devonian) sedimentary rocks.

From the previous locality, drive back 1.25 km to the fork in the road, turn sharp left and follow the road towards the small hamlet of Kirtomy. After 1 km the road turns abruptly right; instead, carry on straight ahead on the grassy track and parking for a minibus or three to four cars is available at the end of the track [NC 7417 6401]. Allow ½-1 hour for this locality. **Do not** take the path to the right down to the small jetty; instead walk to the west side of the headland and descend down the grassy path on a narrow ridge westwards onto a stony beach. On the west side at [NC 7411 6399] are low polished beach outcrops of steeply-dipping banded semi-pelitic gneisses within the Torrisdale Steep Belt showing numerous shear bands and asymmetrically deformed leucosomes and melt pods that demonstrate a dextral sense of shear parallel to a gently-plunging L_4 lineation. Walk back 100 m or so to the east and climb over the bottom of the grassy ridge that you descended into the next small bay [NC 7413 6408] to view the cliffs at the back of the beach that comprise Old Red Sandstone boulder conglomerates with thin sandstone layers. Look across the small bay to the north to see the irregular Moine-Old Red Sandstone faulted landscape unconformity completely exposed where sheets of conglomerate drape steeply-dipping gneisses. Boulders within the conglomerate are mostly of Moine migmatitic gneisses of the Kirtomy migmatites (Burns, 1994) and pink granites typical of the late Caledonian 'Newer Granite' Suite. Their rounded appearance most likely indicates fluvial erosion and transportation prior to deposition.

Faults bounding the basin trend NNW-SSE and N-S (the former reactivating basement fabrics) with a system of minor ENE-WSW faults appearing to accommodate along-strike variation in throw. The sedimentary rocks are also cut by various late, brittle faults which are attributed to dextral reactivation of the basin-bounding faults, probably during the Permian (Wilson *et al.*, 2010). Before leaving this locality, take a moment to look east across the bay to see the classic half-graben structure, with conglomerate units adjacent to the bounding fault grading into more blocky, moderately-dipping sandstone units outcropping across much of the bay, which then onlap basement exposures on the far hillside.

Locality 13.14 [NC 7567 6344]

Cnoc Mor (Fig. 13.10). Moine migmatites.

Turn off the A836 *c.*400m east of the minor road to Kirtomy, onto a small tarred track that leads up to the radio and mobile telephone masts on Cnoc Mor. Park in a large space below the mast at [NC 7567 6344]. There is sufficient space for minibuses and cars. Allocate ½ hour for this locality. Walk up the road and examine the first outcrops on the east side. These are of banded migmatitic gneisses with substantial layers of anatectic melt (Fig. 13.11). Some melt layers are concordant, whereas others are clearly discordant and have probably migrated locally from their source. Similar migmatitic gneisses along strike to the NNW at Kirtomy Point have yielded a U-Pb zircon (SHRIMP) age of 461 ± 13 Ma that is interpreted to date melting during the Grampian phase of the Caledonian orogeny (Kinny *et al.*, 1999). Two sets of folds are present at the present locality: (a) early syn-migmatite folds (D_{1N}?) that are commonly disharmonic and cut by discordant melt layers; and (b) later asymmetric folds (D_4?) associated with the steep foliation and development of the major upright synform located a few hundred metres to the east. Note that the gently-plunging L_4 lineation and its associated dextral shear indicators are now absent. A U-Pb monazite age of 431 ± 10 Ma has been obtained from the migmatites near here, indicating substantial reheating during the Scandian phase of the Caledonian orogeny (Kinny *et al.*, 1999). If weather permits, excellent views can be had from the top of the hill of the entire north coast of Sutherland, the south coast of the Orkney Islands, and various mountains inland.

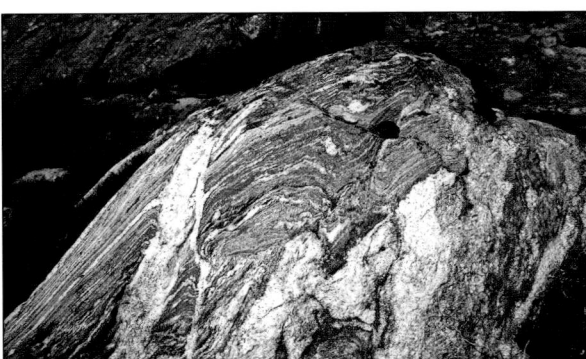

Fig. 13.11
Migmatitic gneisses at Locality 13.14, showing extensive partial melting.

Locality 13.15 [NC 7735 6547]

Port Mor (Fig. 13.12). The boundary between Moine migmatites and the rocks of the Strathy Complex.

If driving from the previous locality, continue east on the A836 to the right-hand bend around Crasbackie Hill [NC 7780 6388]. Stop here briefly to look south and view the escarpment of the Moine gneisses above the lower ground of the Strathy Complex to the east. The Strathy Complex is dominated by siliceous gneisses and amphibolites that may be volcanic in origin (Moorhouse & Moorhouse, 1983; Burns et al., 2004). The age of the complex is uncertain: limited isotopic evidence suggests a late Mesoproterozoic to early Neoproterozoic age, and thus it may represent local basement to the adjacent Moine migmatites (Burns et al., 2004). Turn off the A836 onto the minor road signposted to Armadale. Follow the minor road and continue until the road splits into three at the bus shelter [NC 7861 6466]; coaches should go no further. Smaller vehicles take the left-hand road for 250m and parking is available for a minibus or three to four cars by the house at [NC 7845 6483]. Allow 2 hours for this locality. Take the track signposted to Poulouriscaig westwards across the moor and then follow a narrow valley northwards to reach Port Mor, 15A [NC 7735 6547]. It is best to stay in the valley beside the stream and not to attempt to climb the cliffs on the east side. On the east side of the bay are steeply dipping siliceous gneisses and amphibolites of the Strathy Complex, cut by numerous sheets of discordant granite and pegmatite. At least two sets of folds are present: a tight D_4 fold pair verges east just above high water mark, and later gentle folds with flat-lying to gently inclined axial planes are visible higher in the cliff. The westernmost outcrops before the stream are of a calcite-scapolite-diopside-orange spinel-bearing marble (Harrison & Moorhouse, 1976; Moorhouse & Moorhouse, 1983) that is unlike any Moine lithology in the area. The yellowish-orange marble is interlayered with green calc-silicate bands characterized by abundant diopside with scapolite and plagioclase. A sub-vertical brittle fault located along the western margin of the marble is

Opposite page:
Fig. 13.12 Simplified coastal geology of the Strathy Complex (from Moorhouse et al., 1988 and Burns, 1994) showing Localities 13.15, 13.16A and 13.17.

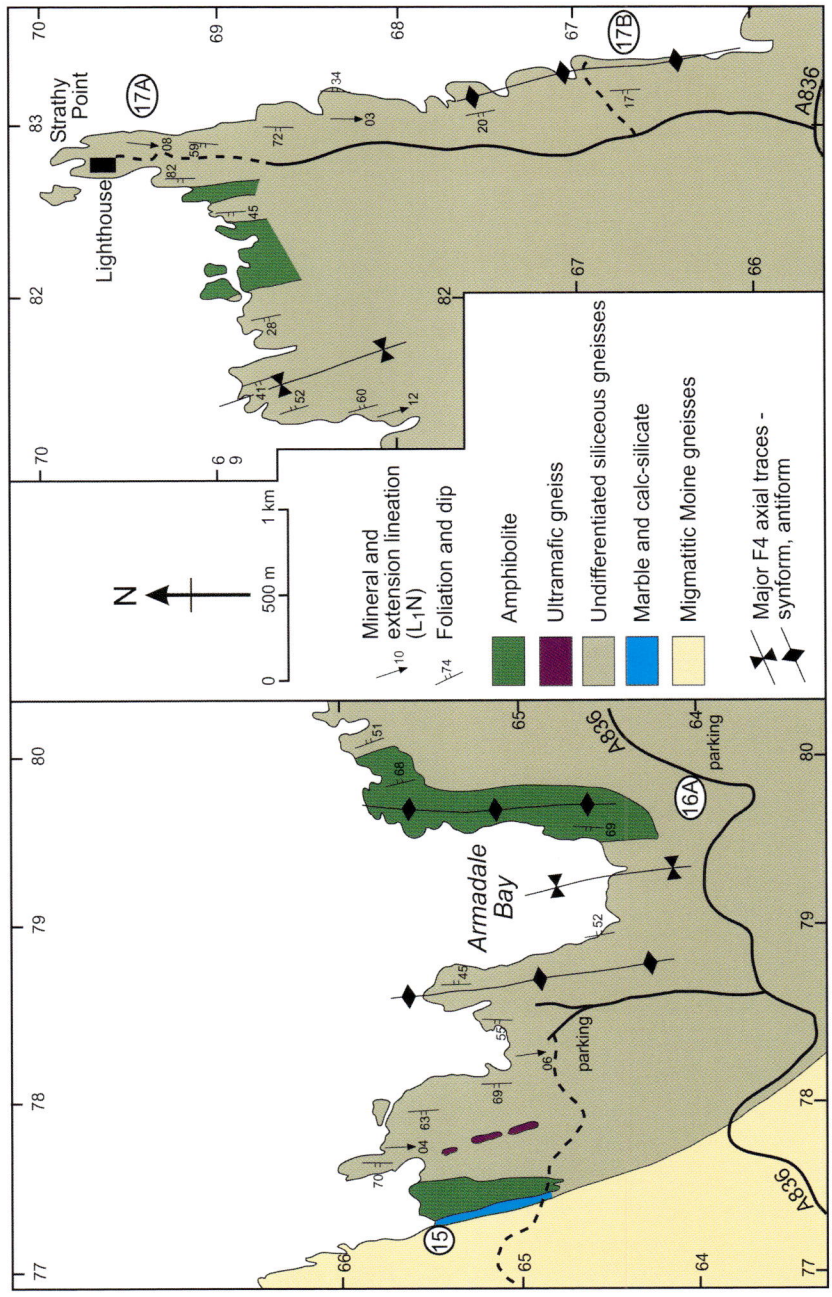

inferred to separate the Strathy Complex from steeply-dipping and strongly deformed Moine pelitic and psammitic gneisses to the west. These gneisses do not carry any penetrative lineations and the high strains appear to be simply related to flattening: there is no evidence that the two different rock units are separated by, for example, a ductile thrust. A 3-4m-wide vertical sheet of pink, unfoliated granite intrudes the Moine gneisses at the back of the bay. Look up to the steep cliffs on the west side of the bay to see pods of amphibolite within the Moine gneisses, engulfed by ramifying late granite sheets. Walk back up the gorge until it is safe to climb out on the east side. Walk east to Locality 13.15B which corresponds to exposures of bright green ultramafic amphibolite, forming grassy knolls at [NC 7784 6532] and [NC 7777 6545]. The rock comprises anthophyllite with relic clinopyroxene and the Cr-rich chemistry has been interpreted to indicate that they represent retrogressed pyroxenites (Moorhouse & Moorhouse, 1983).

Locality 13.16   [NC 7992 6395 to NC 8127 6487]

Strathy road sections (Figs 13.1, 13.12). Strathy Complex siliceous gneisses and amphibolites, fold structures and cross-cutting granitic intrusions.

This locality comprises three separate roadside exposures; about 20 minutes could be allocated to each. If driving from Armadale, take the A836 east and cross the Armadale Burn and, as the hill starts to rise, turn onto the south side of the road and park by a small turning at [NC 7992 6395]. Cross the road to the cutting at 16A that exposes amphibolites and siliceous gneisses of the Strathy Complex, here dipping moderately to the west and cut by late sheets of granitic material and pegmatite. Upright open D_4 folds verge east towards an antiform. Continue driving eastwards for *c.* 800 km and park on the verge on the north side of the road opposite 16B at [NC 8078 6470]. *Care should be taken here – this straight stretch of road is dangerous for unwary pedestrians.* The road cutting shows extensive exposures of layered, siliceous gneisses of the Strathy Complex, mostly dipping east and folded by asymmetric D_4 folds that verge west. The gneisses and the folds are cross-cut by sheets of undeformed white pegmatitic granitoids and later, finer-grained pink granites. Drive a few hundred metres further on to park either on the verge opposite the next large roadside exposure that is 16C at [NC 8127

6487] or a few hundred metres further east in the designated parking area. The road cutting exposes gently undulating interbanded siliceous gneisses and amphibolites. These are intruded by younger metabasic sheets, seen as a series of en-echelon, folded boudins, and ramifying granite sheets that are essentially undeformed. At the east end of the cutting, U-Pb dating of zircons from one of these granite sheets has yielded a Silurian crystallization age (Kinny & Friend, unpublished data).

Locality 13.17   [NC 8282 6964 to NC 8340 6680]

Strathy Point (Fig. 13.12). Rock types and structures in the Strathy Complex.

Continue east along the A836 to the western parts of Strathy and turn north on the road to Strathy Point [NC 8296 6553]. Follow this road to Totegan, where the private road to the lighthouse begins, and park in the extensive car park by the sheep pens [NC 8270 6859]. Allocate two hours for this locality. Walk down the road to the lighthouse, noting various outcrops of siliceous gneiss and fold structures en route. Head for the east side of the lighthouse to 17A at [NC 8282 6964]. In an exposure below the wall, crossed by a drainpipe, a prominent boudin carries an interesting mineral assemblage (garnet-staurolite-sillimanite) and a steep gneissic fabric that is oriented approximately normal to the subhorizontal enveloping S_{1N} fabric within the host siliceous gneisses. A second boudin immediately to the north is less accessible. The margins of the boudins are sheared and retrogressed. The mineral assemblage comprises quartz + plagioclase (An_{47-63}) + garnet + staurolite + sillimanite + anthophyllite + brown and green biotite + spinel. The garnet porphyroblasts may show three growth phases, with sillimanite, staurolite and rutile inclusions found in the second phase (Burns, 1994). The idioblastic outlines of phase 2 garnet are marked by fibrolite, and a second phase of sillimanite may be found outside the third zone, intergrown with biotite. Staurolite is replaced by green spinel. It is clear that the boudins, and presumably also their host siliceous gneisses, have undergone a complex metamorphic history. Metamorphic conditions during the formation of the early mineral assemblage have been estimated by Burns (1994) at $c.$ 700°C and 6kb – substantially lower pressures than those deduced for regional Grampian migmatization in the Moine rocks west and east of the complex. Walk NE from the boudin down the slope onto the rocky headland where there are excellent exposures of interbanded siliceous gneisses and amphibolites cut by discordant granitoids and pegmatites. The dominant S_{1N} fabric dips mostly to the west and is associated with a strong, north-south trending L_{1N} lineation defined by aligned minerals as well as rods and mullions in places. Tight to open D_4 folds plunge gently to the north, more or less parallel to L_{1N}, but can occasionally be seen to fold the lineation.

Return to the vehicles and drive south along the road to the track running east at [NC 8294 6672]. Walk east, through the gate past the bungalow on the south side, then 150m along keep straight on past 'Caberfeidh', through two more gates, and follow the faint track southeast down the cliff to an old fishing slipway and 17B at [NC 8340 6680]. In the exposures behind the winch are flat-lying isoclinal D_{1N} folds refolded by upright, non-cylindrical, tight to open folds, of probable D_4 age. A L_{1N} lineation trends north-south, approximately parallel to the axes of the D_{1N} folds. Ramifying networks of coarse white pegmatite are cut by sheets of finer-grained, pink granite that occupy distinctive tension gashes in the cliffs to the south of the slipway.

Locality 13.18   [NC 8740 6644]

Portskerra (Fig. 13.1). Moine migmatitic gneisses; late Caledonian igneous intrusions; Devonian sedimentary rocks.

Drive east along the A836, turning left onto the minor road by the Melvich Hotel [NC 8767 6507]. Follow the road for 600m and take the left-hand fork. Continue for 850m around a bend and then park at [NC 8749 6620] or further east at the next bend in the road. Parking here is restricted to two minibuses or cars; coaches would have to park further south in Portskerra. Allocate 1½-2 hours for this locality. Low tide is useful but not essential. From the parking place walk 100m west, turn right at the end of the fence and follow a faint path beside the stream to a grassy flat above the beach. The two stones mark the graves of shipwrecked mariners. Climb down the grassy cliff into a stony cove. Devonian sandstones are exposed on both sides of the cove, resting unconformably on Moine gneisses. The unconformity surface is highly irregular and may be examined at 18A [NC 8755

6633]. The underlying Moine rocks are highly migmatized semi-pelitic gneisses showing several generations of melt layers deformed by complex, often disharmonic folds. These are interpreted to represent the same gneisses seen at Cnoc Mor and Port Mor, on the east side of the broad antiform cored by the Strathy Complex. The gneisses are cut by sheets of undeformed pink granite. Walk along the cliffs on the east side of the cove to emerge on the headland where there are extensive exposures. The semi-pelitic gneisses appear to pass transitionally eastwards into psammitic gneisses. At 18B [NC 8740 6644], migmatized psammitic gneisses with numerous concordant melt layers are highly veined and sheeted by late, pink granites that are thought to correlate with the $c.425$ Ma Strath Halladale Granite that intrudes Moine rocks 10km to the SE (Kocks *et al.*, 2006). Some of the late granite sheets contain magmatic fabrics defined by aligned feldpars and micas. Walk east along the cliff line; keep above the next major cove, but look down into it to observe Devonian sandstones again resting unconformably upon Moine gneisses on the west side. Walk around the inlet and head northwards to the next headland. At 18C [NC 8768 6660], psammitic gneisses are intruded by a large, essentially undeformed diorite body which is itself cut by late granite sheets. The diorite is probably the same age as the Reay Diorite 8km to the east. Walk east a few tens of metres into the next bay to examine in detail exposures of the Moine/Devonian unconformity that preserves some 3-4m of relief. The sedimentary rocks are mostly fine to medium-grained sandstones with siltstones; pebble beds are thin and localized. This facies may be contrasted with those seen at Coldbackie and Kirtomy. If tide allows, continue around to the slipway at 18D [NC 8786 6629] (alternatively, return to the previous locality, scramble up the grassy cliff and walk around to the slipway). At the slipway, look west to view the Moine/Devonian unconformity and observe large-scale gentle variations in dip of the lowermost Devonian strata. Whether these variations represent original depositional angles or the effects of compaction and draping over an irregular Moine land surface is uncertain. The rocks at the end of the slipway are psammitic gneisses and late granites as seen before.

Excursion 14
Great Glen

Martyn Stewart

Purpose: To examine deformation that may be related to displacements along the Great Glen Fault and hence to consider late stage events in the Caledonian orogeny.

Aspects covered: Fault-bounded blocks of Moine within the centre of the fault zone; strike-slip related shear fabrics from different crustal levels; interleaved Lewisian basement and Moine adjacent to the fault zone.

Maps: OS: 1:25,000 sheets 399 Loch Arkaig, 400 Loch Lochy & Glen Roy and 432 Black Isle; BGS: 1:50,000 sheets 62E Loch Lochy, 63W Glen Roy, 1:63,360 sheet 94 Cromarty.

Type of terrain: Localities 14.1-14.6 are each within 20 minutes walk from a vehicle on roadsides or established forest tracks and riverside paths. Locality 14.7 involves a steep walk on established paths and a 3km walk along pebble beaches; it is also tide-dependent.

Distance and time: A long distance excursion along the length of the Great Glen. Localities 14.1-14.6 can be completed in a single full day. Locality 14.7 will take most of a day.

Short itinerary: Localities 14.1 and 14.2 can be completed in half a day from Fort William and provide an appreciation of the main aspects of the geology along the Great Glen Fault Zone.

Localities 14.1 and 14.2 are a related pair of exposures that occur in the River Lochy within the valley bottom of the Great Glen north of Fort William. Due to ease of access, it is suggested to visit Locality 14.1 first and then from here proceed to Locality 14.2. Allocate 2-3 hours for these sites.

Great Glen

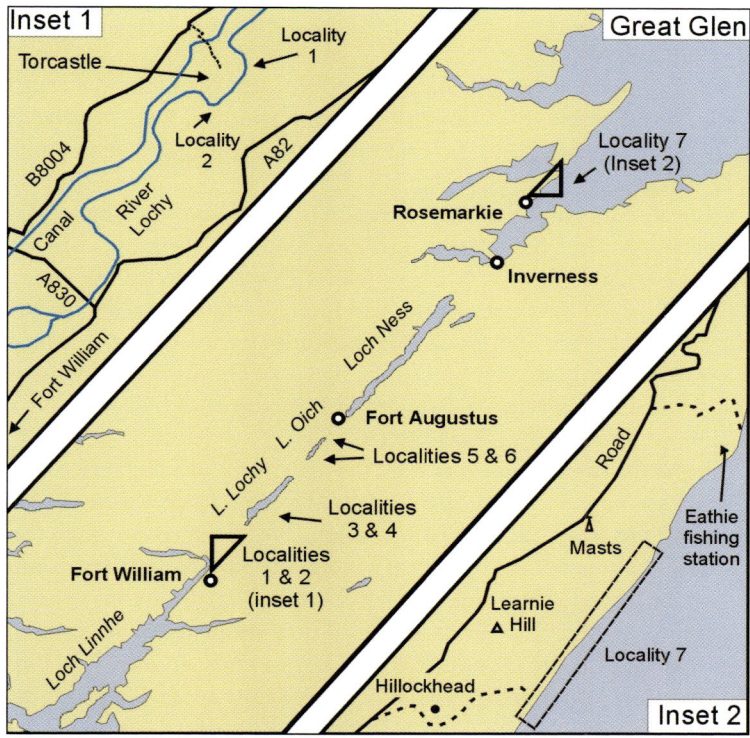

Fig. 14.1 Map of the Great Glen showing localities referred to in the text. The arrows indicate site localities adjacent to the fault.

Locality 14.1 [NN 135 791]

Torcastle, River Lochy: northern outcrop. (Fig. 14.1, inset 1). A highly deformed fault-bounded sliver of probable Glenfinnan Group protolith within the core of the Great Glen Fault Zone, with structural evidence for sinistral shear.

From Fort William, drive NE along the A82 and take the A830 Mallaig road. Just past the Caledonian Canal crossing at Neptune's Staircase, take the B8004 towards Gairlochy. Loch Eil Group psammites are exposed in occasional cuttings. After approximately 4 km, just after the sign for Muirshearlich, turn off to the right at the small restaurant by the side of the road. Continue down this track for 100m, and then take the right-hand fork following the track downslope. At the bottom there is a flat area to park just in front of the tunnel that runs beneath the Caledonian Canal [NN 132 793].

Localities 14.1 and 14.2, referred to here as Torcastle North and South respectively after the nearby ruins, are found at the banks of the River Lochy exposing sheared Moine rocks within the core of the Great Glen fault. These are the closest exposures found to the central axis of the fault anywhere along its length. Examination of exposures at Locality 14.1 provides a good introduction to the macro-scale shear fabrics associated with the fault, whilst meso- to micro- scale textures and shear fabrics can be examined more easily at Locality 14.2.

From the car-parking, bay walk through the tunnel and then turn immediately left up onto a path that leads to the canal towpath. At the towpath turn right to head NE. After a few minutes walking, go through the large metal gate at the side of the path at a junction with a forestry track and walk down the track to the point where it changes from deciduous woodland to coniferous forestry. At this point, head downslope to the River Lochy. At the riverbank is the north end of the large exposure shown in Fig. 14.2 which is Torcastle North ([NN 135 791], Locality 14.1). Exposures here are estimated to lie less than 200m from the interpreted central axis of the Great Glen Fault (Stewart *et al.*, 1999). The rocks are commonly obscured by frequent mud-draping at periods of flooding. The best place to view the detailed lithology and texture is at the SE tip of this outcrop or at Locality 14.2, but a traverse over the outcrop here provides a good impression of the macro-scale distribution of fault-related deformation.

Carefully walking over the rocks towards the SE, it is clear that the outcrop surface comprises high standing areas and flat, low-lying areas. These correspond with, respectively, areas of low and high strain related to fault movements. The high-standing blocky areas comprise psammitic gneiss and quartzite protolith that represent domains of relative low strain during faulting. These rigid areas are typically heavily fractured and dissected by discrete faults. In contrast, the low-lying flat areas, which are often covered by loose boulders or flooded, correspond with mica-rich material that preserves a highly penetrative shear fabric. These high strain belts are between 0.5m to 10m wide.

A good place to view the fabric within the high strain shear zones is to clamber up onto the high-standing ridge (1A on Fig. 14.2) and to look down

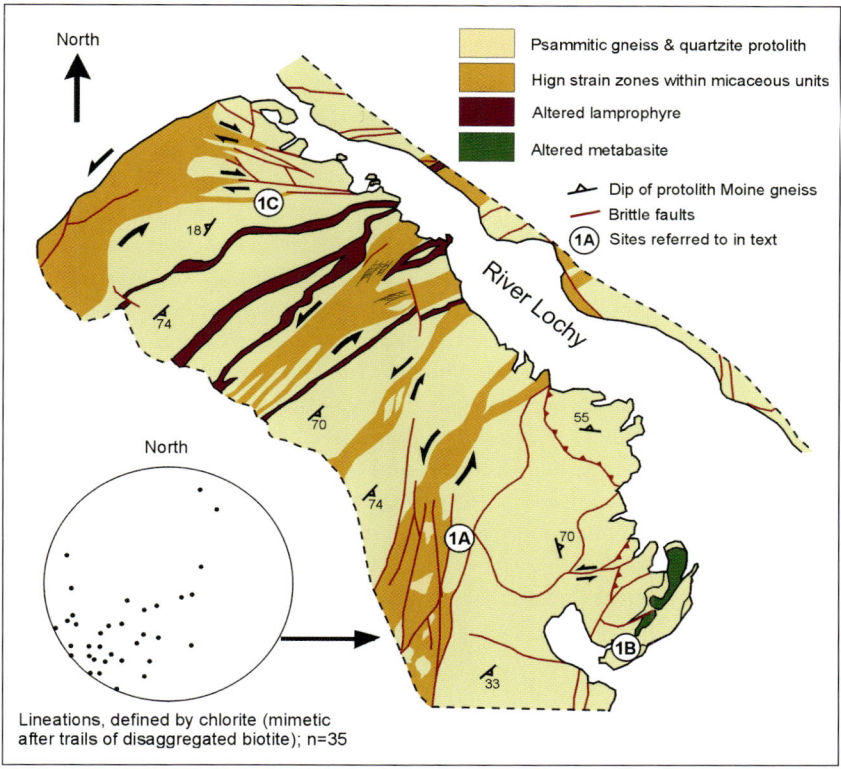

Fig. 14.2 Map of Locality 14.1, on the banks of the River Lochy to the north of Torcastle. The exposures comprise Moine protolith subsequently sheared by late Caledonian displacements on the Great Glen fault. The inset shows trend and plunge of mineral lineations within the shear zones.

onto the flat low-lying area immediately to the SSW; this is usually flooded by a few inches of water which actually enhances the appearance of the shear fabric. The intensely cleaved, shattered and fragmented nature of this fabric should be evident, as should the more prominent faults that cut through it. Use of a hand-lens reveals sub-horizontal slickenfibre lineations (Fig. 14.2 inset) on the surfaces of the augen around which the microshears envelop. This fabric comprises an intense and diffuse connected network of mm-wide shear bands that contain both mechanically fragmented and crystal-plastic deformed muscovite and chlorite with subordinate quartz. These shears anastomose around lenses of undeformed feldspar and quartz that have experienced weak to moderate levels of crystal-plastic strain.

Rare shear-sense indicators, such as mm to cm-scale truncated folds, indicate a sinistral sense of displacement (Stewart *et al.*, 2000).

A good place to study the original protolith is at 1B (Fig. 14.2). Here, the rocks are polished clean by the river water and reveal bands of quartzite and arkosic psammite along with micaceous partitions. Within the latter, the fine mm-scale through-going shears can be seen. Also note the presence of a retrogressed metabasite. These Moine rocks are lithologically most similar to the Glenfinnan Group (Stewart *et al.*, 2000). In the northern third of the exposure, altered lamprophyre dykes are assigned to a regional Permo-Carboniferous swarm (Baxter & Mitchell, 1984). These are locally truncated by minor faults, indicating that at least some of the deformation here relates to later, post-Caledonian movement. In the area by the narrow rapids in the northern corner of the exposure (1C on Fig. 14.2), note how the fractures that trend WNW swing round from the main flat low-lying belt that occurs in the shallow stream bed north of this outcrop. These faults are thought to represent antithetic shears to the main NE-trending high strain belts and is a pattern repeated at the next set of exposures.

Locality 14.2 [NN 132 786]

Torcastle, River Lochy: southern outcrop. (Fig. 14.1, inset 1, and Fig. 14.3). A highly deformed fault-bounded sliver of probable Glenfinnan Group protolith within the core of the Great Glen Fault Zone, with structural evidence for sinistral shear.

Leave the Torcastle North outcrops along the small track along the riverbank and follow this downstream. After a while, the track passes onto the pebbly point-bar and then back onto the riverbank path. Walk upslope above the cliffs to the overgrown ruins marked 'Tor Castle' on the map and then down to the flat-lying exposures ahead which are Locality 14.2, Torcastle South [NN 132 786].

The exposures here are generally cleaner than at the previous locality. Semi-pelitic gneiss, quartzite and feldspathic psammite protoliths are clearly visible, albeit with a strong overprint of fault-related fabrics. A Glenfinnan Group origin is again suggested by the semi-pelitic composition and supported by the presence of garnetiferous amphibolites which are common elsewhere within the group (Johnstone *et al.*, 1969; Holdsworth *et*

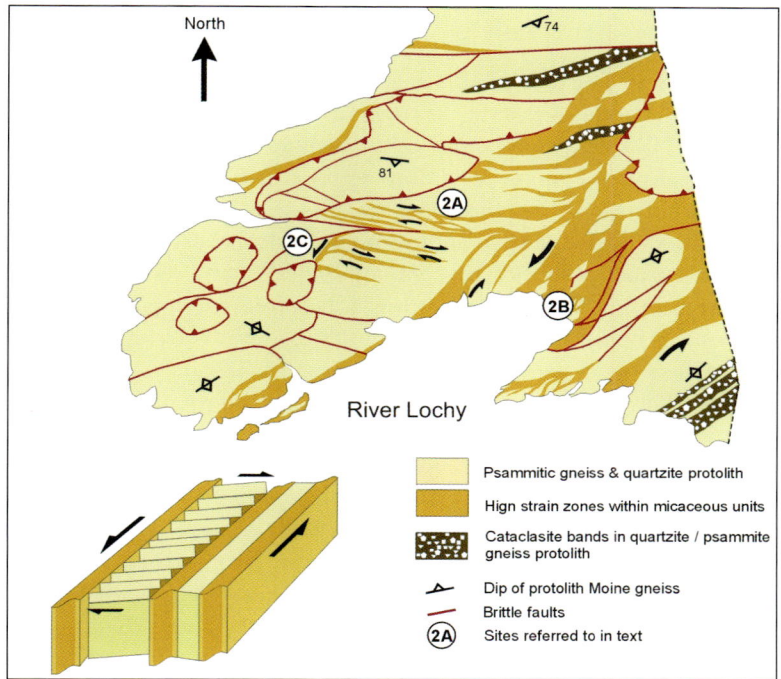

Fig. 14.3 Map of sheared Moine protolith exposed at Locality 14.2, to the south of Torcastle. The inset is a schematic diagram to show the interpreted geometric relationship between the main vertical sinistral shear zones and subordinate dextral antithetic shears. The latter are thought to have formed by 'bookshelf' faulting where low strain domains between the main shear zones have experienced horizontal anticlockwise rotations.

al., 1987). If indeed these correlate with the Glenfinnan Group, it implies that rocks at both exposures are part of a single fault-bound unit that has been either uplifted relative to Loch Eil Group strata to the west (along the road), and/or moved laterally here by strike-slip displacement.

Locate the prominent upstanding blocks and note how at their base they are cut by a low angle fault. Much of the flat surface here is the eroded footwall of this fault. On the surface itself, where it is polished and fabrics are well-exposed (e.g. 2A on Fig. 14.3), note cm- to dcm-wide, WNW-trending, steeply-dipping to vertical micaceous shear-bands forming a pervasive and anastomosing network. Detailed examination of shear-sense indicators here and the overall sense of asymmetry suggests formation in response to localised dextral shearing.

In the low-lying area at 2B, which may be partly submerged depending on the height of the river-level, the rocks have a more fractured, slaty appearance and NE trend, comparable to the high-strain belts seen at Torcastle North. Lineations plunge at shallow to moderate angles to both NE and SW. Examination of the fabric here reveals a few truncated folds and anti-clockwise-deflected fabrics, indicating a sinistral shear sense. It appears that these sinistral fabrics are contemporaneous with the dextral fabrics seen at site 2A. This interpretation is based largely on the geometric relationship between the two sets of fabrics. The WNW trend of the dextral shears suggests an antithetic relationship to the larger NE-trending sinistral belts, accommodating anticlockwise (sinistral) rotation of low strain material between the high strain belts (Fig. 14.3, inset). This is supported by observations of WNW-trending dextral shear-bands that curve into NE-trending shears, rather than truncating them (see 2C on Fig. 14.3). This relationship is thought to explain similar fault patterns seen at site 2C at Torcastle North (Fig. 14.2). Taken together, these macro-scale kinematic indicators, supported by lineation data, meso-scale shear-sense indicators seen in the field and those seen in oriented thin-sections, provide compelling evidence for sinistral strike-slip shearing along the Great Glen Fault (Stewart *et al.*, 1999, 2000).

The evidence at these localities indicates that the Great Glen Fault was initiated at a late stage in the Caledonian orogeny. The style of deformation is characteristic of deformation within the brittle-ductile transition zone, broadly equivalent to depths of 10-15km (Snoke *et al.*, 1998; Holdsworth *et al.*, 2001b). This is consistent with the growth of chlorite and muscovite in shear-zone fabrics, implying shearing at depths equivalent to greenschist facies metamorphic conditions (Stewart *et al.*, 2000). In addition, it would appear that most of this shearing was completed within the mid crust, with no evidence for progressive overprinting by brittle deformation during exhumation to the surface. Retrace the route back to vehicles.

Locality 14.3 [NN 235 892]

View over Loch Lochy towards the Clunes Tonalite (Figs 14.1, 14.4).

From Torcastle, continue northeast towards Gairlochy and take the B8004 towards the A82 and Spean Bridge. Turn left at the Commando Memorial along the A82 towards the northeast. After ~9km pull into the lay-by where the overhead power cables cross the road. Look to the opposing side of the valley towards the southern end of Loch Lochy and different bedrock should be identifiable from different colours of the exposure (Fig. 14.4). The bulk of the hillside comprises feldspathic psammites of the Loch Eil Group Moine (pink-orange colouration), but to the southern half of the hill occurs a tonalite pluton, marked by a darker grey-coloured rock.

A study of this Clunes Tonalite pluton has concluded that intrusion occurred synchronous with sinistral displacements along the Great Glen Fault (Stewart *et al.*, 2001). This is based on observations of a sinistral swing of the magmatic fabric at its northeastern margin, together with intrusion into a ductile sinistral shear zone within the marginal Moine host rock. U-Pb zircon dating of the tonalite has yielded an age of 428 ± 2 Ma which is thus interpreted to date early sinistral displacements (Stewart *et al.*, 2001). Near the shoreline of Loch Lochy the northeastern end of the pluton is displaced by a post-Caledonian brittle fault demonstrating dextral offset.

As you proceed to drive north, note the deeply incised gorges running down the eastern side of the valley. The streams flowing down-slope here are actually very small, and the deep-cutting incision is an indicator as to how intensely fractured and shattered the bedrock is adjacent to the fault.

Fig. 14.4 View southwest across Loch Lochy to Coire Lochain and the Clunes Tonalite.

Locality 14.4 [NN 255 918]

Loch Lochy shoreline (Fig. 14.1). Fault-bounded block of gneiss of probable Moinian affinity on the SE side of the Great Glen Fault.

Continue along the A82 for a few km until the road drops down to run straight alongside the shoreline of Loch Lochy and pull into the parking lay-by on the left-hand side. Cross the road to exposures of steep-dipping psammitic gneisses by the waterfall; this is best seen by following the track up just above the fall. These rocks are part of a narrow fault-bounded block that makes up most of the lower valley side south of the Great Glen Fault in this area and northwards until pinching out into the centre of the fault zone near Fort Augustus. They are unlike rocks of the Grampian and Dalradian groups seen locally, which are more flaggy, finer grained and less gneissic in appearance. Also in thin-section, metamorphic textures of Dalradian and Grampian rocks are almost always polygonal in appearance, suggesting a single major phase of metamorphic growth. In contrast, thin-sections of the rocks exposed here reveal complex textures with sutured grain boundaries and very coarse crystals enclosing subgrains, indicating a secondary grain growth over an earlier metamorphic texture, typical of microtextures seen throughout the Moine. The status of these rocks is uncertain, but they appear Moine-like and it is possible that they represent a slice of Moinian material derived from the NW side of the fault and transplanted SE of the axis during strike-slip displacements. Alternatively they may correlate with gneissic basement material underlying the Grampian terrane (Badenoch Group) that has been uplifted adjacent to the fault in response to transpressional displacements.

Locality 14.5 [NN 304 985]

Loch Oich shoreline (Fig. 14.1). Moinian rocks SE of the Great Glen Fault, severely altered by fault-zone fluids.

Continue NE along the A82 towards North Laggan. Just before the road turns sharp left to cross the swing-bridge at the head of Loch Oich, take the road on the right to the Great Glen Water Park. Follow this until the entrance of the park, marked by 'No Entry' signs, and continue down the forestry track on the right. Park where the ground opens out – this is the old

Laggan Station from the disused Fort William to Fort Augustus railway line [NN 304 985]. Continue NE along this track by foot for about 10-15 minutes until the path crosses a great scree slope.

The bedrock here is again Moine-like and belongs to the same fault-bound sliver visited at Locality 14.4. However, examination of hand specimens from the scree reveals that even in very fresh samples, the rock itself is extremely difficult to identify, quite grubby in appearance, and often with an original coarse crystalline texture replaced by a very fine grained speckled matrix. It is primarily composed of retrogressive and alteration mineral phases formed during intense fluid-enhanced reaction-softening. The protolith, seen occasionally within the scree, is coarse grained psammitic gneiss. In most cases, however, feldspars are either heavily or completely altered to sericite with fine quartz and chlorite within the matrix. Cataclastic deformation of these highly altered rocks has subsequently produced an extremely fine-grained, fractured rock which varies in colour from black, green, grey or cream, and contains 'clasts' of less altered material, occasionally with relict gneissic textures.

Continue along the disused railway line past the concrete retaining wall. About 50m along from here, look for scree amongst the woodland on the slope to your right and examine the textures. Here are a mixture of rocks which include: (1) pale psammitic gneiss in which feldspars have experienced partial to complete sericitisation; (2) 'hydrated' cataclasite comprising clasts of quartz and feldspar within a black, fine-grained, often laminated matrix of alteration product; and (3) various rocks representing intermediate phases of this alteration and cataclasis process.

These highly retrogressed, altered and cataclastically deformed rocks, perhaps best termed 'hydrated cataclasites', dominate most of the outcrop within this sliver of Moine-like rock and suggest deformation occurred within a seismogenic upper-crustal environment characterised by brittle fragmentation and dilatancy leading to high fluid influx. This deformation is interpreted as pre-Devonian in age on the basis that a fault-bounded sliver of Old Red Sandstone is seen to lie unconformably upon these cataclasites and is comparatively free of fracturing and evidence of fluid flow. Similar highly altered and shattered cataclasites may well underlie the whole core of the fault zone, explaining why the fault zone was so prone to excavation during glaciation. The Moinian protolith seen here (and at the previous locality) is similar to that seen at Torcastle, but the type of fault rock produced is quite different, a product of shearing in the upper crust, in contrast to inferred mid-crustal shearing seen at Torcastle. Such comparisons suggest the present configuration of fault-bound units is a consequence of late-Caledonian or later differential uplift along the length of the structure.

Locality 14.6 [NN 277 957]

Kilfinnan Burn (Fig. 14.1). Granite veining within Loch Eil Group in the Great Glen Fault Zone.

Drive back onto the A82, cross the Laggan Swing Bridge at the head of Loch Oich, and then after *c*. 200 m turn left onto the single track road running SW towards Kilfinnan along the northern side of the valley. By the farm at Kilfinnan, park near the bridge over the stream Kilfinnan Burn. Walk about 40m upstream to exposures in the stream bed of dcm-scale microgranite and microdiorite veins of the Glen Garry Vein Complex (Fettes & McDonald, 1978) intruding Loch Eil Group psammitic gneiss. This vein complex is defined by intense granodiorite veining over an area of over 300km². Although typically randomly oriented, a NE trend is more common in the vicinity of the Great Glen Fault. At this locality, the ENE-trending gneissic foliation is rotated towards these veins so that a sinistral sigmoidal fabric results. This relationship is seen elsewhere in the Moine along the fault zone and suggests that vein emplacement was synchronous with sinistral shearing along the fault.

Before leaving, look across Loch Lochy to the huge gulleys incised into the valleysides on the SE of the Great Glen, indicating the weak nature of the highly shattered bedrock. Note also how the gulleys are very wide upslope, but then narrow abruptly. Thus sudden change marks the boundary between, upslope, shattered Moine (e.g. Locality 14.4) and Grampian Group bedrock and, downslope, less fractured Old Red Sandstone which occurs as a local fault-bounded sliver. Such evidence suggests that the majority of fracturing observed along the Great Glen Fault Zone relates to pre-Old Red Sandstone sinistral displacements. From here continue on to Inverness and then to Locality 14.7.

Locality 14.7 [NH 773 627 to NH 765 615]

Rosemarkie (Fig. 14.1, inset 2, and Fig. 14.5). Interleaved Lewisian and Moine rocks with intrusive granites, showing evidence for ductile deformation that may be related to displacement along the Great Glen Fault.

This locality exposes psammites, semi-pelites and hornblendic gneisses that lie adjacent to the Great Glen Fault and are interpreted as interleaved Moine and Lewisianoid basement (Rathbone & Harris, 1980; Mendum *et al.*, 2010). These outcrops represent the most southeastern outcrop of basement north of the Great Glen, implying that Lewisian-like basement underlies the whole of the Moines. Amphibolite facies fabrics contain a strong sub-horizontal lineation that may record ductile, strike-slip movements along the adjacent Great Glen Fault. Microgranites and pegmatites appear to have been injected synchronous with this deformation. These metamorphic rocks occur as two inliers either side of the Cromarty Firth, surrounded by Old Red Sandstone sediments.

The 3km section described here is the best-preserved and most accessible section to have survived later brittle brecciation. Exposures are tide-dependent and occur along a narrow shoreline backed by steep grassy and wooded cliffs. The section described is a circular route that traverses the whole 3km shoreline section and will take most of a day. It requires either two cars, parked at either end, or a 5km walk back along the road to the start point. The route as described below approaches the shore at the NE end of the section, but if a shorter excursion is required, or only one car is available, it is best accessed and returned to from the SW end.

From Inverness take the A9 north and then at Tore take the A832 for Rosemarkie. Drive through Rosemarkie and after approximately 1.5km turn right onto the single-track road to Eathie (Fig. 14.1, inset 2). Approximately 3km along the Eathie road take the right-hand turn at Hillockhead [NH 742 604]. This is where the excursion ends, so if two vehicles are available, one can be left off here. There is room for parking in the small cutting 50m down the track. To reach the start (4.5km from here), continue along the Eathie road, 1.5km past the prominent TV Mast Station at Eathie Hill, and then pull into the parking area [NH 769 635]. Follow the track here down to the shoreline to an old fishing station. From here walk SW for 1km along the boulder and cobble shoreline.

Fig. 14.5

(a) Plan view of shear fabrics within hornblende gneisses at Locality 14.7A [NH 773 627]. Syn-metamorphic and later brittle shear fabrics demonstrate sinistral offsets.

(b) Plan view of amphibolite-grade shear fabrics. A sinistral sense of shear is indicated by geometry of the sigmoidal vein in the lower middle section of photograph.

(c) Foliated microgranite at Locality 14.7B [NH 772 626] showing well-developed feldspar augen indicating sinistral shearing.

(d) Shear banding within granite sheet at Locality 14.7B. Geometry of shear fabrics is highlighted in right-hand image and is suggestive of sinistral shear.

(e) White foliated leucogranite at Locality 14.7C [NH 769 621] with sheared muscovite rich core and development of quartz & feldspar augen where muscovite envelopes porphyroclasts.

(f) Detail of leucogranite vein at Locality 14.7C. Geometry of synthetic shears within sheared core indicate dextral shearing. [Inset: Evidence of pressure solution (impacting of muscovite) in top left and bottom right quadrants implies dextral shear.]

The first exposures at [NH 773 627] are Locality 14.7A. These are of coarse-grained hornblende gneisses with a foliation trending SW, parallel to the trace of the Great Glen Fault, and dipping 60° to the SE. This is interpreted by Rathbone & Harris (1980) to be composite S_1/S_2. Hornblende and stretched quartz define a prominent lineation that plunges at 0-20° to the NE. This foliation trend and lineation plunge is largely typical of the whole section. It is this fabric that is thought likely to be the result of ductile deformation during displacement along the Great Glen Fault, the trace of which is only 1 km offshore. The foliation encloses micro- to meso-scale lenses of quartzo-feldspathic material (Fig. 14.5A). Shear-sense indicators include the sense of overturning of ptygmatically folded veins, sigmoidal quartzo-feldspathic lenses (Fig. 14.5B), asymmetric pressure shadows accompanying rigid grains, asymmetric micas, and the rotation of foliation from lower- to higher-strained horizons. Both sinistral and dextral shear-sense indicators are commonplace.

To the SW the gneiss is intruded by mm- to dcm-scale veins and sheets of pink microgranite. The microgranites often have transitional boundaries and appear as migmatitic segregations concordant with the gneissic foliation. These usually carry a foliation which is concordant with the composite S_1/S_2 foliation in the gneiss. Many contain feldspar augen that may represent deformed phenocrysts. The thicker veins more typically have sharp boundaries that cross-cut the foliation at shallow angles, suggesting they are post D_1, although they too carry a foliation parallel to the gneiss, suggesting a syn-D_2 age of emplacement.

Locality 14.7B is ~40m from the first outcrops. Beneath a large Lewisian-like erratic [NH 772 626], a granite vein contains a foliation and an internal shear-band fabric. A sinistral sense of asymmetry is shown by porphyroclasts and synthetic shear bands that cut an earlier foliation (see Figs 14.5C, 14.5D). There is also evidence of cataclasis in the core of the vein, which might be consistent with a continuum of ductile to brittle deformation, presumably late- to post-D_2, as the granite sheet cooled and crystallized. The logic behind this arises from the tendency of cataclasis to nucleate within weak horizons or along boundaries with contrasting rheological properties, rather than within the core of a rigid, crystalline granite. If veins like this were emplaced post-D_2, the internal foliation may be a high strain magmatic fabric rather than the S_1/S_2 foliation present within the host gneisses. The comparable orientation might simply reflect the fact that the vein intruded along the grain of the gneissic fabric.

Various horizons have experienced retrogression, probably as a result of the migration of fault-related fluids. These are finer-grained, schistose and pale green, reflecting the growth of muscovite, biotite and chlorite as reaction products of hornblende and biotite. Microgranite veins occur within some of these horizons, again suggesting that vein intrusion was not a single event but occurred at multiple stages during uplift and cooling. One such example is a leucogranite within a retrogressed band exposed at the headland just north of an upstanding stack which is Locality 14.7C ([NH 769 621], Figs 14.5E, 14.5F). Lineations within such retrogressive horizons are harder to identify due to finer grain sizes and predominance of platy minerals, but slickenfibres and elongate hornblende reveal a sub-horizontal to shallow plunge. Surfaces perpendicular to this lineation contain dextral shear bands (Figs 14.5E, 14.5F).

Near the headland and stack, prominent late stage folding is likely related to late movements along the Great Glen Fault. Earlier, small and medium-scale similar folds (i.e. thickened hinges) fold the main gneissic fabric and are accompanied by an axial-planar schistosity which appears retrogressive as it breaks down the earlier coarse fabric. This may be coeval with retrogression described above, as it is associated with new growth of finer biotite and muscovite. Rathbone & Harris (1980) interpret these folds as D_3 structures. Pre-D_3 folds are rare but typically isoclinal and carry the main (S_1/S_2) gneissic foliation within the axial plane. This S_1/S_2 foliation appears to be the result of progressive D_1 and D_2 events as indicated by the complex deformation history of microgranite veins intruded at different stages.

The boundary between the Moine and Lewisian is exposed at Locality 14.7D [NH 765 615], and to the SW of here exposures comprise Moinian psammite and micaceous schist, interleaved with Lewisian acidic and hornblendic gneisses. The Moine-Lewisian boundary is parallel to S_1, and is folded by a small-scale isoclinal D_2 fold which is itself affected by a D_3 fold. The relationship between folding and the numerous granite veins present is complex and intrusion was probably long-lived or multi-phase. Some, if not most, granite veins are strongly deformed and carry the S_1/S_2 high-grade metamorphic foliation and associated low-angle lineation. Monazite and zircon data from one of these deformed veins shows that it was intruded ~399 Ma (Mendum & Noble, 2010). Other veins however, are post-D_2, because they cut these folds (e.g. [NH 766 616]) and many seem either synchronous with or post-D_3 (Rathbone & Harris, 1980), intruding lower-grade retrogressive horizons and experiencing cataclastic deformation at late stages of crystallisation.

In summary, the rocks of the Rosemarkie locality record a protracted history, from the D_1/D_2 development of amphibolite facies tectonites, through subsequent phases of down-temperature refolding with accompanying retrogressive fabrics and semi-brittle shear plane development. Later warping, cataclasis, brecciation and pervasive fracturing reflects upper-crustal deformation. The shallow-plunging lineation within the S_1/S_2 gneissic fabric appears to record ductile displacement along the Great Glen Fault during the Devonian and thus the Rosemarkie section could represent an exhumed portion of the fault zone from the ductile lower crust. The kinematic history of these early ductile movements is, however, difficult to decipher because of the lack of any consistency in the shear-sense indicators.

The path back to the road is signposted [NH 754 604] and steps should be visible up the hillside. Continue uphill along the track to Brown Hill with its radio mast at the summit [NH 747 604]; to the SW are visible the houses and track of Hillockhead that leads back up to the Eathie road.

References

AFTALION, M. and VAN BREEMEN, O. (1980): U-Pb zircon, monazite and Rb-Sr whole-rock systematics of granitic gneiss and psammitic to semipelitic host gneiss from Glenfinnan, northwestern Scotland, *Contributions to Mineralogy and Petrology*, **72**, pp. 87-98.

ALSOP, G. I. (1992): Late Caledonian sinistral strike-slip displacement across the Leannan Fault system, northwest Ireland, *Geological Journal*, **27**, pp. 119-25.

ALSOP, G. I. and HOLDSWORTH, R. E. (1999): Vergence and facing patterns in large-scale sheath folds, *Journal of Structural Geology*, **21**, pp. 1335-49.

ALSOP, G. I. and HOLDSWORTH, R. E. (2002): The geometry and kinematics of flow perturbation folds, *Tectonophysics*, **350**, pp. 99-125.

ALSOP, G. I. and HOLDSWORTH, R. E. (2004a): The geometry and topology of natural sheath folds: a new tool for structural analysis, *Journal of Structural Geology*, **26**, pp. 1561-89.

ALSOP, G. I. and HOLDSWORTH, R. E. (2004b): Shear zone folds: records of flow perturbation or structural inheritance?, in ALSOP, G. I., HOLDSWORTH, R. E., McCAFFREY, K. J. W. and HAND, M. (eds): *Flow Processes in Faults and Shear Zones*, Geological Society, London, Special Publications, **224**, pp. 177-99.

ALSOP, G. I. and HOLDSWORTH, R. E. (2007): Flow perturbation folding in shear zones, in REES, A. C., BUTLER, R. W. H. and GRAHAM, R. H. (eds): *Deformation of the Continental Crust: The Legacy of Mike Coward*, Geological Society, London, Special Publications, **272**, pp. 75-101.

ANDERSON, D. E. and OLYMPIO, J. C. (1977): Progressive homogenization of metamorphic garnets, S. Morar, *Canadian Mineralogist*, **15**, pp. 205-16.

ANDERTON, R. (1976): Tidal-shelf sedimentation: an example from the Scottish Dalradian, *Sedimentology*, **23**, pp. 429-58.

ANDERTON, R. (1985): Sedimentation and tectonics in the Scottish Dalradian, *Scottish Journal of Geology*, **21**, pp. 407-36.

ATHERTON, M. P. and GHANI, A. A. (2002): Slab breakoff: a model for Caledonian, late granite syn-collisional magmatism in the orthotectonic (metamorphic) zone of Scotland and Donegal, Ireland, *Lithos*, **62**, pp. 65-85.

BARBER, A. J. and SOPER, N. J. (1973): Summer field meeting in the North-West of Scotland, *Proceedings of the Geologists' Association*, **84**, pp. 207-35.

BAIRD, A. W. (1982): The Sgurr Beag Slide within Moine rocks at Loch Eilt, Inverness-shire, *Journal of the Geological Society, London*, **139**, pp. 647-54.

BAIRD, A. W. (1985): Discussion of the structural setting of the Glen Dessary syenite, Inverness-shire, *Journal of the Geological Society, London*, **142**, p. 713.

BARR, D. (1983): *Genesis and structural relations of Moine migmatites*, University of Liverpool PhD thesis.

BARR, D. (1985): Migmatites in the Moine, in ASHWORTH, J. R. (ed): *Migmatites* (Glasgow & London: Blackie), pp. 226-64.

BARR, D., HOLDSWORTH, R. E. and ROBERTS, A. M.

(1986): Caledonian ductile thrusting in a Precambrian metamorphic complex: the Moine of north-western Scotland, *Geological Society of America Bulletin*, **97**, pp. 754-64.

BARR, D., ROBERTS, A. M., HIGHTON, A. J., PARSON, L. M. and HARRIS, A. L. (1985): Structural setting and geochronological significance of the West Highland Granitic Gneiss, a deformed early granite within the Proterozoic, Moine rocks of NW Scotland, *Journal of the Geological Society, London*, **142**, pp. 663-75.

BAXTER, A. N. and MITCHELL, J. G. (1984): Camptonite-monchiquite dyke swarms of northern Scotland: age relationships and their implications, *Scottish Journal of Geology*, **20**, pp. 297-308.

BEACOM, L. E. (1999): *The kinematic evolution of reactivated and non-reactivated faults in basement rocks*, PhD thesis, Queen's University of Belfast.

BINGEN, B., DEMAIFFE, D. and VAN BREEMEN, O. (1998): The 616 Ma Old Egursund Basaltic Dike Swarm, SW-Norway, and Late Neoproterozoic Opening of the Iapetus Ocean, *Journal of Geology*, **106**, pp. 565-74.

BLACKBOURN, G. (1981): Probable Old Red Sandstone conglomerates around Tongue and adjacent areas, north Sutherland, *Scottish Journal of Geology*, **17**, pp. 103-18.

BONSOR, H. C. and PRAVE, A. R. (2008): The Upper Morar Psammite of the Moine Supergroup, Ardnamurchan Peninsular, Scotland: depositional setting, tectonic implications, *Scottish Journal of Geology*, **44**, pp. 111-22.

BOWES, D. R. (1968): The absolute time scale and the subdivision of Precambrian rocks in Scotland, *Geologiska Föreningens i Stockholm Förhandlingar*, **90**, pp. 175-88.

BOYER, S. and ELLIOTT, D. (1982): Thrust systems, *Bulletin of the American Association of Petroleum Geologists*, **66**, pp. 1196-230.

BREWER, J. A. and SMYTHE, D. K. (1984): MOIST and the continuity of crustal reflector geometry along the Appalachian-Caledonian orogen, *Journal of the Geological Society, London*, **141**, pp. 105-20.

BREWER, M. S., BROOK, M. and POWELL, D. (1979): Dating of the tectono-metamorphic history of the southwestern Moine, in HARRIS, A. L., HOLLAND, C. H. and LEAKE, B. E. (eds): *The Caledonides of the British Isles – Reviewed*, Geological Society, London, Special Publications, **8**, pp. 129-37.

BREWER, T. S., STOREY, C. D., PARRISH, R. R., TEMPERLEY, S. and WINDLEY, B. F. (2003): Grenvillian age decompression of eclogites in the Glenelg-Attadale Inlier, NW Scotland, *Journal of the Geological Society, London*, **160**, pp. 565-74.

BRIDEN, J. C., TURNELL, H. B. and WATTS, D. R. (1984): British paleomagnetism, Iapetus Ocean and the Great Glen Fault, *Geology*, **12**, pp. 136-39.

BRITISH GEOLOGICAL SURVEY (1997): *Tongue, Scotland Sheet 114E, Solid Geology, 1:50,000*, British Geological Survey, Keyworth, Nottingham.

BRITISH GEOLOGICAL SURVEY (2002): *Loch Eriboll, Scotland Sheet 114W, Solid Geology, 1:50,000*, British Geological Survey, Keyworth, Nottingham.

BROWN, R. L., DALZIEL, I. W. D. and JOHNSON, M. R. W. (1970): A review of the structure and stratigraphy of the Moinian of Ardgour, Moidart and Sunart-Argyll and Inverness-shire, *Scottish Journal of Geology*, **6**, pp. 309-35.

BURNS, I. M. (1994): *Tectonothermal evolution and petrogenesis of the Naver and Kirtomy nappes, north Sutherland*, PhD thesis, Oxford Brookes University.

BURNS, I. M., FOWLER, M. B., STRACHAN R. A. and GREENWOOD, P. B. (2004): Geochemistry, petrogenesis and structural setting of the meta-igneous Strathy Complex: a unique basement block within the Scottish Caledonides?, *Geological Magazine*, **141**, pp. 209-23.

BUTLER, R. W. H. (1982): A structural analysis

of the Moine Thrust zone between Loch Eriboll and Foinaven, NW Scotland, *Journal of Structural Geology*, **4**, pp. 19-29.

BUTLER, R. W. H. (1987): Thrust sequences, *Journal of the Geological Society, London*, **144**, pp. 619-34.

BUTLER, R. W. H. (2004a): Mountain Building with Henry Cadell, *Geoscientist*, **14**(6), pp. 4-7.

BUTLER, R. W. H. (2004b): The nature of 'roof thrusts' in the Moine Thrust Belt, NW Scotland: implications for the structural evolution of thrust belts, *Journal of the Geological Society, London*, **161**, pp. 849-59.

BUTLER, R. W. H. (2009): Eriboll in MENDUM, J R., BARBER, A. J., BUTLER, R. W. H., FLINN, D., GOODENOUGH, K. M., KRABBENDAM, M., PARK, R.G. and STEWART, A.D. (eds): *Lewisian Torridonian and Moine Rocks of Scotland*, Geological Conservation Review Series, Joint Nature Committee, Peterborough, **34**, pp. 242-49.

BUTLER, R. W. H. and COWARD, M. P. (1984): Geological constraints, structural evolution and the deep geology of the NW Scottish Caledonides, *Tectonics*, **3**, pp. 347-65.

BUTLER, R. W. H., HOLDSWORTH, R. E. and MATTHEWS, S. J. (2006): Styles of basement involvement in the Moine thrust Belt, NW Scotland, in MAZZOLI, S. and BUTLER, R. W. H. (eds): *Styles of continental contraction*, Geological Society of America Special Paper, **414**, pp. 133-50.

CADELL, H. M. (1888): Experimental researches in mountain building, *Transactions of the Royal Society of Edinburgh*, **35**, pp. 337-57.

CANNING, J. C., HENNEY, P. J., MORRISON, M. A. and GASKARTH, J. W. (1996): Geochemistry of late Caledonian minettes from Northern Britain: Implications for the Caledonian sub-continental lithospheric mantle, *Mineralogical Magazine*, **128**, pp. 385-88.

CANNING, J. C., HENNEY, P. J., MORRISON, M. A., VAN CALSTEREN, P. W., GASKARTH, J. W. and SWARBRICK, A. (1998): The Great Glen Fault: A major lithospheric boundary, *Journal of the Geological Society, London*, **155**, pp. 424-27.

CAWOOD, P. A., NEMCHIN, A. A., STRACHAN, R. A., KINNY, P. D. and LOEWY, S. (2004): Laurentian provenance and tectonic setting for the upper Moine Supergroup, Scotland, constrained by detrital zircons from the Loch Eil and Glen Urquhart successions, *Journal of the Geological Society, London*, **161**, pp. 863-74.

CAWOOD, P. A., NEMCHIN, A. A., STRACHAN, R. A., PRAVE, A. R. and KRABBENDAM, M. (2007): Sedimentary basin and detrital zircon record along east Laurentia and Baltica during assembly and breakup of Rodinia, *Journal of the Geological Society, London*, **164**, pp. 257-75.

CAWOOD, P. A., STRACHAN, R. A., CUTTS, K., KINNY, P. D., HAND, M. and PISAREVSKY, S. (2010): Neoproterozoic orogeny along the margin of Rodinia: development of the Valhalla Orogen, North Atlantic, *Geology*, **38**, pp. 99-102.

CHENG, Y. C. (1942): A hornblendic complex, including appinitic types, in the migmatite area of northern Sutherland, *Proceedings of the Geologists' Association*, **53**, pp. 67-85.

CHENG, Y. C. (1943): The migmatite area around Bettyhill, Sutherland, *Quarterly Journal of the Geological Society of London*, **99**, pp. 107-54.

CLIFFORD, T. N. (1957): The stratigraphy and structure of part of the Kintail district of southern Ross-shire – its relationship to the Northern Highlands, *Quarterly Journal of the Geological Society of London*, **113**, pp. 57-92.

COCKS, L. R. M. and TORSVIK, T. H. (2002): Earth geography from 500 to 400 million years ago: a faunal and palaeomagnetic review, *Journal of the Geological Society, London*, **159**, pp. 631-44.

COWARD, M. P. (1980): The Caledonian thrust and shear zones of N.W. Scotland, *Journal of Structural Geology*, **2**, pp. 11-17.

References

COWARD, M. P. (1984): A geometrical study of the Arnaboll and Heilam thrust sheets, NW of Ben Arnaboll, Sutherland, *Scottish Journal of Geology*, **20**, pp. 87-106.

COWARD, M. P. (1988): The Moine Thrust and the Scottish Caledonides, in *Geometries and Mechanics of Thrusting, with special reference to the Appalachians*, in MITRA, G. and WOJTAL, S. (eds): Geological Society of America Special Paper, **222**, pp. 1-16.

COWARD, M. P. (1990): The Precambrian, Caledonian and Variscan framework to NW Europe, in HARDMAN, R. F. P. and BROOKS, J. (eds): *Tectonic Events Responsible for Britain's Oil and Gas Reserves*, Geological Society, London, Special Publications, **55**, pp. 1-34.

COWARD, M. P. and ENFIELD, M. A. (1987): The structure of the West Orkney Basin and adjacent basins, in BROOKS, J. and GLENNIE, K. (eds): *Petroleum Geology of North West Europe: proceedings from the 3rd Petroleum Geology Conference*, (UK: Graham & Titman), pp. 687-96.

CUTTS, K. A., HAND, M., KELSEY, D. E. and STRACHAN, R. A. (2009): Orogenic versus extensional settings for regional metamorphism: Knoydartian events in the Moine Supergroup revisited, *Journal of the Geological Society, London*, **166**, pp. 201-204.

CUTTS, K. A., KINNY, P. D., STRACHAN, R. A., HAND, M., KELSEY, D. E., EMERY, M., FRIEND, C. R. L. and LESLIE, A. G. (2010): Three metamorphic events recorded in a single garnet: coupled phase modelling with in situ LA-ICPMS, and SIMS geochronology from the Moine Supergroup, NW Scotland, *Journal of Metamorphic Geology*, **28**, pp. 249-67.

DALLMEYER, R. D., STRACHAN, R. A., ROGERS, G., WATT, G. R. and FRIEND, C. R. L. (2001): Dating deformation and cooling in the Caledonian thrust nappes of north Sutherland, Scotland: insights from $^{40}Ar/^{39}Ar$ and Rb-Sr chronology, *Journal of the Geological Society, London*, **158**, pp. 501-12.

DALZIEL, I. W. D. (1966): A structural study of the granitic gneiss of Western Ardgour, Argyll and Inverness-shire, *Scottish Journal of Geology*, **2**, pp. 125-52.

DALZIEL, I. W. D. and SOPER, N. J. (2001): Neoproterozoic extension on the Scottish promontory of Laurentia: paleogeographic and tectonic implications, *Journal of Geology*, **109**, pp. 299-317.

DEWEY, J. F. and SHACKLETON, R. J. (1984): A model for the evolution of the Grampian tract in the early Caledonides and Appalachians, *Nature*, **312**, pp. 115-20.

DEWEY, J. F. and RYAN, P. D. (1990): The Ordovician evolution of the South Mayo Trough, western Ireland, *Tectonics*, **9**, pp. 887-903.

DEWEY, J. F. and MANGE, M. (1999): Petrography of Ordovician and Silurian sediments in the western Irish Caledonides: tracers of a short-lived Ordovician continent-arc collision orogeny and the evolution of the Laurentian Appalachian-Caledonian margin, in MACNIOCAILL, C. and RYAN, P. D. (eds): *Continental Tectonics*, Geological Society, London, Special Publications, **164**, pp. 55-107.

DEWEY, J. F. and STRACHAN, R. A. (2003): Changing Silurian-Devonian relative plate motion in the Caledonides: sinistral transpression to sinistral transtension, *Journal of the Geological Society, London*, **160**, pp. 219-29.

ELLIOTT, D. and JOHNSON, M. R. W. (1980): Structural evolution in the northern part of the Moine thrust belt, NW Scotland, *Transactions of the Royal Society of Edinburgh: Earth Sciences*, **71**, pp. 69-96.

ENFIELD, M. A. and COWARD, M. P. (1987): The structure of the West Orkney Basin, northern Scotland, *Journal of the Geological Society, London*, **144**, pp. 871-84.

EVANS, D. J. and WHITE, S. H. (1984): Microstructural and fabric studies from the Moine rocks of the Moine Nappe, Eriboll, NW Scotland, *Journal of Structural Geology*, **6**, pp. 369-89.

FETTES, D. J. (1979): A metamorphic map of the British and Irish Caledonides, in HARRIS, A.

References

L., HOLLAND, C. H. and LEAKE, B. E. (eds): *The Caledonides of the British Isles – Reviewed*, Geological Society, London, Special Publications, **8**, pp. 307-21.

FETTES, D. J. and MACDONALD, R. (1978): The Glen Garry vein complex, *Scottish Journal of Geology*, **14**, pp. 335-58.

FETTES, D. J., LONG, C. B., MAX, M. D. and YARDLEY, B. W. D. (1985): Grade and time of metamorphism in the Caledonide Orogen of Britain and Ireland, in HARRIS, A. L. (ed.): *The Nature and Timing of Orogenic Activity in the Caledonian Rocks of the British Isles*, Geological Society, London, Memoirs, **9**, pp. 41-53.

FISCHER, M. W. and COWARD, M. P. (1982): Strains and folds within thrust sheets: the Heilam sheet NW Scotland, *Tectonophysics*, **88**, pp. 291-312.

FLETT, J. S. (1905): *On the petrographic characters of the inliers of Lewisian rocks among the Moine gneisses of the north of Scotland*, Memoirs of the Geological Survey, Summary of Progress for 1905, pp. 155-67.

FLINN, D. (1961): Continuation of the Great Glen Fault beyond the Moray Firth, *Nature*, **191**, pp. 589-91.

FREEMAN, S. R., BUTLER, R. W. H., CLIFF, R. A. and REX, D. C. (1998): Dating mylonite evolution: an Rb-Sr and K-Ar study of the Moine mylonites, NW Scotland, *Journal of the Geological Society, London*, **155**, pp. 745-58.

FRIEND, C. R. L., JONES, K. A. and BURNS, I. M. (2000): New high-pressure granulite facies event in the Moine Supergroup, northern Scotland: implications for Taconic (early Caledonian) crustal evolution, *Geology*, **28**, pp. 543-46.

FRIEND, C. R. L., KINNY, P. D., ROGERS, G., STRACHAN, R. A. and PATERSON, B. A. (1997): U-Pb zircon geochronological evidence for Neoproterozoic events in the Glenfinnan Group (Moine Supergroup): the formation of the Ardgour granite gneiss, north-west Scotland, *Contributions to Mineralogy and Petrology*, **128**, pp. 101-13.

FRIEND, C. R. L., STRACHAN, R. A. and KINNY, P. D. (2008): U-Pb zircon dating of basement inliers within the Moine Supergroup, Scottish Caledonides: implications of Archaean protolith ages, *Journal of the Geological Society, London*, **165**, pp. 807-15.

FRIEND, C. R. L., STRACHAN, R. A., KINNY, P. D. and WATT, G. R. (2003): Provenance of the Moine Supergroup of NW Scotland: evidence from geochronology of detrital and inherited zircons from sediments, granites and migmatites, *Journal of the Geological Society, London*, **160**, pp. 247-57.

GEIKIE, A. (1884): The crystalline schists of the Scottish Highlands, *Nature*, **31**, pp. 29-31.

GILETTI, B. J., MOORBATH, S. and LAMBERT, R. ST. J. (1961): A geochronological study of the metamorphic complexes of the Scottish Highlands, *Quarterly Journal of the Geological Society of London*, **117**, pp. 233-72.

GLENDINNING, N. R. W. (1988): Sedimentary structures and sequences within a late Proterozoic tidal shelf deposit: the Upper Morar Psammite Formation of north-western Scotland, in WINCHESTER, J. A. (ed.): *Later Proterozoic Stratigraphy of the Northern Atlantic Regions* (Glasgow: Blackie), pp. 14-31.

GRANT, C. J. and HARRIS, A. L. (2000): The kinematic and metamorphic history of the Sgurr Beag Thrust, Ross-shire, Scotland, *Journal of Structural Geology*, **22**, pp. 191-205.

HALL, J., BREWER, J. A., MATTHEWS, D. H. and WARNER, M. (1984): Crustal structure across the Caledonides from the WINCH seismic reflection profile: Influences on the evolution of the Midland Valley of Scotland, *Transactions of the Royal Society of Edinburgh: Earth Sciences*, **75**, pp. 97-109.

HARRISON, V. E. and MOORHOUSE, S. J. (1976): A possible early Scourian supracrustal assemblage within the Moine, *Journal of* the *Geological Society, London*, **132**, pp. 461-66.

HIPPLER, S. J. and KNIPE, R. J. (1990): The evolution of cataclastic rocks from a pre-existing mylonite, in KNIPE, R. J. and RUTTER, E. H. (eds): *Deformation Mechanisms, Rheology and Tectonics,* Geological Society, London, Special Publications, **54**, pp. 71-79.

HOLDSWORTH, R. E. (1987): *Basement/cover relationships, reworking and Caledonian ductile thrusts of the Northern Moine, NW Scotland,* University of Liverpool PhD thesis.

HOLDSWORTH, R. E. (1988): The stereographic analysis of facing, *Journal of Structural Geology*, **10**, pp. 219-23.

HOLDSWORTH, R. E. (1989a): The geology and structural evolution of a Caledonian fold and ductile thrust zone, Kyle of Tongue region, Sutherland, northern Scotland, *Journal of the Geological Society, London*, **146**, pp. 809-23.

HOLDSWORTH, R. E. (1989b): Late brittle deformation in a Caledonian ductile thrust wedge: new evidence for gravitational collapse in the Moine Thrust sheet, Sutherland, Scotland, *Tectonophysics*, **170**, pp. 17-28.

HOLDSWORTH, R. E. (1990): Progressive deformation structures associated with ductile thrusts in the Moine Nappe, Sutherland, N. Scotland, *Journal of Structural Geology*, **12**, pp. 443-52.

HOLDSWORTH, R. E. and ROBERTS, A. M. (1984): A study of early curvilinear fold structures and strain in the Moine of the Glen Garry region, Inverness-shire, *Journal of the Geological Society, London*, **141**, pp. 327-38.

HOLDSWORTH, R. E. and STRACHAN, R. A. (1988): The structural age and possible origin of the Vagastie Bridge granite and associated intrusions, central Sutherland, *Geological Magazine*, **125**, pp. 613-20.

HOLDSWORTH, R. E. and GRANT, C. J. (1990): Convergence-related 'dynamic spreading' in a mid-crustal ductile thrust zone: a possible orogenic wedge model, in KNIPE, R. J. and RUTTER, E. H. (eds): *Deformation Mechanisms, Rheology and Tectonics,* Geological Society, London, Special Publications, **54**, pp. 491-500.

HOLDSWORTH, R. E., HARRIS, A. L. and ROBERTS, A. M. (1987): The stratigraphy, structure and regional significance of the Moine rocks of Mull, Argyllshire, W Scotland, *Geological Journal*, **22**, pp. 83-107.

HOLDSWORTH, R. E., STRACHAN, R. A. and HARRIS, A. L. (1994): Precambrian rocks in northern Scotland east of the Moine Thrust: the Moine Supergroup, in GIBBONS, W. and HARRIS, A. L. (eds): *A revised correlation of Precambrian rocks in the British Isles,* Geological Society, London, Special Report, **22**, pp. 23-32.

HOLDSWORTH, R. E., McERLEAN, M. A. and STRACHAN, R. A. (1999): The influence of country rock structural architecture during pluton emplacement: the Loch Loyal syenites, Scotland, *Journal of the Geological Society, London*, **156**, pp. 163-75.

HOLDSWORTH, R. E., STRACHAN, R. A. and ALSOP, G. I. (2001a): *Geology of the Tongue District,* Memoir of the British Geological Survey (UK: HMSO).

HOLDSWORTH, R. E., STEWART, M., IMBER, J. and STRACHAN, R. A. (2001b): The structure and rheological evolution of reactivated continental fault zones: a review and case study, in MILLER, J. A., HOLDSWORTH, R. E., BUICK, I. and HAND, M. (eds): *Continental Reworking and Reactivation,* Geological Society, Special Publication, **184**, pp. 115-37.

HOLDSWORTH, R. E., STRACHAN, R. A., ALSOP, G. I., GRANT, C. J. and WILSON, R. W. (2006): Thrust sequences and the significance of low-angle, out-of-sequence faults in the northernmost Moine Nappe and Moine Thrust Zone, NW Scotland, *Journal of the Geological Society, London*, **163**, pp. 801-14.

References

HOLDSWORTH, R. E., ALSOP, G. I. and STRACHAN, R. A. (2007): Tectonic stratigraphy and structural continuity of the northernmost Moine Thrust Zone and Moine Nappe, Scottish Caledonides, in RIES, A. C., BUTLER, R. W. H. and GRAHAM, R.H. (eds): *Deformation of the Continental Crust: The Legacy of Mike Coward*, Geological Society, London, Special Publication, **272**, pp. 121-42.

HOWARTH, R. J. and LEAKE, B. E. (2002): *The Life of Frank Cole Phillips* (London: Geological Society).

HUTTON, D. H. W. and McERLEAN, M. (1991): Silurian and Early Devonian sinistral deformation of the Ratagain granite, Scotland: constraints on the age of Caledonian movements on the Great Glen fault system, *Journal of the Geological Society, London*, **148**, pp. 1-4.

HYSLOP, E. K. (1992): *Strain-induced metamorphism and pegmatite development in the Moine rocks of Scotland*, PhD thesis, University of Hull.

JACQUES, J. M. and REAVY, R. J. (1994): Caledonian plutonism and major lineaments in the SW Scottish Highlands, *Journal of the Geological Society, London*, **151**, pp. 955-69.

JOHNSON, H. D. (1978): Shallow siliciclastic seas, in READING, H. (ed): *Sedimentary Environments and Facies* (Oxford: Blackwell Scientific Publications), pp. 209-58.

JOHNSON, M. R. W. and DALZIEL, I. W. D. (1963): Metamorphosed lamprophyres and the late thermal history of the Moine, *Geological Magazine*, **103**, pp. 240-49.

JOHNSON, M .R. W., KELLEY, S. P., OLIVER, G. J. H. and WINTER, D. A. (1985): Thermal effects and timing of thrusting in the Moine Thrust zone, *Journal of the Geological Society, London*, **142**, pp. 863-74.

JOHNSTONE, G. S. (1975): The Moine Succession, in HARRIS, A. L., SHACKLETON, R. M., WATSON, R. M., WATSON, J. V., DOWNIE, C., HARLAND, W. B. and MOORBATH, S. (eds): *A Correlation of Precambrian rocks in the British Isles*, Geological Society, London, Special Report, **6**, pp. 30-42.

JOHNSTONE, G. S., SMITH, D. I. and HARRIS, A. L. (1969): The Moinian Assemblage of Scotland, in KAY, M. (ed.): *North Atlantic geology and continental drift*, American Association of Petroleum Geologists, Memoirs, **12**, pp. 159-80.

KELLEY, S. P. (1988): The relationship between K-Ar mineral ages, mica grainsizes and movement on the Moine Thrust Zone, NW Highlands, Scotland, *Journal of the Geological Society, London*, **145**, pp. 1-10.

KELLEY, S. P. and POWELL, D. (1985): Relationships between marginal thrusting and movement on major, internal shear zones in the N. Highland Caledonides, Scotland, *Journal of Structural Geology*, **7**, pp. 43-56.

KENNEDY, W. Q. (1946): The Great Glen Fault, *Quarterly Journal of the Geological Society of London*, **102**, pp. 41-76.

KENNEDY, W. Q. (1949): Zones of progressive metamorphism in the Moine Schists of the Western Highlands of Scotland, *Mineralogical Magazine*, **86**, pp. 43-56.

KENNEDY, W. Q. (1955): The tectonics of the Morar anticline and the problems of the north-west Caledonian front, *Quarterly Journal of the Geological Society of London*, **110**, pp. 375-90.

KINNY, P. D. and FRIEND, C. R. L. (1997): U-Pb isotopic evidence for the accretion of different crustal blocks to form the Lewisian Complex of northwestern Scotland, *Contributions to Mineralogy & Petrology*, **129**, pp. 26-340.

KINNY, P. D., FRIEND, C. R. L. and LOVE, G. J. (2005): Proposal for a terrane-based nomenclature for the Lewisian Gneiss Complex of NW Scotland, *Journal of the Geological Society, London*, **162**, pp. 175-86.

KINNY, P. D., FRIEND, C. R. L., STRACHAN, R. A., WATT, G. R. and BURNS, I. M. (1999): U-Pb geochronology of regional migmatites, East Sutherland, Scotland: evidence for crustal melting during the Caledonian orogeny, *Journal of the Geological Society, London*, **156**, pp. 1143-52.

KINNY, P. D., STRACHAN, R. A., ROGERS, G. R., FRIEND, C. R. L. and KOCKS, H. (2003a): U-Pb geochronology of deformed metagranites in central Sutherland, Scotland: evidence for widespread Silurian metamorphism and ductile deformation of the Moine Supergroup during the Caledonian orogeny, *Journal of the Geological Society, London*, **160**, pp. 259-69.

KINNY, P. D., STRACHAN, R. A., KOCKS, H. and FRIEND, C. R. L. (2003b): U-Pb geochronology of late Neoproterozoic augen granites in the Moine Supergroup, NW Scotland: dating of rift-related, felsic magmatism during supercontinent break-up?, *Journal of the Geological Society of London*, **160**, pp. 925-34.

KIRKLAND, C. L., STRACHAN, R. A. and PRAVE, A. R. (2008): Detrital zircon signature of the Moine Supergroup, Scotland: contrasts and comparisons with other Neoproterozoic successions within the circum-North Atlantic region, *Precambrian Research*, **163**, pp. 332-50.

KOCKS, H., STRACHAN, R. A. and EVANS, J. A. (2006): Heterogeneous reworking of Grampian metamorphic complexes during Scandian thrusting in the Scottish Caledonides: insights from the structural setting and U-Pb geochronology of the Strath Halladale Granite, *Journal of the Geological Society, London*, **163**, pp. 525-38.

KRABBENDAM, M, PRAVE, A. R. and CHEER, D. (2008): A fluvial origin for the Neoproterozoic Morar Group, NW Scotland: implications for Torridon-Morar group correlations and the Grenville Orogen Foreland Basin, *Journal of the Geological Society, London*, **165**, pp. 379-94.

LAMBERT, R. ST. J. (1959): The mineralogy and metamorphism of the Moine schists of the Morar and Knoydart districts of Inverness-shire, *Transactions of the Royal Society of Edinburgh*, **63**, pp. 553-86.

LAMBERT, R. ST. J. (1969): Isotopic studies relating to the Pre-Cambrian history of the Moinian of Scotland, *Proceedings of the Geological Society, London*, **1652**, pp. 243-45.

LAMBERT, R. ST. J. and McKERROW, W. S. (1976): The Grampian Orogeny, *Scottish Journal of Geology*, **12**, pp. 271-92.

LAPWORTH, C. (1883): On the structure and metamorphism of the rocks of the Durness-Eriboll district, *Proceedings of the Geological Association*, **8**, pp. 438-42.

LAPWORTH, C. (1885): The Highland Controversy in British geology: its causes, course and consequence, *Nature*, **32**, pp. 558-59.

LAW, R. D. and JOHNSON, M. R. W. (2010) Microstructures and crystal fabrics of the Moine thrust zone and Moine nappe: history of research and changing tectonic interpretations in: LAW, R. D., BUTLER, R. W. H., HOLDSWORTH, R. E., KRABBENDAM, M. and STRACHAN R. A. (eds): *Continental Tectonics and Mountain Building – The Legacy of Peach and Horne*, Geological Society, London, Special Publication, **335**, pp. 441-501.

LAW, R. K., CASEY, M. and KNIPE, R. J. (1986): Kinematic and tectonic significance of microstructural and crystallographic fabrics within quartz mylonites from the Assynt and Eriboll regions of the Moine thrust zone, NW Scotland, *Transactions of the Royal Society of Edinburgh, Earth Sciences*, **77**, pp. 99-126.

LEEDAL, G. P. (1952): The Cluanie igneous intrusion, Inverness-shire and Ross-shire, *Quarterly Journal of the Geological Society of London*, **108**, pp. 35-63.

LESLIE, A. G., KRABBENDAM, M., KIMBALL, G. S. and STRACHAN, R. A. (2010): Regional-scale lateral variation and linkage in ductile thrust architecture: the Oykell Traverse Zone, and mullions, in the Moine Nappe, NW Scotland in Scotland in LAW, R. D., BUTLER, R. W. H., HOLDSWORTH, R. E., KRABBENDAM, M. and STRACHAN R. A. (eds): *Continental Tectonics and Mountain*

Building – *The Legacy of Peach and Horne*, Geological Society, London, Special Publication, **335**, pp. 357-79.

LONG, L. E. and LAMBERT, R. ST. J. (1963): Rb-Sr isotopic ages from the Moine series, in JOHNSON, M. R. W. and STEWART, F. H. (eds): *The British Caledonides* (Edinburgh: Oliver & Boyd), pp. 217-46.

McBRIDE, J. H. (1994) Investigating the crustal structure of a strike-slip 'step-over' zone along the Great Glen Fault, *Tectonics*, **13**, pp. 1150-60.

McCLAY, K. R. and COWARD, M. P. (1981): The Moine Thrust zone: an overview, in McCLAY, K. R. and PRICE, N. J. (eds): *Thrust and Nappe Tectonics*, Geological Society, London, Special Publications, **9**, pp. 241-60.

MACKENZIE, W. S. (1949): Kyanite gneiss within a thermal aureole, *Geological Magazine*, **86**, pp. 251-55.

MacQUEEN, J. A. and POWELL, D. (1977): Relationships between deformation and garnet growth in Moine (Precambrian) rocks of western Scotland, *Geological Society of America Bulletin*, **88**, pp. 235-40.

MACKIE, D. W. (1975): The Brochs of Scotland, in FOWLER, P. J. (ed): *Recent Work in Rural Archaeology* (Bradford: Moonraker Press), pp. 72-92.

MAY, F., PEACOCK, J. D., SMITH, D. I. and BARBER, A. J. (1993): *Geology of the Kintail District*, Memoir of the British Geological Survey.

MENDUM, J. R. (1976): A strain study of the Strathan Conglomerate, North Sutherland, *Scottish Journal of Geology*, **12**, pp. 135-46.

MENDUM, J. R., BARBER, A. J., BUTLER, R. W. H., FLINN, D., GOODENOUGH, K. M., KRABBENDAM, M., PARK, R. G. and STEWART, A. D. (eds) (2009): *Lewisian, Torridonian and Moine Rocks of Scotland*, Geological Conservation Review Series, Joint Nature Committee, Peterborough, **34**, 722 pp.

MENDUM, J. R. and NOBLE, S. R. (2010): Mid-Devonian sinistral transpressional movements on the Great Glen fault: the rise of the Rosemarkie Inlier and the Acadian Event in Scotland in LAW, R. D., BUTLER, R. W. H., HOLDSWORTH, R. E., KRABBENDAM, M. and STRACHAN R. A. (eds): *Continental Tectonics and Mountain Building – The Legacy of Peach and Horne*, Geological Society, London, Special Publication, **335**, pp. 159-85.

MILLAR, I. L. (1999): Neoproterozoic extensional basic magmatism associated with emplacement of the West Highland granite gneiss in the Moine Supergroup of NW Scotland, *Journal of the Geological Society, London*, **156**, pp. 1153-62.

MOORHOUSE, S. J. (1976): The geochemistry of the Lewisian and the Moinian of the Borgie area, north Sutherland, *Scottish Journal of Geology*, **12**, pp. 159-67.

MOORHOUSE, S. J. and MOORHOUSE, V. E. (1979): The Moine amphibolite suites of central and northern Sutherland, *Mineralogical Magazine*, **43**, 211-25.

MOORHOUSE, S. J. and MOORHOUSE, V. E. (1988): The Moine Assemblage in Sutherland, in WINCHESTER, J. A. (ed.): *Later Proterozoic Stratigraphy in the Northern Atlantic Regions* (Glasgow: Blackie), pp. 54-73.

MOORHOUSE, V. E. (1979): *The geology and geochemistry of the Bettyhill-Strathy area of north-east Sutherland*, PhD thesis, University of Hull.

MOORHOUSE, V. E. and MOORHOUSE, S. J. (1983): The geology and geochemistry of the Strathy complex of north-east Sutherland, Scotland, Mineralogical Magazine, **47**, pp. 123-37.

MOORHOUSE, S. J., MOORHOUSE, V. E. and HOLDSWORTH, R. E. (1988): Excursion 12: North Sutherland, in ALLISON, I., MAY, F. and STRACHAN, R. A. (eds): *An Excursion Guide to the Moine Geology of the Scottish Highlands* (Edinburgh: Scottish Academic Press), pp. 216-48.

MYKURA, W. (1982): The Old Red Sandstone east of Loch Ness, Inverness-shire, *Institute of Geological Sciences, Report*, **82-13**, Edinburgh.

OLIVER, G. J. H. (2001): Reconstruction of the Grampian episode in Scotland: its place in the Caledonian Orogeny, *Tectonophysics*, **332**, pp. 23-49.

OLIVER, G. J. H., WILDE, S. and YUSHENG, W. (2008): Geochronology of Scottish granitoids between ~600 and ~390 Ma: from break-up of Rodinia to collisions with a Iapetan arc, Avalonia, Baltica and Armorica, *Journal of the Geological Society, London*, **165**, pp. 661-74.

OPEN UNIVERSITY (2003): *Mountain Building in Scotland* (Milton Keynes: Open University).

PANKHURST, R. J. (1982): Geochronological tables for British igneous rocks, in SUTHERLAND, D.S. (ed): *Igneous Rocks of the British Isles* (Chichester: John Wiley & Sons), pp. 575-81.

PARK, R. G., STEWART, A. D. and WRIGHT, D. T. (2002): The Hebridean Terrane, in TREWIN, N. (ed): *Geology of Scotland* (4th edition) (London: Geological Society), pp. 45-80.

PASSCHIER, C. W. (2001): Flanking structures, *Journal of Structural Geology*, **23**, pp. 951-62.

PEACH, B. N., HORNE, J., GUNN, W., CLOUGH, C. T. and HINXMAN, L. W. (1907): *The Geological Structure of the Northwest Highlands of Scotland*, Memoirs of the Geological Survey of Great Britain.

PEACH, B. N., HORNE, J., WOODWARD, H. B., CLOUGH, C. T., HARKER, A. and WEDD, C. D. (1910): *The Geology of Glenelg, Lochalsh and south-east part of Skye*, Memoirs of the Geological Survey of Great Britain.

PEACH, B. N., GUNN, W., CLOUGH, C. T., HINXMAN, L. W., CRAMPTON, C. B., ANDERSON, E. M. and FLETT, J. S. (1912): *The Geology of Ben Wyvis, Carn Chuinneag, Inchbae and the surrounding country*, Memoirs of the Geological Survey of Great Britain.

PEACH, B. N. *et al.* (1913): *The geology of central Ross-shire*, Memoirs of the Geological Survey of Great Britain.

PEACOCK, J. D. (1975): 'Slide' rocks in the Moine of the Loch Shin area, Northern Scotland, *Bulletin of the Geological Survey, Great Britain*, **49**, pp. 23-30.

PEACOCK, J. D. (1977): Metagabbros in granitic gneiss, Inverness-shire, and their significance in the structural history of the Moines, *Institute of Geological Sciences Report*, **77/20**.

PIASECKI, M. A. J. (1984): Ductile thrusts as time markers in orogenic evolution: an example from the Scottish Caledonides, in GALSON, D. and MUELLER, S. E. (eds): *First European Geotraverse Workshop: the northeastern segment* (Strasbourg: Publication of the European Science Foundation), pp. 109-14.

PIASECKI, M. A. J. and VAN BREEMEN, O. (1983): Field and isotopic evidence for a *c.*750 Ma tectonothermal event in the Moine rocks of the central Highland region of the Scottish Caledonides, *Transactions of the Royal Society of Edinburgh: Earth Sciences*, **73**, pp. 119-34.

PICKERING, K. T., BASSETT, M. G. and SIVETER, D. J. (1988): Late Ordovician-Early Silurian destruction of the Iapetus Ocean: Newfoundland, British Isles and Scandinavia – a discussion, *Transactions of the Royal Society of Edinburgh: Earth Sciences*, **79**, pp. 361-82.

PITCHER, W. S., ELWELL, R. W. D., TOZER, C. F. and CAMBRAY, F. W. (1964): The Leannan Fault, *Quarterly Journal of the Geological Society of London*, **120**, pp. 241-73.

POOLE, A. B. and SPRING, J. S. (1974): Major structures in Morar and Knoydart, NW Scotland, *Journal of the Geological Society, London*, **130**, pp. 43-53.

POTTS, G. J., HUNTER, R. H., HARRIS, A. L. and FRASER, F. M. (1995): Late-orogenic extension tectonics at the NW margin of the Caledonides in Scotland, *Journal*

of the Geological Society, London, **152**, pp. 907-10.
POWELL, D. (1964): The stratigraphical succession of the Moine schists around Lochailort (Inverness-shire) and its regional significance, *Proceedings of the Geologists' Association*, **75**, pp. 223-50.
POWELL, D. (1966): The Structure of the South-Eastern Part of the Morar Antiform, Inverness-shire, *Proceedings of the Geologists' Association*, **77**, pp. 79-100.
POWELL, D. (1974): Stratigraphy and structure of the western Moine and the problem of Moine orogenesis, *Journal of the Geological Society, London*, **130**, pp. 575-93.
POWELL, D. and MacQUEEN, J. A. (1976): Relationships between garnet shape, rotational inclusion fabrics and strain in some Moine metamorphic rocks of Skye, Scotland, *Tectonophysics*, **35**, pp. 391-402.
POWELL, D., BROOK, M. and BAIRD, A. W. (1983): Structural dating of a Precambrian pegmatite in Moine rocks of northern Scotland and its bearing on the status of the 'Morarian Orogeny', *Journal of the Geological Society, London*, **140**, pp. 813-23.
POWELL, D., BAIRD, A. W., CHARNLEY, N. R. and JORDAN, P. J. (1981): The metamorphic environment of the Sgurr Beag Slide: a major crustal displacement zone in Proterozoic, Moine rocks of Scotland, *Journal of the Geological Society, London*, **138**, pp. 661-73.
RAMSAY, J. G. (1956): The supposed Moinian basal conglomerate at Glen Strathfarrar, Inverness-shire, *Geological Magazine*, **93**, pp. 32-40.
RAMSAY, J. G. (1957): Superimposed folding at Loch Monar, Inverness-shire and Ross-shire, *Quarterly Journal of the Geological Society of London*, **113**, pp. 271-307.
RAMSAY, J. G. (1958): Moine-Lewisian relations at Glenelg, Inverness-shire, *Quarterly Journal of the Geological Society of London*, **113**, pp. 487-523.
RAMSAY, J. G. (1960): The deformation of early linear structures in areas of repeated folding, *Journal of Geology*, **68**, pp. 75-93.
RAMSAY, J. G. (1962): The geometry and mechanics of 'similar type folds, *Journal of Geology*, **70**, pp. 309-27.
RAMSAY, J. G. (1967): *Folding and fracturing of rocks* (New York: McGraw Hill Book Co.).
RAMSAY, J. G. (1997): The geometry of a deformed unconformity in the Caledonides of NW Scotland, in SENGUPTA, S. (ed): *Evolution of Geological Structures in Micro- to Macro-scales* (London: Chapman and Hall), pp. 445-72.
RAMSAY, J. G. and SPRING, J. S. (1962): Moinian stratigraphy in the western Highlands of Scotland, *Proceedings of the Geologists' Association*, **73**, pp. 295-326.
RAMSAY, J. G. and HUBER, M. I. (1987): *The Techniques of Modern Structural Geology 2: Folds and Fractures* (London: Academic Press).
RAMSAY, J. G. and LISLE, R. J. (2000): *The Techniques of Modern Structural Geology 3: Applications of Continuum Mechanics in Structural Geology* (London: Academic Press).
RATHBONE, P. A. and HARRIS, A. L. (1979): Basement-cover relationships at Lewisian inliers in the Moine rocks, in HARRIS, A. L., HOLLAND, C. H. and LEAKE, B. E. (eds): *The Caledonides of the British Isles – Reviewed*, Geological Society, London, Special Publications, **8**, pp. 101-107.
RATHBONE, P. A. and HARRIS, A. L. (1980): Moine and Lewisian near the Great Glen Fault in Easter Ross, *Scottish Journal of Geology*, **16**, pp. 51-64.
RATHBONE, P. A., COWARD, M. P. and HARRIS, A. L. (1983): Cover and basement: A contrast in style and fabrics, in HARRIS, L. D. and WILLIAMS, H. (eds): *Tectonics and Geophysics of Mountain Chains*, Geological Society of America Memoir, **158**, pp. 213-23.

RAWSON, J. R. (2004): *Origin and metamorphic evolution of high pressure rocks in the Glenelg-Attadale inlier of the Caledonian Moine Thrust Nappe, Northwest Scotland*, University of Sheffield PhD thesis.

RAWSON, J. R., CARSWELL, D. A. and SMALLWOOD, D. (2001): Garnet-bearing olivine-websterite within the Eastern Glenelg Lewisian of the Glenelg Inlier, NW Highlands, *Scottish Journal of Geology*, **37**, pp. 27-34.

READ, H. H. (1931): *The Geology of Central Sutherland*, Memoir of the Geological Survey of Great Britain.

READ, H. H. (1961): Aspects of the Caledonian magmatism in Britain, *Liverpool and Manchester Geological Journal*, **2**, pp. 653-83.

RICHEY, J. E. and KENNEDY, W. Q. (1939): The Moine and Sub-Moine Series of Morar, Inverness-shire, *Geological Survey of Great Britain Bulletin*, **2**, pp. 26-45.

RICKWOOD, P. C. (1990): The anatomy of a dyke and the determination of propogation and magma flow directions, in RICKWOOD, P. C. and TUCKER, R. (eds): *Mafic Dykes and Emplacement Mechanisms* (Rotterdam: Balkema), pp. 81-100.

ROBERTS, A. M. (1984): *Stratigraphy and structure in Moine rocks along the Loch Quoich Line, Inverness-shire*, PhD thesis, University of Liverpool.

ROBERTS, A. M. and HARRIS, A. L. (1983): The Loch Quoich Line – a limit of early Palaeozoic crustal reworking in the Moine of the northern Highlands of Scotland, *Journal of the Geological Society, London*, **140**, pp. 883-92.

ROBERTS, A. M., SMITH, D. I. and HARRIS, A. L. (1984): The structural setting and tectonic significance of the Glen Dessary syenite, Inverness-shire, *Journal of the Geological Society, London*, **141**, pp. 1033-42.

ROBERTS, A. M., STRACHAN, R. A., HARRIS, A. L., BARR, D. and HOLDSWORTH, R. E. (1987): The Sgurr Beag nappe: a reassessment of the northern Highland Moine, *Bulletin of the Geological Society of America*, **98**, pp. 497-506.

ROCK, N. M. S., MACDONALD, R., WALKER, B. H., MAY, F., PEACOCK, J. D. and SCOTT, P. (1985): Intrusive metabasite belts within the Moine assemblage, west of Loch Ness, Scotland: evidence for metabasite modification by country rock interactions, *Journal of the Geological Society, London*, **142**, pp. 643-61.

ROGERS, G. and DUNNING, G. R. (1991): Geochronology of appinitic and related granitic magmatism in the W Highlands of Scotland: constraints on the timing of transcurrent fault movement, *Journal of the Geological Society, London*, **148**, pp. 17-27.

ROGERS, G., HYSLOP, E. K., STRACHAN, R. A., PATERSON, B. A. and HOLDSWORTH, R. E. (1998): The structural setting and U-Pb geochronology of Knoydartian pegmatites in W. Inverness-shire: evidence for Neoproterozoic tectonothermal events in the Moine of NW Scotland, *Journal of the Geological Society, London*, **155**, pp. 685-96.

ROGERS, G., KINNY, P. D., STRACHAN, R. A., FRIEND, C. R. L. and PATERSON, B. A. (2001): U-Pb geochronology of the Fort Augustus granite gneiss: constraints on the timing of Neoproterozoic and Palaeozoic tectonothermal events in the NW Highlands of Scotland, *Journal of the Geological Society, London*, **158**, pp. 7-14.

RYAN, P. D. and SOPER, N. J. (2001): Modelling anatexis in intra-cratonic rift basins: an example from the Neoproterozoic rocks of the Scottish Highlands, *Geological Magazine*, **138**, pp. 577-88.

SANDERS, I. S. (1979): Observations on eclogite- and granulite-facies rocks in the basement of the Caledonides, in HARRIS, A. L., HOLLAND, C. H. and LEAKE, B. E. (eds): *The Caledonides of the British Isles – Reviewed*, Geological Society, London, Special Publication, **8**, pp. 97-101.

SANDERS, I. S. (1988): Plagioclase breakdown

and regeneration reactions in Grenville kyanite eclogite at Glenelg, NW Scotland *Contributions to Mineralogy and Petrology*, **98**, pp. 33-39.

SANDERS, I. S., VAN CALSTEREN, P. W. C. and HAWKESWORTH, C. J. (1984): A Grenville Sm-Nd age for the Glenelg eclogite in northwest Scotland, *Nature*, **312**, pp. 439-40.

SMITH, D. I. (1979): Caledonian minor intrusions of the N Highlands of Scotland, in HARRIS, A. L., HOLLAND, C. H. and LEAKE, B. E. (eds): *The Caledonides of the British Isles – Reviewed*, Geological Society, London, Special Publications, **8**, pp. 129-37.

SMITH, D. I. and WATSON, J. V. (1983): Scale and timing of movements on the Great Glen Fault, Scotland, *Geology*, **11**, pp. 523-26.

SMITH, M., ROBERTSON, S. and ROLLIN, K. E. (1999): Rift basin architecture and stratigraphical implications for basement-cover relationships in the Neoproterozoic Grampian Group of the Scottish Caledonides, *Journal of the Geological Society, London*, **156**, pp. 1163-73.

SNOKE, A. W. and TULLIS, J. (1998): An overview of fault rocks, in SNOKE, A. W., TULLIS, J. and TODD, V. R. (eds): *Fault-related rocks: a photographic atlas* (Princeton, New Jersey: Princeton University Press), pp. 3-18.

SNOKE, A. W., TULLIS, J. and TODD, V. R. (1998): *Fault-related rocks: a photographic atlas* (Princeton, NJ: Princeton University Press).

SOPER, N. J. (1971): The earliest Caledonian structures in the Moine Thrust belt, *Scottish Journal of Geology*, **7**, pp. 241-47.

SOPER, N. J. and BROWN, P. E. (1971): Relationship between metamorphism and migmatisation in the northern part of the Moine Nappe, *Scottish Journal of Geology*, **7**, p. 305-25.

SOPER, N. J. and WILKINSON, P. (1975): The Moine Thrust and Moine Nappe at Loch Eriboll, Scotland, *Scottish Journal of Geology*, **11**, pp. 239-59.

SOPER, N. J. and BARBER, A. J. (1982): A model for the deep structure of the Moine thrust zone, *Journal of the Geological Society, London*, **139**, pp. 127-38.

SOPER, N. J. and ANDERTON, R. (1984): Did the Dalradian slides originate as extensional faults?, *Nature*, **307**, pp. 357-60.

SOPER, N. J. and HUTTON, D. H. W. (1984): Late Caledonian sinistral displacements in Britain: Implications for a three-plate model, *Tectonics*, **3**, pp. 781-94.

SOPER, N. J. and HARRIS, A. L. (1997): Report: Highland field workshops 1995-1996, *Scottish Journal of Geology*, **33**, pp. 187-90.

SOPER, N. J., HARRIS, A. L. and STRACHAN, R. A. (1998): Tectonostratigraphy of the Moine Supergroup: a synthesis, *Journal of the Geological Society, London*, **155**, pp. 13-24.

SOPER, N. J., RYAN, P. D. and DEWEY, J. F. (1999): Age of the Grampian orogeny in Scotland and Ireland, *Journal of the Geological Society, London*, **156**, pp. 1231-36.

SOPER, N. J., STRACHAN, R. A., HOLDSWORTH, R. E., GAYER, R. A. and GREILING, R. O. (1992): Sinistral transpression and the Silurian closure of Iapetus, *Journal of the Geological Society, London*, **149**, pp. 871-80.

STEPHENS, W. E. and HALLIDAY, A. N. (1984): Geochemical contrasts between late Caledonian granitoid plutons of northern, central and southern Scotland, *Transactions of the Royal Society of Edinburgh: Earth Sciences*, **75**, pp. 259-73.

STEPHENSON, D., BEVINS, R. E., MILLWARD, D., HIGHTON, A. J., PARSONS, I., STONE, P. and WADSWORTH, W. J. (1999): *Caledonian Igneous Rocks of Great Britain*, Geological Conservation Review Series, **17**, (Peterborough: Joint Nature Conservation Committee).

STEWART, A. D. (2002): *The Later Proterozoic Torridonian Rocks of Scotland: their Sedimentology, Geochemistry and Origin*, Geological Society, London, Memoir, **24**, 130 pp.

STEWART, M., STRACHAN, R. A. and HOLDSWORTH, R. E. (1997): Direct field evidence for sinistral displacements along the Great Glen Fault Zone: late Caledonian reactivation of a regional basement structure? *Journal of the Geological Society, London*, **154**, pp. 135-39.

STEWART, M., STRACHAN, R. A. and HOLDSWORTH, R. E. (1999): Structure and early kinematic history of the Great Glen Fault Zone, Scotland, *Tectonics*, **18**, pp. 326-42.

STEWART, M., HOLDSWORTH, R. E. and STRACHAN, R. A. (2000): Deformation processes and weakening mechanisms within the frictional-viscous transition zone of major crustal-scale faults: insights from the Great Glen Fault Zone, Scotland, *Journal of Structural Geology*, **22**, pp. 543-60.

STEWART, M., STRACHAN, R. A., MARTIN, M. W. and HOLDSWORTH, R. E. (2001): Dating early sinistral displacements along the Great Glen Fault Zone, Scotland: structural setting, emplacement and U-Pb geochronology of the syn-tectonic Clunes Tonalite, *Journal of the Geological Society, London*, **158**, pp. 821-30.

STOKER, M. S. (1982): Old Red Sandstone sedimentation and deformation in the Great Glen Fault Zone NW of Loch Linnhe, *Scottish Journal of Geology*, **19**, pp. 369-86.

STOKER, M. S. (1983): The stratigraphy and structure of the Moine rocks of eastern Ardgour, *Scottish Journal of Geology*, **19**, pp. 369-85.

STOKER, M. S., HITCHEN, K. and GRAHAM, C. C. (1993): *The geology of the Hebrides and West Shetland shelves and adjacent deep-water areas*, United Kingdom offshore regional report, British Geological Survey.

STOREY, C. D. (2002): *Tectono-metamorphic evolution of the Glenelg-Attadale Inlier, northwest Scotland*, University of Leicester PhD thesis.

STOREY, C. D., BREWER, T. S. and PARRISH, R. R. (2004): Late-Proterozoic tectonics in northwest Scotland: one contractional orogeny or several? *Precambrian Research*, **134**, pp. 227-47.

STOREY, C. D., BREWER, T. S. and TEMPERLEY, S. (2005): P-T conditions of Grenville-age eclogite facies metamorphism and amphibolite facies retrogression of the Glenelg-Attadale Inlier, NW Scotland, *Geological Magazine*, **142**, pp. 605-15.

STOREY, C. D., BREWER, T. S., ANCZKIEWICZ, R., PARRISH, R. R. and THIRLWALL, M. F. (2010): Multiple high-pressure metamorphic events and crustal telescoping in the northwest Highlands of Scotland, *Journal of the Geological Society, London* (in press).

STRACHAN, R. A. (1985): The stratigraphy and structure of the Moine rocks of the Loch Eil area, West Inverness-shire, *Scottish Journal of Geology*, **21**, pp. 9-22.

STRACHAN, R. A. (1986): Shallow marine sedimentation in the Proterozoic Moine succession, northern Scotland, *Precambrian Research*, **32**, pp. 17-33.

STRACHAN, R. A. and EVANS, J. A. (2008): Structural setting and U-Pb zircon geochronology of the Glen Scaddle Metagabbro: evidence for polyphase Scandian ductile deformation in the Caledonides of northern Scotland, *Geological Magazine*, **145**, pp. 361-71.

STRACHAN, R. A. and HOLDSWORTH, R. E. (1988): Basement-cover relationships and structure within the Moine rocks of central and southeast Sutherland, *Journal of the Geological Society, London*, **145**, pp. 23-36.

STRACHAN, R. A., MAY, F. and BARR, D. (1988): The Glenfinnan and Loch Eil Divisions of the Moine Assemblage, in WINCHESTER, J. A. (ed.): *Later Proterozoic Stratigraphy of the Northern Atlantic Regions* (Glasgow: Blackie), pp. 32-45.

STRACHAN, R. A., SMITH, M., HARRIS, A. L. and FETTES, D. J. (2002): The Northern Highland and Grampian terranes, in TREWIN, N. (ed): *Geology of Scotland* (4th edition), (London: Geological Society), pp. 81-147.

STRACHAN, R. A., HOLDSWORTH, R. E., KRABBENDAM, M. and ALSOP, G. I. (2010): The Moine Supergroup of NW Scotland: insights into the analysis of polyorogenic supracrustal sequences, in LAW, R. D., BUTLER, R. W. H., HOLDSWORTH, R. E., KRABBENDAM, M. and STRACHAN R. A. (eds): *Continental Tectonics and Mountain Building – The Legacy of Peach and Horne*, Geological Society, London, Special Publication, **335**, pp. 231-52.

SUTTON, J. and WATSON, J. V. (1954): The structural and stratigraphic succession of Fannich Forest and Strath Bran, Ross-shire, *Quarterly Journal of the Geological Society of London*, **110**, pp. 21-54.

SUTTON, J. and WATSON, J. V. (1959): Structures in the Caledonides between Loch Duich and Glenelg, North-West Highlands, *Quarterly Journal of the Geological Society of London*, **114**, pp. 231-54.

SWETT, K. (1969): Interpretation of depositional and diagenetic history of Cambrian-Ordovician Succession of northwest Scotland, in KAY, M. (ed.): *North Atlantic Geology and Continental Drift*, American Association of Petroleum Geologists, Memoirs, **12**, pp. 630-46.

TALBOT, C. J. (1983): Microdiorite sheet intrusions as incompetent time- and strain-markers in the Moine assemblage NW of the Great Glen Fault, Scotland, *Transactions of the Royal Society of Edinburgh: Earth Sciences*, **74**, pp. 137-52.

TANNER, P. W. G. (1965): *Structural and metamorphic history of the Kinloch Hourn area, Inverness-shire*, Scotland, PhD thesis, University of London.

TANNER, P. W. G. (1971): The Sgurr Beag Slide – a major tectonic break within the Moinian of the western Highlands of Scotland, *Quarterly Journal of the Geological Society of London*, **126**, pp. 435-63.

TANNER, P. W. G. (1976): Progressive regional metamorphism of thin calcareous bands from the Moinian rocks of NW Scotland, *Journal of Petrology*, **17**, pp. 100-34.

TANNER, P. W. G. and MILLER, R. G. (1980): Geochemical evidence for loss of Na and K from Moinian calc-silicate pods during prograde metamorphism, *Geological Magazine*, **117**, pp. 267-75.

TANNER, P. W. G. and EVANS, J. A. (2003): Late Precambrian U-Pb titanite age for peak regional metamorphism and deformation (Knoydartian orogeny) in the western Moine, Scotland, *Journal of the Geological Society, London*, **160**, pp. 555-64.

TANNER, P. W. G., JOHNSTONE, G. S., SMITH, D. I. and HARRIS, A. L. (1970): Moinian Stratigraphy and the problem of the Central Ross-shire Inliers, *Bulletin of the Geological Society of America*, **81**, pp. 299-306.

TEMPERLEY, S. and WINDLEY, B. F. (1997): Grenvillian extensional tectonics in northwest Scotland, *Geology*, **25**, pp. 53-56.

THIRLWALL, M. F. (1988): Geochronology of late Caledonian magmatism in northern Britain, *Journal of the Geological Society, London*, **145**, pp. 951-67.

TILLEY, C. E. (1936): Eulysites and related rock types from Loch Duich, Ross-shire, *Mineralogical Magazine*, **154**, pp. 331-42.

VAN BREEMEN, O., PIDGEON, R.T. and JOHNSON, M. R. W. (1974): Precambrian and Palaeozoic pegmatites in the Moines of northern Scotland, *Journal of the Geological Society, London*, **130**, pp. 493-507.

VAN BREEMEN, O., HALLIDAY, A. N., JOHNSON, M. R. W. and BOWES, D. R. (1978): Crustal additions in late Precambrian times, in BOWES, D. R. and LEAKE, B. E. (eds): Crustal Evolution in Northwestern Britain and Adjacent Regions, *Geological Journal Special Issue*, **10**, pp. 81-106.

VAN BREEMEN, O., AFTALION, M. and JOHNSON, M. R. W. (1979a): Age of the Loch Borrolan complex, Assynt, and late movements along the Moine Thrust zone, *Journal of the Geological Society, London*, **136**, pp. 489-96.

VAN BREEMEN, O., AFTALION, M., PANKHURST, R. J. and RICHARDSON, S. W. (1979b): Age of the Glen Dessary syenite, Inverness-shire: diachronous Palaeozoic metamorphism across the Great Glen, *Scottish Journal of Geology*, **15**, pp. 49-62.

VANCE, D., STRACHAN, R. A. and JONES, K. A.

(1998): Extensional versus compressional settings for metamorphism: garnet chronometry and pressure-temperature-time histories in the Moine Supergroup, northwest Scotland, *Geology*, **26**, pp. 927-30.

WALKER, R. G. (1979): Shallow marine sands, in WALKER, R. G. (ed): *Facies Models, Geosciences Canada*, reprint series 1, pp. 75-89.

WHEELER, J., MORGAN, L. S. and PRIOR, D. J. (2004): Disequilibrium in the Ross of Mull contact metamorphic aureole, Scotland: a consequence of polymetamorphism, *Journal of Petrology*, **45**, pp. 835-53.

WHITE, S. H. (1998): Fault rocks from Ben Arnaboll, Moine Thrust Zone, Northwest Scotland, in SNOKE, A. W., TULLIS, J. and TODD, V. R. (eds): *Fault-related rocks: a photographic atlas* (Princeton, NJ: Princeton University Press), pp. 382-91.

WIBBERLEY, C. (2005): Initiation of basement thrust detachments by fault-zone reaction weakening, in BRUHN, D. and BURLINI, L. (eds): *High Strain Zones: Structure and Physical Properties*, Geological Society, London, Special Publications, **245**, pp. 347-72.

WIBBERLEY, C. A. J. and BUTLER, R. W. H. (2010): Structure and internal deformation of the Arnaboll Thrust Sheet, north-west Scotland: implications for strain localization in thrust belts in LAW, R. D., BUTLER, R. W. H., HOLDSWORTH, R. E., KRABBENDAM, M. and STRACHAN R. A. (eds): *Continental Tectonics and Mountain Building – The Legacy of Peach and Horne*, Geological Society, London, Special Publication (in press).

WILSON, G. (1953): Mullion and rodding structures in the Moine Series of Scotland, *Proceedings of the Geologists' Association*, **64**, pp. 118-51.

WILSON, D. and SHEPHERD, J. (1979): The Carn Chuinneag granite and its aureole, in HARRIS, A. L., HOLLAND, C. H. and LEAKE, B. E. (eds): *The Caledonides of the British Isles – Reviewed*, Geological Society, London, Special Publications, **8**, pp. 669-75.

WILSON, R. W., HOLDSWORTH, R. E., WILD, L. E.,

McCAFFREY, K. J. W., ENGLAND, R. W., IMBER, J. and STRACHAN, R. A. (2010): Basement-influenced rifting and basin development: a reappraisal of post-Caledonian faulting patterns in the North Coast Transfer Zone, Scotland in LAW, R. D., BUTLER, R. W. H., HOLDSWORTH, R. E., KRABBENDAM, M. and STRACHAN R. A. (eds): *Continental Tectonics and Mountain Building – The Legacy of Peach and Horne*, Geological Society, London, Special Publication, **335**, pp. 785-816.

WINCHESTER, J. A. (1971): Some geochemical distinctions between Moinian and Lewisian rocks and their use in establishing the identity of supposed inliers in the Moinian, *Scottish Journal of Geology*, **7**, pp. 327-44.

WINCHESTER, J. A. (1974): The zonal pattern of regional metamorphism in the Scottish Caledonides, *Journal of the Geological Society, London*, **130**, pp. 509-24.

WINCHESTER, J. A. (1976): Different Moine amphibolite suites in northern Ross-shire, *Scottish Journal of Geology*, **12**, pp. 187-204.

WINCHESTER, J. A. (1984): The geochemistry of the Strathconon amphibolites, Northern Scotland, *Scottish Journal of Geology*, **20**, 37-51.

WINCHESTER, J. A. and LAMBERT, R. ST. J. (1970): Geochemical distinctions between the Lewisian of Cassley, Durcha and Loch Shin, Sutherland and the surrounding Moinian, *Proceedings of the Geologists' Association*, **81**, pp. 275-301.

WINCHESTER, J. A. and FLOYD, P. A. (1983). The geochemistry of the Ben Hope sill suite, northern Scotland, UK, *Chemical Geology*, **43**, pp. 49-75.

WOODCOCK, N. H. and STRACHAN, R. A. (eds) (2000): Geological History of Britain and Ireland (Oxford: Blackwell).

ZANIEWSKI, A., REAVY, R. J. and HARRIS, A. L. (2006): Field relationships and emplacement of the Caledonian Ross of Mull Granite, Argyllshire, *Scottish Journal of Geology*, **42**, pp. 179-89.

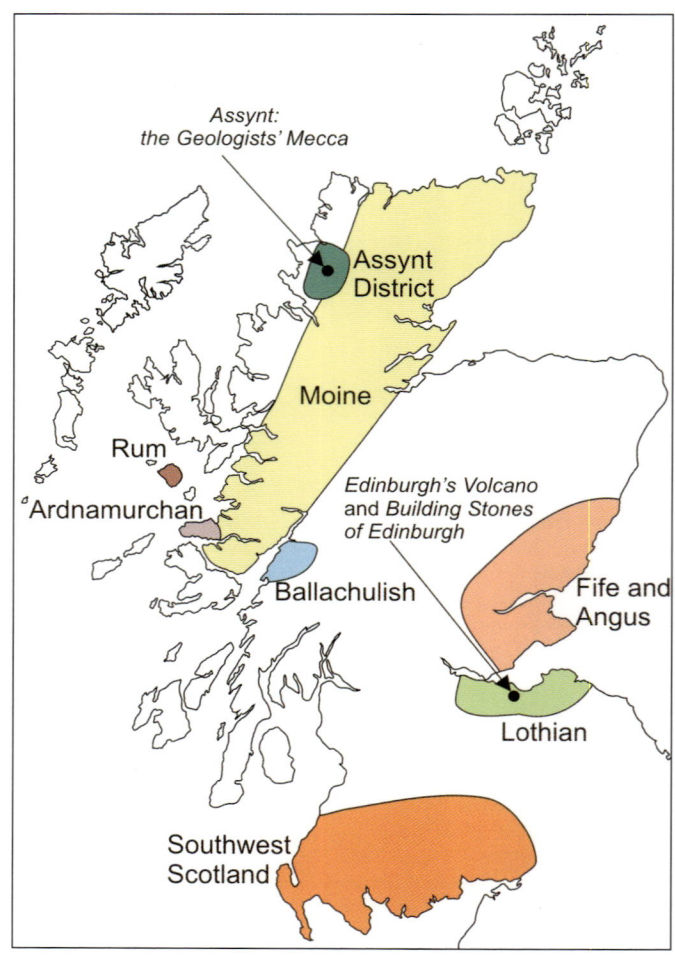

Map of Scotland showing the areas covered by Excursion Guides of the Edinburgh Geological Society. Leaflets are indicated in italics. For more information:

www.edinburghgeolsoc.org

1. NW Highlands to be published 2010; 2. SW Scotland is published jointly with the British Geological Survey; 3. Fife and Angus is published by the Pentland Press with financial support from the Society.